Happy Projects!

ROLAND GAREIS

Happy Projects!

- ⋯⋗ **Project and programme management**
- ⋯⋗ **Project portfolio management**
- ⋯⋗ **Management of the project-oriented organization**
- ⋯⋗ **Management in the project-oriented society**

MANZ

The author: Roland Gareis
Professor for project management at the University of Economics and Business Administration, Vienna, Director of the post-graduate university programme "International Project Management", managing partner of ROLAND GAREIS CONSULTING GmbH.

Printed in Austria

ISBN 3-214-08268-X

© 2005 MANZ´sche Verlags- und Universitätsbuchhandlung GmbH, Vienna
Telephone: (01) 531 61-0
E-mail: verlag@MANZ.at
World Wide Web: www.MANZ.at
Translated by: Caroline Wellner
Page Layout: BuX. Verlagsservice, www.bux.cc
Cover and Interior Designer: designby frank scheikl · www.frankscheikl.at
Print: MANZ CROSSMEDIA, 1051 Vienna

For Haldis, my "muse", and
Luisa and Lorenz, who give my life meaning.

Introduction

"Happy Projects!" is a wish.

"Happy Projects!" is not a promise, but a wish from me to all those thinking and acting in projects. The chance to experience "Happy Projects!" increases dramatically when applying the models and methods described in the present book. Just do it!

The target group

The target group for "Happy Projects!" is you, "Happy Projects!" is the handbook

- for project owners, project managers and programme managers as well as
- for managers of project-oriented organizations, such as members of a Project Portfolio Group, managers of Expert Pools and PM Offices,

it is the textbook

- for management trainers and consultants,
- for teachers at universities, at universities of applied science and at colleges,
- for project management students as well as
- for candidates for project management certifications,
- and serves as motivation
- for organizers in small municipalities, in churches and in associations,
- for organizers in schools and in families to use project management for their projects, too.

The basis

"Happy Projects!" is based on

- *ROLAND GAREIS Project and Programme Management®,*
- *ROLAND GAREIS Management of the Project-oriented Company®,*
- *ROLAND GAREIS Management in the Project-oriented Society®.*

These management approaches are characterized by a future-oriented management paradigm.

The contents

Project orientation is a management strategy, not only for organizations but also for societies. Competencies for the professional management of projects,

programmes and project portfolios create competitive advantages for organizations and for societies. I will describe "Management by Projects" as a vision becoming reality in Chapter A. In the following chapters I attempt to supply evidence concerning this matter in the form of examples and numerous case studies.

Even though it may seem trivial, I cannot spare you an organizational definition of the construct "project" in Chapter B. This is done by differentiating projects and business processes, programmes, project portfolios, etc. It is also important for me to work out the specifics of my approach to project and programme management.

The hurried project manager, who needs instructions for designing his/her project organization straight away, can, of course, skip the first two chapters and start with Chapters C and D. These chapters serve to clarify that "empowerment" is not a buzzword in projects and that the design of the project organization presents a central

success factor in projects.

In Chapter E project management as a business process of the project-oriented organization is described. The corresponding methods of project and programme management are depicted in detail in Chapter F. As a continuous example for exemplifying the methods the project "Realization e-application" is used. In

Chapter G: Programme management, this business process and specifics in the use of methods in programme management are described. Programme management is the integration instrument of this decade!

The use of relatively new business processes of the project-oriented organization, i.e. management consulting and management auditing of projects and programmes, is recommended in Chapter H.

The specific business processes of the project-oriented organization, i.e. project portfolio management, organizational design and personnel management, are described in Chapters I, J and K. In these chapters the importance of Expert Pools, of a Project Portfolio Group and of a PM Office is elaborated and advice is offered, for example on how to establish a project management career path.

For all those, who, like me, believe in it, the model of the project-oriented society is introduced in Chapter L.

Non-contents of this book are methods which I do not consider important for my work as project owner or project manager, such as methods of CPM-based resource optimization, for example.

General leadership methods, which should also be applied in projects, such as creativity techniques and moderation techniques, are not treated. Descriptions of other project management approaches, such as Prince 2, PROPS, PMBoK®, ICB, etc., were not made either. However, I generally consider these as interesting. Maybe there will be a corresponding addition in the next edition?In the next edition I will also gladly consider your feedback, which I hope will be plenty.

History of origins and acknowledgements

In 1990, I already published "Management by Projects" with Manz publishers, which was followed in 1991 with "Project management in engineering and engineering construction". Due to the brisk sales of the project management book Manz concluded a contract with me in 1993 for publishing a revised edition. Intensive research work on crisis management in projects and in the project-oriented organization as well as on management in the project-oriented society has delayed this publication for several years. Even the use of project management and several new starts of this publication project did not advance the process. My sincere thanks therefore to the publishers for their patience and trust.

I would also like to thank the students of the programme "Project management" at the University of Economics and Business Administration, Vienna, who promote the further development of project management as a "discipline" by analytical questions and creative contributions. I would like to express my thanks to my consulting clients for joint development of practical solutions and for providing case studies for this book.

I would like to thank my colleagues at the PROJEKTMANAGEMENT GROUP from the University of Economics and Business Administration, Vienna, and ROLAND GAREIS CONSULTING, on the one hand for never giving up hope on completing "Happy Projects!", and on the other hand, for providing me with lots of ideas in great discussions. This is especially true for the co-authors of individual chapters, i.e. Mag. Dr. Martina Huemann and Mag. Michael Stummer.

Thank you, thank you, thank you to my family, for bearing with me and for accepting my love for my projects.

Happy Projects![1]

Roland Gareis

---

1) There exists no trademark for "Happy Projects!", i.e. the greeting, the wish, the "order" is freely disposable and can be used by everybody.

Contents

F   Methods of project and programme management

I Project portfolio management

J Organizational design of the project-oriented organization

K   Personnel management in the project-oriented organization

L   The project-oriented society

Contents

A

Project orientation as a management strategy

Projects and programmes are temporary organizations for the performance of complex, relatively unique business processes. The objectives of using projects and programmes are to establish organizational flexibility within an organization and to assure the quality of the results of the performed business processes. By that, competitive advantages can be developed.

In the project-oriented society projects and programmes are becoming more important, not only in profit and non-profit organizations, but also in new areas of application, such as in small communities, associations, schools and even in families.

The definition of a project requires the use of project management. A project is "a difference that makes a difference". A project requires management attention and the design of an adequate project organization, the creation and controlling of project plans and the establishment of project context relationships. The definition of a programme requires the use of programme management.

A Project orientation as a management strategy

Contents

A 1 Projects and programmes, project and programme management

A 1.1 The history of project and programme management

The first projects for which formal project management documentation was created were military projects and projects in the US space programme.[1] The first project management documentation was created about 1941 for the development of the atomic bomb for the Manhattan Engineering District Project. In 1956, the US Air Force published the C/SCSC (Cost/Schedule Control System Criteria) specification and the Air Force project management concept with the designation AFSCM 375 (Air Force System Command Manual). These publications were the standard works for modern project management. In 1958, the scheduling method PERT (Programme Evaluation and Review Technique) was presented by the US Navy as a part of the Polaris Missile Program.

This first phase in the use of projects and project management was characterized by projects with technical objectives in the military and the space program, as well as by a small number of projects of high complexity, with high project costs and long project duration.

In the 1960s and 70s the experiences of the US-Air Force, US-Navy and NASA projects were used as a basis for projects in construction, engineering and information technology. Furthermore, international organizations which support worldwide research and development projects, such as the World Bank, the UNIDO (United Nations Industrial Development Organization) or the ILO (International Labour Organization), published their own standards for project management.

This second phase of development in project management was also aimed at projects of high complexity with technical objectives, but they were to be found in different industries.

The third phase of project management development was heralded with the theme "Management by Projects" during the Project Management World Congress of the IPMA-International Project Management Association in Vienna in 1990.[2] "Management by Projects" was presented as a new organizational strategy. It was based on the assumption that projects as temporary organizations present a strategic option for the organizational structure of an organization.

The perception of projects as temporary organizations dramatically increased the importance of project management. It was acknowledged that projects with different objectives (such as contracting projects, offer projects, marketing projects, organizational projects and personnel development projects, etc.) in all branches of industry, in the public sector and in non-profit organizations[3] could increase an organization's efficiency

---

1) Cp. *Schelle, H.*, (Projekte zum Erfolg) 2001, p. 12
2) Cp. *Gareis, R.*, (Handbook) 1990, p. 35 f
3) Cp. *Gareis, R.*, (Non-Profit) 1997, p. 299 ff

and chances of survival. In addition to projects of high complexity, small projects and projects of medium complexity increased in importance.

Meanwhile the organizational strategy "Management by Projects", which was formulated in 1990, is being applied in more and more companies. Well-known national and international companies in practically every branch of industry have project portfolios of different types of projects. So projects, whether consciously or unconsciously, are an instrument of organizational differentiation. As opposed to the permanent organizational structures of companies (such as divisions, profit centers, departments, etc.) which fulfill the routine business processes, projects are used to perform more complex and relatively unique business processes.

A relatively new development[4] is the organizational differentiation between projects and programmes. The performance of processes of different durations and different scopes requires different types of organizations. A programme is a temporary organization established to fulfill a one-time business process of high complexity with a medium to long duration. The projects which are linked to the programme serve to realize the programme objectives. A programme which is perceived as an organization is, for example, the development of a product family in a service organization, the implementation of a complex IT-solution in an international organization or the reorganization of multiple companies into a holding.

CHAPTER

A

characteristic of business process	scale		
frequency	continuous	unique	unique
duration	short-term	short-term – medium-term	medium-term – long-term
importance	low	medium – high	high
scope	small	medium – large	large
resource demand	low	medium	high
cost	low – medium	medium – high	high
organizations involved	few	several – many	many
	↓	↓	↓
organizational form	permanent organization or working group	project	programme

Fig. A1.1: Appropriate organizations for fulfilling different business processes

Although the term "programme" has existed a long time in reference works, the use of programmes as temporary organizations has emerged within the last few years.

---

4) Cp. *Gareis, R.*, (Programmmanagement) 2001, p. 4 ff

A 1.2 The benefits of projects and programmes

The performance of projects and programmes serves to ensure the competitiveness of companies which should be ensured by providing an appropriate organizational complexity. Different organizations to fulfill different business processes are required.

The social environments of companies can be seen as being increasingly complex. Due to the globalization of markets, new technological developments, new cooperative relationships with customers and suppliers as well as changing values in society, the complexity in the environment is increasing.

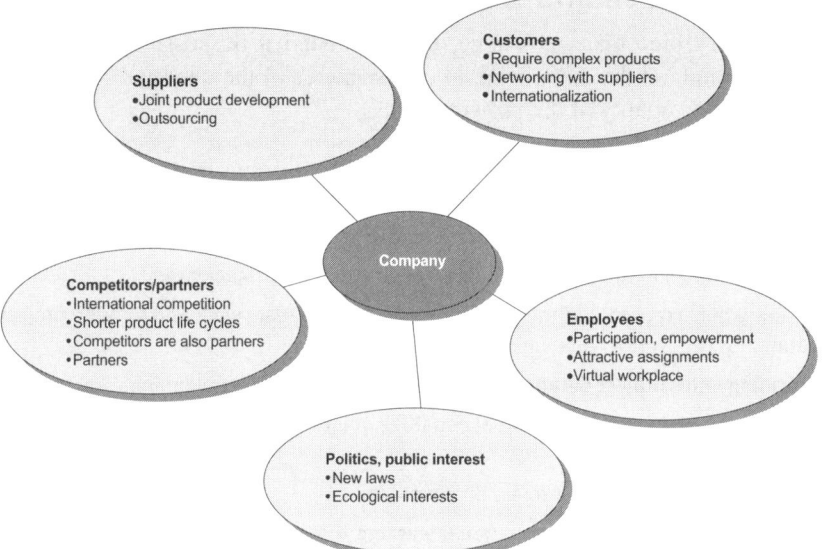

Fig. A1.2: Examples of the increasing complexity of the environment of an organization

The complexity of the social environment of an organization can be measured on the basis of variety. Variety can be defined by the number of possible states which a social system can take on. A competitor can at the same time, for example, be a partner and a supplier of an organization. Ashby laid down the law of "required variety".[5] This states that "...only variety can absorb variety". Companies must therefore build up a certain amount of complexity within their organizations in order to be able to match the complexity of their environments. The differentiation which results from the use of projects and programmes contributes to the building up of complexity of the organization.

Organizational design creates competitive advantages for companies![6] Through the use of projects and programmes temporary organizations are established which can be dissolved again after performing a relatively unique business process. Project team members with the competencies required to meet the objectives are recruited and released

---

5) Cp. *Ashby, W.R.*, (Process of model-building) 1970, p. 94 ff
6) Cp. *Senge, P.*, (Fieldbook) 1994, p. 10 ff

again after project close-down. Organizations are created according to needs and used temporarily.

To borrow a phrase from Bateson[7], a project represents "a difference that makes a difference". By using the term "project", a certain amount of management attention is granted. Through the definition of a "project", project management has to be applied! Thereby the fulfillment of the project objectives and the quality of the result should be assured.

A 1.3   The benefits of project and programme management

The benefits of project management lie, on the one hand, in the possibility to realize complex projects and, on the other hand, in the assurance of the quality of the project performance and the quality of the project results.

benefits of project management
• making complex projects feasible
• assuring quality in the project results through a holistic project view
• assuring the acceptance of the project results through team work and through project marketing
• providing short project durations and high accuracy in project planning
• optimizing costs by saving eventual penalties or interest payments, or through the optimization of interest yields
• transparency by providing project documentation
• assuring individual and organizational learning through reflections within the project organization
• constructive relationships between customers, suppliers and partners

Fig. A1.3: Benefits of project management

The threats of the inadequate use of projects and project management lie in both the inflationary and undifferentiated use of the term "project" and the inflated expectations of the integration functions of project management.

When the term "project" is used in an inflationary manner, in other words for anything which is relatively unique and for which boundaries can be defined, and when there is no clear difference between what is a project and what is not, then there will be projects for which the use of project management is not necessary. On the other hand, it often happens that unidentified programmes are managed as projects, although the integration functions which are necessary for the programme cannot be fulfilled by project management. Both situations are dysfunctional and damage the acceptance of project and programme management.

---

7) Cp. *Bateson, G.,* (Geist und Natur) 1990, p. 274

Programme management is more that the management of a project of high complexity. Programme management tasks are additional to the management of the individual projects of a programme. Ensuring the "big programme picture", planning and controlling of programme objectives, programme schedule and the programme budget, the establishment of the programme environment relationships, programme organization and programme culture, as well as programme marketing are the instruments for integrating the different projects of a programme.

Sometimes it is expected that the coordinating tasks of the permanent organization, such as taking over the tasks of further technical development or of personnel development, should be fulfilled by project management within the framework of individual projects. These expectations are usually the result of expecting too much of project management. Project management is not a substitute for weak management in the permanent organization!

CHAPTER

A 2   The project-oriented organization

A

Companies and parts of companies, such as divisions, business units or profit centers, which use projects and programmes to fulfill complex and relatively unique business processes, can be defined as "project-oriented companies".

Project-oriented companies have specific strategies, specific organizational structures and specific cultures for managing projects, programmes and project portfolios.

A 2.1   Construct: Project-oriented organization

In order to identify an organization as a project-oriented organization, the management of projects and programmes within the organization and the strategic, structural and cultural prerequisites for their performance must be considered.

Of course, an organization can also be viewed from other perspectives: The project-oriented organization only provides a new possibility for constructing organizational reality. By viewing an organization as a "project-oriented organization", new possibilities for management intervention can be created by which the potential for successful performance of projects and programmes can be increased.

Project-oriented organizations are characterized by projects and programmes. At any given time a number of projects can be started, performed, closed down or even stopped. In this way, a state of balance is created which should assure the development of the organization and its survival.

The more varied the projects and programmes of a project-oriented organization are, the more complex the management of the organization will be. This depends on the dynamics and the complexity of the individual projects as well as the relationships between the projects.

The project-oriented organization has the following characteristics:

- Management by projects is an explicit organizational strategy,
- projects and programmes are used as temporary organizations,

- networks of projects, chains of projects and project portfolios are objects of consideration for the management,
- project management, programme management and project portfolio management are specific business processes,
- know-how assurance takes place in Expert Pools,
- project management competence is assured by a PM Office and a Project Portfolio Group, and
- a "new management paradigm" is applied, which is characterized by team work, process orientation and empowerment.

Not only companies but also subsystems of companies, such as divisions, business units and profit centers, can be viewed as project-oriented companies. Therefore, the terms "project-oriented company" and "project-oriented organization" can be used synonymously.

A project-oriented company is a company which has at its disposal specific strategies, structures and cultures for the professional management of projects, programmes and project portfolios. The creation of the requisite "organizational fit" between these strategies, structures and cultures poses a particular management challenge. "Structure follows strategy" is a linear management approach which does not take into account the reciprocal relationships between the three dimensions.

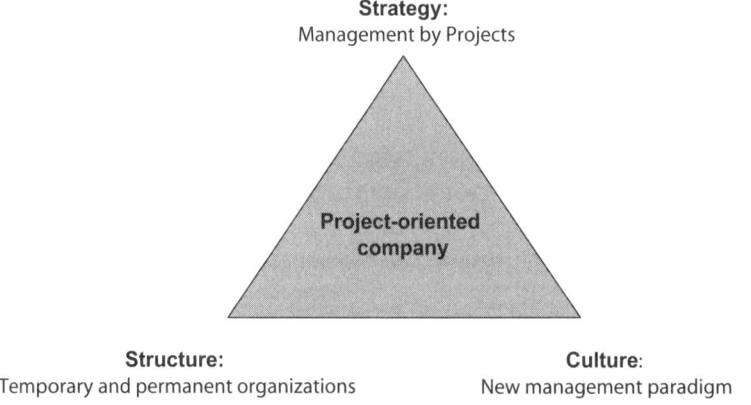

Fig. A2.1: Requirement for an "organizational fit" in the project-oriented organization

A 2.2   Management by Projects as an organizational strategy

The engineering organization Fluor Daniel views projects "[...] as a way of conducting its business and, of late, as a way of improving its internal tasks [...] Fluor Daniel is able to conduct its business effectively in a decentralized, networked organizational atmosphere."[8]

The organization simultaneously performs projects for internal and external customers, small projects as well as projects of medium or high complexity, and projects with different objectives.

from	→	to
• contracting, research and development		• contracting, research and development, offer development, marketing, public relations, personnel development, organizational development, infrastructure development
• few projects of high complexity		• many projects of different complexities, and programmes
• primarily projects for external customers		• projects with external and projects with internal customers

Fig. A2.2: Trends to more project orientation

CHAPTER

A

Project-oriented companies view projects and programmes as a strategic option for designing the structure of the organization. Companies have the choice of designing their organizations with or without projects and programmes. By applying Management by Projects as a strategy it is possible to realize the following organizational objectives:

- creating organizational flexibility by using temporary organizations in addition to the permanent organization,
- delegation of management responsibility to projects and programmes,
- goal-oriented work by defining project and programme objectives and
- assuring organizational learning through the monitoring potential of projects and programmes.

Personnel management goals can also be realized through Management by Projects. In projects it is possible to operationalize and integrate the often isolated or competing leadership models of Management by Objectives, Management by Delegation, Management by Motivation, etc. Management by Projects uses as leadership strategy the motivational and personnel development functions of projects.

Working in projects is attractive for many employees because projects

- usually contain new business processes,
- are socially challenging and integrative because of team work,
- provide freedom of movement and promote creativity and
- are limited in time, and therefore require feedback and new options after the end of the project.

---

8) *Thatcher, J R* , (New Age Managers for Projects) 1990, p. 9

Individual learning in projects is promoted through the complexity of the problem situations.

The marketing relevance of Management by Projects results from

- its effects of projects on relevant environments,
- the application of project management as a sales instrument and
- the possible marketing of project management as a service for organization-internal and/or organization-external markets.

Through the stronger definition and performance of projects many companies apply "Management by Projects" implicitly. A specific benefit of the above described possibilities requires an explicit application of "Management by Projects" and the corresponding structural and cultural prerequisites in the organization.

CHAPTER
A

A 2.3   Organizational structure of the project-oriented organization

The permanent organization of an organization is designed to perform repetitive business processes. The organizational structure and the business processes should provide orientation for employees through clear task definitions and responsibilities. Furthermore, the organization should guarantee continuity in the relationships of the organization to its relevant environments.

For the most part, these organizational objectives can be met by a stable, hierarchical line organization. However, an organization which continuously performs new projects of different contents and degrees of complexity requires a flexible, networked organizational structure.

Simply put, organizations can be positioned on a continuum between extreme "steep hierarchic organizations" and "flat, network-type organizations". The amount of routine work in relation to the amount of project work determines the positioning of an organization on the continuum. There is no clear optimum position for an organization, but a trend toward flatter, networked structures is observable.

In companies with little project orientation, projects are used in addition to the hierarchic line organization. Through the use of projects, this type of organization becomes flatter and more flexible. Flattening comes about through an enlargement of the communications span and a (partial) reduction in the number of levels in the hierarchy. Flexibility is achieved through the possibility of using (project) organizations and then dissolving them.

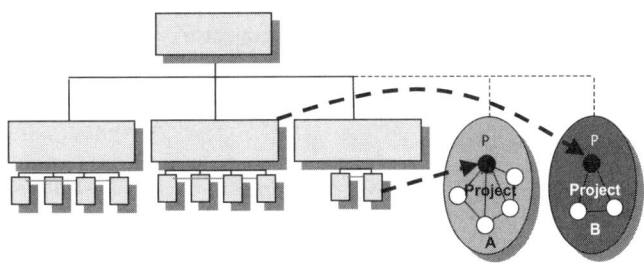

Fig. A2.3: Flattening of the organization through projects

In flat, networked organizations the most important business processes will be performed within projects. An example of a flexible, networked structure is depicted in the
following organizational chart:

CHAPTER

A

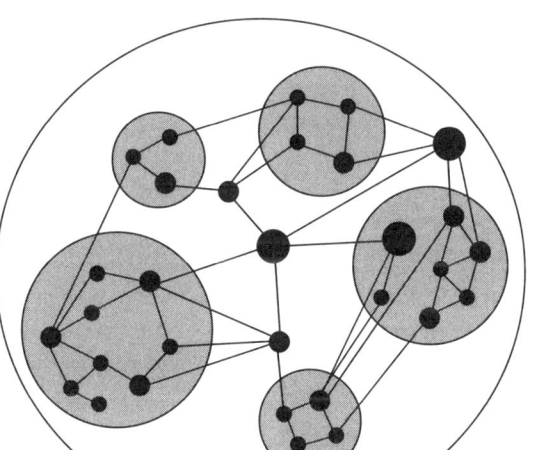

Fig. A2.4: Organizational chart of GORE GmbH

The main characteristic of this organizational structure is that roles, lines of communication and borders between teams, projects and departments are depicted, but not the
hierarchical relationships.

Project-oriented companies perform a number of different projects and programmes at
the same time. This quantity of temporary organizations requires a high level of syn-
chronization. In order to fulfill this integration function projects (and programmes) can
be clustered into chains of projects, project portfolios and networks of projects.

relationships between projects		
a set of sequential projects	all projects of a project-oriented organization	a set of closely-coupled projects
over a period of time	at a point in time	at a point in time
⬇	⬇	⬇
chain of projects	project portfolio	network of projects
cluster		

Fig. A2.5: Clustering of projects

A chain of projects is a number of sequential projects for the performance of several business processes. A chain of projects is viewed over a period of time. Examples of chains of projects are the projects "Conception of an EDP application" and "Realization of an EDP application", or the projects "Offer for customer contract A" and "Performing customer contract A".

A project portfolio is the set of all the projects of a project-oriented organization. A project portfolio represents the set of projects at a particular point in time.

A network of projects is a set of several closely-coupled projects. Grouping projects into networks requires different criteria, such as the use of a particular technology in all the projects, performance of the projects in the same geographic region or performance of the projects for a common customer. The relevant criteria are to be chosen for the construction of a network of projects. A network of projects is to be constructed on a set date. An example of a network of projects is the network of all offer, contracting and joint venture projects for an engineering organization in China.

For the integration of projects and programmes, as well as for the management of chains of projects, of project portfolios and of networks of projects, specific permanent organizational units are required: The PM Office, the Project Portfolio Group and Expert Pools.

From the viewpoint of organizational theory, projects can be understood as instruments of differentiation. To assure that the general objectives and rules of the organization are followed within the projects, project-oriented companies require these new integrating structures.

A 2.4 Business processes of the project-oriented organization

The project-oriented organization is characterized by the following specific business processes:

- project management,
- programme management,
- assuring management quality in a project or a programme,
- assignment of a project or a programme,

- project portfolio coordination,
- networking between projects,
- personnel management in the project-oriented organization and
- organizational design of the project-oriented organization.

In the process of assigning, the decision of whether or not a project or a programme is to be performed is taken. When the decision to perform the project has been taken the project can be assigned to a project team.

The project management process begins with the performed project assignment and ends with the final approval of the project by the project owner. The process is made up of the sub-processes project start, project coordination, project controlling, possibly resolving a project discontinuity, and, project close-down.

Analogous to these are the programme management sub-processes programme start, programme coordination, programme controlling, possibly resolving a programme discontinuity, and programme close-down.

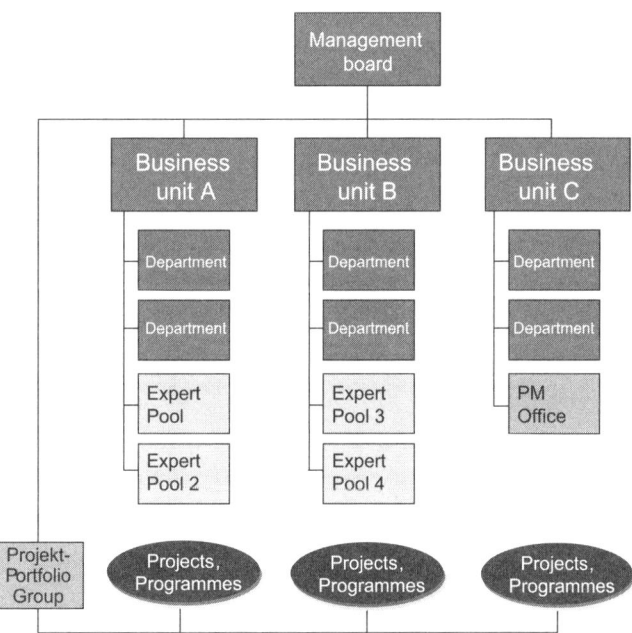

Fig. A2.6: Organizational chart of the project-oriented organization (according to ROLAND GAREIS Management of the Project-oriented Company®)

To assure management quality in a project or a programme the processes of management consulting and management auditing can be performed.

In the process of project portfolio coordination the priorities between projects are set, and internal and external resources are coordinated. The process of networking of projects contributes to the assurance of synergies in the network.

Personnel management in a project-oriented organization includes the recruiting, disposition and (continuous) development of project personnel. Those employees of the project-oriented organization who have the role of project or programme owner, project or programme manager, project team member, project contributor, etc., are to be understood as project personnel.

The organizational design of a project-oriented organization includes the establishment of a PM Office, a Project Portfolio Group and Expert Pools, the establishment of guidelines for project and programme management and for project portfolio management, as well as the development of standard project plans.

These specific processes can be depicted graphically in a maturity model of a project-oriented organization.

CHAPTER

A

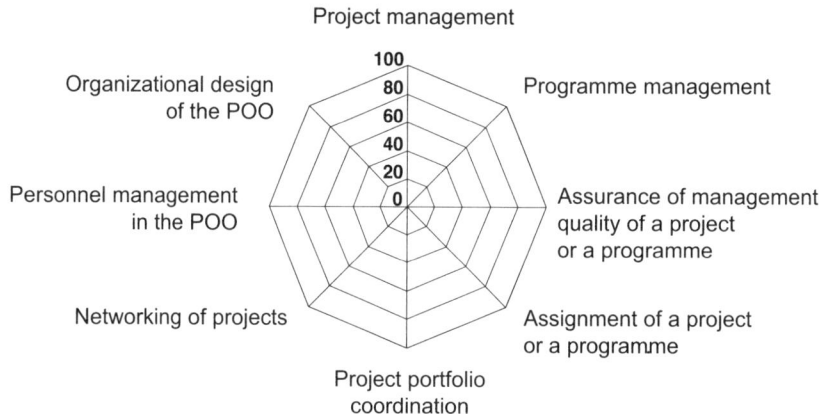

Fig. A2.7: Maturity model of the project-oriented organization (according to ROLAND GAREIS Management in the Project-oriented Society®)

A 2.5   Cultures of the project-oriented organization

The use of a "new management paradigm" in the organization is a prerequisite for the successful and efficient performance of projects and programmes. A traditional management paradigm, in which the hierarchy is seen as the central instrument of integration, where cooperation is organized on the basis of interfaces, and where business processes are structured according to functional units; does not exhaust the organizational potentials of projects and programmes. This can only be achieved through the implementation of a "new management paradigm".

Due to the following characteristics which are derived from management approaches such as lean management, organizational learning, knowledge management, total quality management, etc., this "new management paradigm" can be characterized as:

- customer orientation,
- organization as competitive advantage,

- process orientation,

- team orientation,

- empowerment of employees and of (temporary) organizations,

- promotion of networks between employees and between (temporary) organizations and

- encouraging (dis-)continuous change.

In project-oriented organizations cultural differentiation is promoted through the development of project- and programme-specific cultures.

A 3 The project-oriented society

A society in which profit and non-profit organizations often use projects and programmes as temporary organizations to perform relatively unique business processes of medium to high complexity, can be perceived as a project-oriented society.

In the project-oriented society institutions provide project management-related training, research and marketing services. The project-oriented society has competencies in project and programme management, in project portfolio management, in the management of project personnel and in the organizational design of project-oriented organizations.

A 3.1 Project orientation as a macro-economic phenomenon

In many national societies projects and programmes are being used more and more often in companies, but also in other organizations such as in (small) communities, in associations, in schools and even in families. "Management by Projects" is becoming an organizational strategy of society in order to better handle the increasing complexity and dynamics of society and its environments.

Not only in industry but also in non-profit organizations projects and programmes are being used as the appropriate organizational form for performing relatively unique business processes of medium to high complexity. New areas of application and new types of projects are becoming more important.

The importance of projects and programmes in society, the structure of the society, its history and its expectations regarding the future influence the development of the project-oriented society.

A 3.2 Construct: The project-oriented society

The perception of a society as a project-oriented society is a construct. It requires one to look at the society through a "special pair of glasses" – the glasses of project orientation. In doing so, the communications within the society related to projects, programmes and project portfolios can be considered.

A society which often uses projects and programmes as temporary organizations for performing relatively unique business processes of medium to high complexity, can be viewed as a project-oriented society.

On the one hand, the project-oriented society model takes into account the practices of project-oriented organizations in project management, programme management, project portfolio management, management of project personnel and in organizational design.[9] On the other hand, the model describes project management-related services from institutions which promote the practice of project, programme and project portfolio management. Project management-related services are fulfilled by training, research and marketing institutions.

The maturity model of the project-oriented society can be visualized in the form of a spider web. Each axis of the web represents a dimension of practice in project-oriented organizations or the project management-related services of the institutions. The "maturity" of a project-oriented society can be evaluated on the basis of the level of achievement along each of the individual axes.

CHAPTER

A

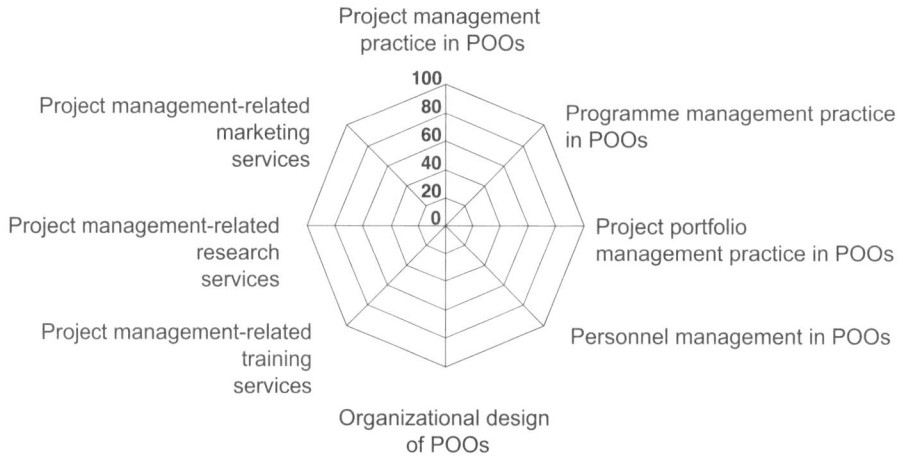

Fig. A3.1: Maturity model of the project-oriented society (according to ROLAND GAREIS Management of the Project-oriented Company®)

The "practice" axes of the project-oriented society model are partially equivalent to the axes of the maturity model of the project-oriented organization described above, whereby the processes "assignment of a project or a programme", "project portfolio coordination" and "networking of projects" are grouped together under "project portfolio man-

---

9)   The term "project-oriented organization" has been used instead of "project-oriented company" since the project orientation refers not necessarily to the entire company, but to one or several parts, such as a division or a profit center, but also organizations, such as schools, communities, non-profit organizations, etc.

agement". Therefore, only the axes of the project-oriented society model which have to do with project management-related services must be described.

Formal project management training programmes can be offered from private and public training institutions. They can lead to an academic degree in project management (for example a master's in project management) or not. Project management training programmes can differ according to the number of courses offered and the project management approach represented.

Services which may be offered by project management research institutions are project management research projects and project management research programmes, project management publications and project management research events. There can also be project management-related research financing.

Project management marketing tasks in project-oriented societies are mostly performed by universities, colleges and national project management associations. The services of national project management associations are membership, certification of project management personnel, performing project management events, etc. The application of project management is also promoted through the creation of project management-related standards and the definition of formal minimum requirements for project management for performing public contracts.

References

Ashby, W. R., (Process of model-building) The process of model-building in the behavioral sciences, Ohio State Univ. Press, Columbia, Ohio, 1970

Bateson, G., (Geist und Natur) Geist und Natur, Suhrkamp, Frankfurt am Main, 1990

Gareis, R., (Non-Profit) Projekte und Projektmanagement in Non-Profit-Organisationen, in: *Badelt, C.,* (Ed.), Handbuch der Nonprofit Organisationen, Schäffer-Poeschel-Verlag, Stuttgart, 1997

Gareis, R., (Handbook) Handbook of Management by Projects, Manz Verlag, Wien, 1990

Gareis, R., (Programmmanagement) Programmmanagement und Projektportfolio-Management, in: Deutsche Gesellschaft für Projektmanagement (Ed.), Projekt Management, TÜV-Verlag, Köln, 1/2001

Schelle, H., (Projekte zum Erfolg) Projekte zum Erfolg führen, Dt. Taschenbuchverlag, 2001

Senge, P., (Fieldbook) The fifth discipline fieldbook: strategies and tools for building a learning organization, Doubleday, New York, 1994

Thatcher J. R., (New Age Managers for Projects) New Age Managers for Projects – Dealing with a „World Turned Upside Down", in: Project Management Network, Volume IV, No 4, Project Management Institute, Drexel Hill, PA, May 1990

B

Construct: Project and project management approaches

Projects are temporary organizations which are used for the performance of relatively unique, short to medium term, strategically important business processes which are medium to large in complexity.

To ensure the advantages of projects the business processes for which projects are to be performed must be differentiated from those business processes which are not project-worthy.

Project management is a business process of the project-oriented organization which includes the sub-processes project start, continuous project coordination, project controlling, project close-down and, possibly, the resolution of a project discontinuity. In the project management process project objectives, project objects of consideration, project schedules, project costs and project income, project resources and project risks as well as the project organization, the project culture and the project context are considered.

B Construct: Project and project management approaches

Contents

B 1   Construct: Project

In project management research, as well as in the practice of project management, various project definitions are used. This is important inasmuch as different perceptions of projects lead to different project management approaches.

The definition of projects as tasks (with special characteristics), rather than the definition of projects as temporary organizations and as social systems, results in a different understanding of the objectives of project management, of the project management tasks, of the objects of consideration of project management and of the project management methods used.

B 1.1   Perceptions of projects

The perception of projects as tasks with special characteristics

Traditionally, projects are defined as tasks with special characteristics. The special characteristics of projects are the "complexity" of the content, the relative uniqueness, the high risk and the high strategic importance for the project-oriented organization. Projects are understood as goal-oriented tasks since the objectives in terms of the scope, the schedule, the required resources, and the costs are planned, agreed on and controlled.

Typical representatives of this project understanding are, for example, PMI, the American Project Management Institute, or GPM, the German Society for Project Management, whose project definitions are cited below. The GPM defines a project as "... an undertaking which is basically characterized by the uniqueness of conditions, e.g. objective, temporal, financial, personnel and other limitations, boundaries against other undertakings, project specific organization". Project management is "...the entirety of management tasks, management organization, management techniques and tools for the performance of a project".[1]

PMI defines a project as "a temporary endeavor undertaken to create a unique product or service. [...] Projects are undertaken at all levels of the organization. They may involve a single person or many thousands. Their duration ranges from a few weeks to more than five years."[2] Project management is "the application of knowledge, skills, tools and techniques to project activities to meet project requirements".[3]

The perception of projects as temporary organizations

According to organizational theory, projects can be perceived as temporary organizations for the performance of business processes which are limited in time.

---

1) *Motzel, E.* et al., (Projektmanagement Kanon) 1998, p. 12
2) *Duncan, W.R.*, (Body of Knowledge) 2000, p. 6
3) *Duncan, W.R.*, (Body of Knowledge) 2000, p. 6 f

As with other organizations, a project has a specific identity, which is characterized by its specific project objectives, project organization, project values and project environment relationships.

A project is a temporary organization. Through this temporary character the establishment of the project in the project start process, as well as the dissolution of the project in the project close-down process, attain a special meaning.

The perception of projects as social systems

The perception of projects as temporary organizations also makes it possible to view them as social systems. According to social systems theory organizations, and therefore also projects, can be viewed as social systems which have clear boundaries to differentiate themselves from their environments. But they are also related to those environments. The specific characteristics of social systems, such as their social complexity, their dynamics and self-reference, are management topics in projects as well.

Therefore, according to ROLAND GAREIS Project and Programme Management®, projects are understood as temporary organizations and social systems. This understanding of projects results in a systemic project management approach which matches the complexity and dynamics of projects.

⋯⟶ EXCURSUS B1: SOCIAL SYSTEMS

Organizations, such as companies, divisions, profit centers and also projects and programmes, can be viewed as social systems. For Luhmann, the "benefit of establishing boundaries is to create systems which are less complex than their environments".[a] Anything can be called a system when it is possible to distinguish between inside and outside. "The inside-outside difference signifies that an order is established, which does not expand in an arbitrary fashion, but sets boundaries through its inner structure and through the characteristics of its relationships."[b]

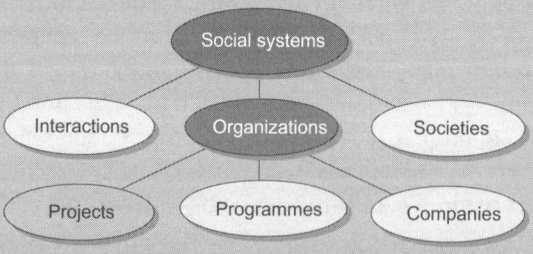

Fig. B1.1: Social systems

The elements of a social system are communications. The structures created by communications have a central meaning for the system. They determine the identity of the system along with the context in which the system operates. A social system is only to be understood in its context, upon which it also depends. The dimensions of the context are the relevant environments, the history of the system and the expectations regarding the future of the system.

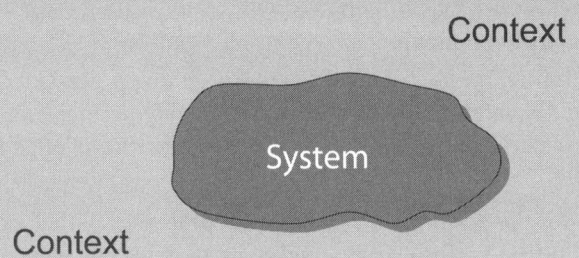

Fig. B1.2: : Structures and context of a social system

Relevant social environments of an organization are, for example, customers, suppliers, competitors and the media, but there are also "internal environments", such as employees. Every social system is shaped by its history. Many "peculiarities" of a system can only be understood and interpreted through its past. On the other hand, the expectations regarding the future determine the current decisions.

Social systems are complex, self-referencing and dynamic. Luhmann defines the grade of complexity of systems in terms of the following influence factors:[c]

- the number of elements in the system,
- the number of possible relationships between these elements,
- the diversification of the relationships and
- the development of these three factors over time.

Complexity is not only reduced through the formation of social systems, it is also built up by it. The ability of a social system to survive is determined by its ability to develop an appropriate complexity of its own and to use it to deal with the complexity of the environment.[d]

[a] Cp. *Kasper, H.,* (Organisierte Sozialsysteme) 1990, p. 156
[b] *Luhmann, N.,* (Funktionen) 1964, p. 24
[c] Cp. *Luhmann, N.,* (Komplexität) 1980, p. 1064 ff
[d] Cp. *Kasper, H.,* (Organisierte Sozialsysteme) 1990, p. 376

B 1.2   Definition: Project

A project is a temporary organization of a project-oriented organization for the performance of a relatively unique, short to medium term, strategically important business process of medium or large complexity.

Projects are, therefore, appropriate for the performance of business processes with the following characteristics:

- relatively unique,
- short to medium duration,
- medium to high strategic importance and
- medium to large complexity.

Projects are used for the performance of relatively unique processes. The more unique the objectives and deliverables to be fulfilled are, the higher the associated risk. Information from past experiences which can be used as reference is often only available to a limited extent.

Projects are used for the performance of business processes with short to medium duration. These should be performed as quickly as possible, in other words in several months. One exception is the performance of infrastructure projects (construction or engineering projects) which can have a longer duration.

Business processes for which projects are used have a medium to high strategic importance for the company performing them. The performance of contracts contributes to, for example, the short- to medium-term survival of the company. The development of new products and the establishment of a new strategic alliance have, however, long-term consequences and are, therefore, strategically more important.

Projects are used for business processes of medium to large complexity. The complexity of a business process can be described by the tasks, the resources required, the costs occurring and the organizations involved.

To operationalize the definition of "what is a project and what is not" the characteristics of a business process are used, i.e. strategic importance, duration, organizations involved, resources required, and costs occurring. The scaling of these characteristics is to be defined by each organization. Following is an example of the definitions from an Austrian bank. In other organizations some aspects (for example costs) will be higher or lower.

criteria	small project	project	programme
strategic importance	net present value: at least EUR 50,000	net present value: at least EUR 50,000	net present value: at least EUR 250,000
duration	at least 2 months	at least 3 months	at least 12 months
organizations involved	at least 3 organizations	at least 5 organizations	at least 7 organizations
resources	at least 100 person-days	at least 200 person-days	at least 500 person-days
costs	at least EUR 0.1 Mio	at least EUR 0.5 Mio	at least EUR 1 Mio

Fig. B1.3: Project and programme definitions of a large Austrian bank

B 1.3 Differentiation between projects and small projects

One possibility for organizational differentiation in project-oriented organizations is the differentiation between projects and small projects. Projects with a lower level of complexity, such as the performance of an event, the development of a brochure or the completion of a small contract, can be defined as small projects. Fewer project management methods are used for small projects than for projects and there is less detail in the project plans than there is with projects. It is usually sufficient to segment the work breakdown structure only to the third level.

For small projects a less differentiated design of the project organization will suffice than for projects. The role of the "project owner" will be filled by one person instead of a team. Sub-teams will probably not be necessary. The project marketing is less extensive for small projects than for projects.

An example of the organizational consequences of this differentiation is shown in Figure B1.4. This is an extract from the checklist "Application of Project Management Methods" from a large Austrian bank.

Chapter B

methods for the project start process	small project	project
methods for project planning		
project scope planning		
• project objectives plan	must	must
• objects of consideration plan	can	can
• work breakdown structure	must	must
• work package specifications	can	must
project scheduling		
• project milestone plan	must	must
• project bar chart	can	must
• CPM schedule	can	can
project resource and cost planning		
• project resource plan	can	can
• project cost plan	must	must
• project income plan	can	can
designing the project context relationships		
• project environment analysis	must	must
• business case analysis	can	must
• project other projects analysis	can	must
• pre- and post-project phase analysis	can	must
• project presentations, project vernissage	can	Kann
designing the project organization		
• project assignment	must	must
• project partial assignment	can	can

methods for the project start process	small project	project
• project organizational chart	must	must
• project role descriptions	must	must
• project responsibility matrix	can	can
• project communications structures	must	must
• project rules	can	must

Fig. B1.4: Extract from the checklist from a large Austrian bank: Application of Project Management Methods

B 1.4   Organization for routine business processes

The differentiation of business processes of different scopes in project-oriented organizations makes it possible to use the appropriate organizations for their performance. Projects are not suitable for the performance of all business processes. Routine business processes are to be taken over either by units of the permanent organization or by work groups.

In order to ensure the organizational advantage of projects, business processes which are fulfilled as projects should be differentiated from business processes which are not project-worthy. Organizations for the fulfillment of routine business processes are the permanent organization as well as work groups.

Permanent organizations of the project-oriented organization are, for example, the departments of profit centers and service centers. Routine processes are, for example, procurement processes, production processes and accounting processes.

Work groups are groups of three to eight persons, are limited in time and are formed to fulfill special assignments. Typical assignments for work groups are, for example, the analysis of weak points in a business process, the creation of a marketing concept or the improvement of the quality in a business process. Work groups are usually used short-term and work in a less formal manner than projects.

Through differentiation in the organization for the performance of business processes of different characteristics, it is possible to have a differentiated personnel management. For projects and programmes on the one hand, and for the permanent organization and work groups on the other, it is necessary to have different roles with different career paths, training and also payment and bonus systems.

The professional performance of routine business processes can be ensured through business process management (see Excursus B2: Business process management).

⋯⫶ EXCURSUS B2: Business process management

Definition: Business process

"By process, in its organizational meaning, one understands a collection of integrated, functionally overlapping tasks with

- measurable input,
- measurable added value and
- measurable output

with which a product is created and/or a service is provided, which meets the requirements of the internal and/or external customers."[a]

A business process is a sequence of tasks with clear boundaries which are performed by several organizations. A business is a social construct. The elements of business processes are tasks and decisions and their relationships.

CHAPTER

B

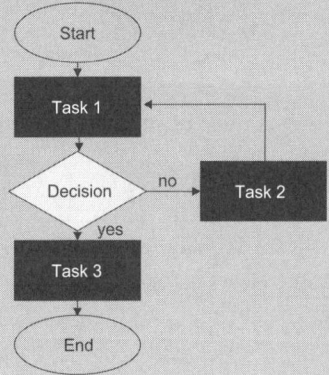

Fig. B1.5: Elements of business processes

A process runs "horizontally" through one or more organizations.

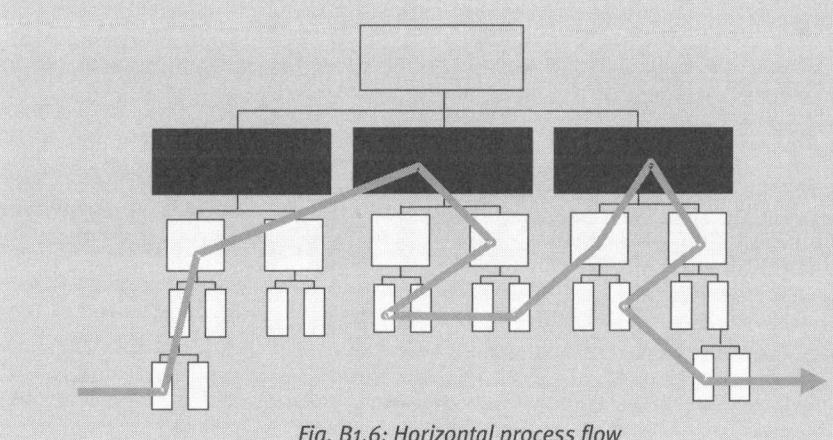

Fig. B1.6: Horizontal process flow

Chains of business processes

There are relationships between business processes. A business process can be performed before or after other processes; a business process can be performed in parallel with other business processes. These relationships between processes can be illustrated in chains of business processes.

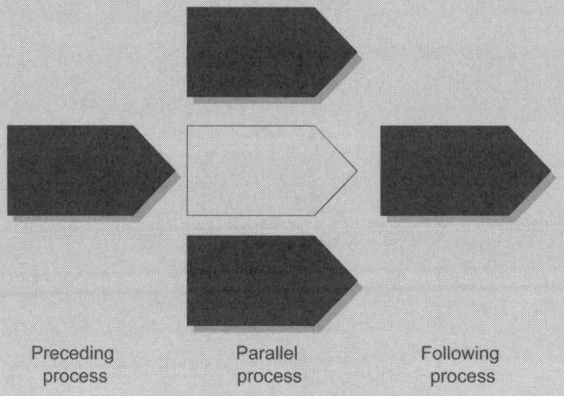

| Preceding process | Parallel process | Following process |

Fig. B1.7: Chain of business processes

Objectives of business process management

Business process management includes the planning, organization and controlling of processes. An integrative process consideration, team orientation, concentration on core competencies, elimination of non-value producing tasks and a minimization of the process costs should take place.[(b)]

Approaches to a process-oriented management can be found in lean management, total quality management and business process re-engineering. The change of paradigm in management, which is characterized by customer orientation, teamwork and networking with suppliers and partners, promotes thinking in business processes. The ISO 9000 standards also promote a process-oriented management approach.

The objectives of business process management are the definition, description and optimization of business processes, in order to increase the efficiency of the organization. Business process management should dissolve unnecessary interfaces and reduce the requirements for supervision.

Business process management leads through a dynamic and team-oriented approach to a new understanding of management. Business process management strives not only for decision support in the process optimization, but also for influencing the behavior of those performing the process. Business process management also promotes organizational learning in an organization.

The tasks of business process management

Business process management tasks have to be performed at a macro- and at a micro-level. The macro-process management tasks are:

- the identification of business processes to be considered and the selection of process names,
- the dissolving of obsolete business processes and the construction of new business processes, and
- the establishment of standards for business process analysis, business process optimization and business process documentation.

The identification of relevant business processes is a social construct. It can take place intuitively and/or analytically. When using the analytical method one can orient oneself on the "identity model" of organizations (see Chapter B1.6). The process identification should include the definition of

- the process boundaries (process objectives, process start and process end) and
- the process context (for example, the other business processes in a chain of processes).

The micro-process management tasks are:

- the process structuring, the planning of the process costs,
- the design of the process organization and

Hammer and *Champy* stress the importance of the definition of the start and end events of processes[c]. The establishment of process boundaries includes the definition of the start and end events as well as the definition of the objectives and the non-objectives of a process. The duration of a process is called its cycle-time.

In process structuring the business process is segmented into tasks and decisions and the logical sequence of the process is depicted. Selected tasks are specified in more detail. The process structure can be shown with the help of flow-charts or responsibility matrices. These can be created in different levels of detail. The symbols in the diagrams are to be used consistently.

Cost planning is a central method of business process optimization. Cost planning takes a prominent place in reference works related to process management. Particularly in earlier works, process management is understood from the aspect of cost planning.[d] The process of cost planning is used to improve transparency, and also to make organizations more efficient.

When designing the process organization the roles responsible for the tasks and decisions and their relationships will be defined. To do so, role descriptions and responsibility matrices can be used. For each task organizational tools, such as checklists and forms, can be defined.

For process controlling, the key values, such as process costs, process duration and qualitative measures, such as customer satisfaction or employee satisfaction, can be used. The objective of controlling is process optimization through the reduction of interfaces, the minimization of costs, the outsourcing of services, etc.

> **Organization of business process management**
>
> For the management of individual processes a "process owner" and possibly a "process management team" can be defined. The work of the process owner and a process management team relates to the reflection and optimization of a process and not to its performance.
>
> For process management meetings of the process team members or workshops can be performed in which the process management team and representatives of relevant environments participate.

[a] Cp. *Fischer, T.M.*, (Sicherung unternehmerischer Wettbewerbsvorteile) 1993, p. 312
[b] Cp. *Gaitanides, M.*, (Prozessmanagement) 1994, p. 3
[c] Cp. *Krüger, W.*, (Organisation der Unternehmung) 1993, p. 126
[d] Cp. e.g. *Witt, F.* (Aktivitätscontrolling) 1991, or Eschenbach, R. et al., (Prozessmanagement) 1993, p. 70 ff

CHAPTER

B

B 1.5 Relationships between projects, business processes, investments and objects

To clarify the definition of what a project is the differentiation between projects, business processes, investments and objects is helpful. The relationships between projects, business processes, investments and objects can then be analyzed.

The relationships between business processes and investments

A business process is a clearly defined sequence of tasks in which several roles of one or more organizations are involved.

Elements of a business process are the tasks to be fulfilled, decisions to be made, and the relationship between the tasks and the decisions. Business processes can be differentiated into primary processes, secondary processes and tertiary processes.

This differentiation is made according to the extent of customer orientation of the business process, whereby customer orientation is largest in primary processes. Primary processes are business processes of the performance of deliverables for customers. Secondary processes support the primary processes directly, while tertiary processes do not support them. Typical primary processes of an ICT company, for example, are developing an offer, contracting and after sales service. Typical secondary processes are product development, performing a marketing campaign or organizing a marketing event. Typical tertiary processes are strategic planning, periodical controlling or the annual balance sheet.

Investments are long-term employment of capital in assets, for example in machinery or buildings, but also in customer relationships, in products, in the organization or in personnel.

In accordance with the identity model of organizations (see Excursus B3: Identity of an organization) investments can be divided into

- customer relationship-related investments,
- product and/or market-related investments,
- infrastructure investments,
- organizational investments,
- personnel-related investments and
- environment relationship-related investments.

⋯⋯⋗ EXCURSUS B3: Identity of an organization

The identity of an organization is determined by:

- the strategies and objectives of the organization,
- the services of the organization and its markets,
- the organizational structure, the business processes, and the organizational culture,
- the personnel (quantitative and qualitative) of the organization,
- the infrastructure (buildings, plants, IT, telecommunications) of the organization,
- the budget and the financing of the organization and
- the context of the organization.

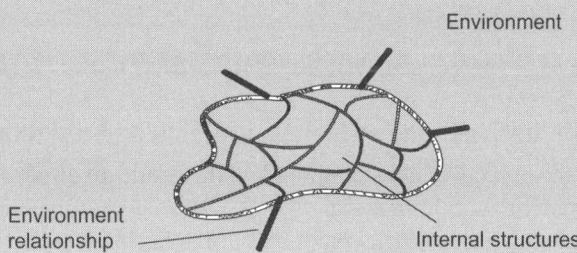

Fig. B1.8: Internal structures and environment of an organization

Organizations differentiate themselves from their environment through their boundaries. On the other hand, they are only to be understood within their context, upon which they are dependant. Context dimensions are, for example, the relevant social environments, the history and the expectations regarding the future of the organization.

The relevant social environments of an organization are, for example, customers, suppliers, competitors and the media, but also "internal environments", such as employees, union representatives or the board of directors. Every organization is shaped by its history. Many "peculiarities" of an organization can only be understood and interpreted through the events of the past. On the other hand, the expectations regarding the future determine the current decisions.

The relationships between investments and projects

Several business processes are combined in an investment process. The investment in an industrial plant combines, for example, the business processes of developing a feasibility study, planning the plant, construction and commissioning the plant, use, maintenance and decommissioning of the plant.

Investments can be initialized by a project and/or a programme. Projects can segment the investment process.

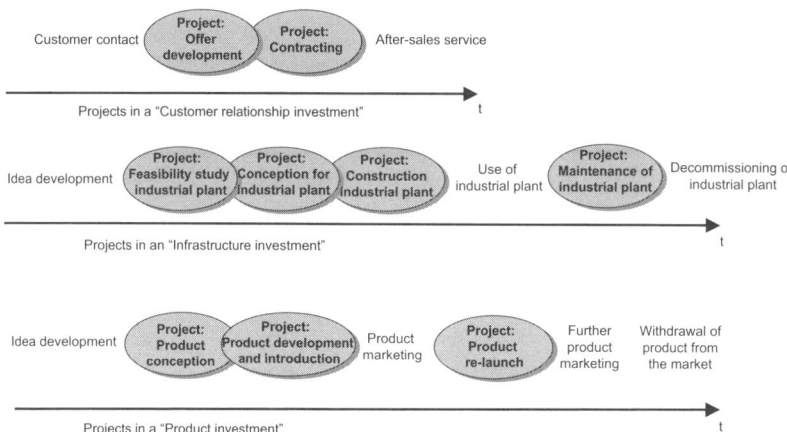

Fig. B1.9: Segmentation of investment processes through projects

The relationship between investments, projects and objects

Objects are both the objects of consideration and the results of an investment. They can be divided into material and immaterial objects. The material object of the investment in an industrial plant is the industrial plant. All of the business processes to be fulfilled in the framework of this investment are related to this object.

A project is to be differentiated from the object which results from the project. Projects are not only to be labeled with the object name (e.g. "Product XY") but also with their function (e.g. "Development of product XY").

B 1.6   Continuous and discontinuous development of projects and programmes

The development of projects and programmes can take place continuously and discontinuously.

Levy and *Merry* offer a development model for organizations which differentiates between "first order change" and "second order change".[4] The first order change leads to quantitative, content-related and gradual change, the second order change, however, is characterized by qualitative, sudden changes. This leads to a change of identity in the affected organization..

First Order Change	Second Order Change
• a change in few dimensions	• a change in many dimensions
• a change at one or few levels	• a change at many levels
• a change in one or two behavioral aspects	• changes in many behavioral aspects
• a quantitative change	• a qualitative change
• continuity	• discontinuity
• does not change the identity	• changes the identity, the paradigm

Fig. B1.10: First order change and second order change according to Levy and Merry

A continuous development of projects takes place, for example, because of a change in the project objectives, a change in the scope, a requirement for additional resources, etc. This requirement for continuous development is taken into account in project controlling and results in the adaptation of the existing project plans.

A discontinuous development of a project, for example in the case of a crisis, occurs by considering new technological developments, the loss of a strategic partner or the loss of the project owner. These situations can lead to a change in the project identity through the creation of completely new project structures.

Discontinuities of projects and programmes

A discontinuity of an organization can be understood as a bifurcation, that is, a phase of instability in which the organization's possible paths of development branch out. Discontinuities offer the possibility of a higher development, but they also offer the possibility of a catastrophe. In the end, only one of these paths of development can be realized.[5]

Discontinuities of organizations can occur due to various causes. They can occur due to changes in the relationships of the organization to its relevant environments or through the dynamics of the organization itself.

When the organization interprets a situation as an existential threat, then it is referred to as a crisis.

A crisis is defined as a phase of instability in the development of an organization, which is characterized by an existential threat to the organization in question and by the ambivalence of the outcome. The definition of a crisis is a social construction process.

---

4) Cp. *Klimecki, R.* et al., (Systementwicklung als Managementproblem) 1991, p. 103-162
5) Cp. *Paslack, R.,* (Urgeschichte der Selbstorganisation) 1991, p. 91

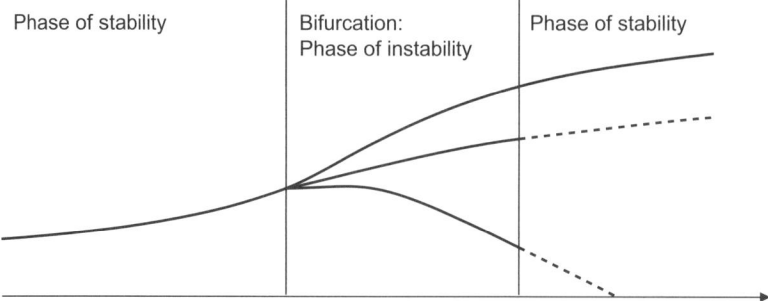

| Phase of stability | Bifurcation: Phase of instability | Phase of stability |

Fig. B1.11: Discontinuity as bifurcation in the development of organizations

To provide orientation in a project-oriented organization it is possible to add the definition of a project crisis to the guidelines for project and programme management.

Definitions: Project crisis and project chance

- A project crisis is characterized by the following features:
  - negative business case for the investments initiated by the project
  - project costs exceeded by 50%
  - customer unable to pay
  - a strategic partner drops out
- A project chance is characterized by the following features:
  - customer contract expanded by more than 50% of the project turnover
  - possible use of substantially more efficient methods/technologies/resources

Fig. B1.12: : Criteria for defining a project crisis (from the guidelines for project and programme management of an international chemical company based in Austria)

A project crisis leads to a change in the identity of the project. After the crisis the structures of the project and the relationships to its relevant environments are different.

Catastrophes are different from crises in that during a catastrophe the structures of the organization in question are destroyed, it is no longer possible to achieve objectives and the outcome is not ambivalent anymore.[6] When the strategies and actions for resolving a crisis do not work this can lead to a catastrophe. Thus, a project crisis which has not

---

6) Hundreds of passengers died in a crash of a Lauda-Air jet due to a technical problem with the thrust reverser of the plane's engines. The crash was a catastrophe for the passengers who died. For Lauda-Air the crash led to a crisis in the company.

been resolved can, for example, lead to the cancellation of the project, which is a catastrophe for the project.

Positive causes can also lead to a discontinuity, which is considered a chance. A chance is also a phase of instability for an organization which can lead to a change in the identity. The difference between a chance and a crisis is the positive cause and existing potential for the further development of the organization, as opposed to existential threat. Reasons for a project chance could be, for example, technological innovations which would have a positive effect on the project, the availability of additional highly qualified resources, etc.

Structurally determined changes in identity can also occur in projects and programmes. A requirement for a structurally determined identity change exists, for example, in an IT-project at the beginning of the roll-out phase, after the end of product development or prototyping. In an engineering project, too, at the beginning of the construction phase after engineering, procurement and production a structurally determined discontinuity exists. In both situations, fundamentally new structures in the project organization and project culture are required.

As opposed to project crises or project chances, a structurally determined identity change in a project is predictable. A structurally determined identity change can already be planned in the project start process. No additional project roles are required to manage the structurally determined identity change. It can be performed by the existing members of the project organization.

CHAPTER B

B 2   Project management approaches

Project management approaches can be differentiated by the way in which projects are perceived. Traditional, method-oriented project management approaches are based on the perception of projects as tasks with special characteristics.

The systemic and process-oriented project management approach described here, ROLAND GAREIS Project and Programme Management®, is based on the perception of projects as temporary organizations and as social systems.

B 2.1   Traditional project management

The traditional perception of projects as tasks with special characteristics promotes the planning orientation in project management.[7] The main focus is on how an assignment is to be performed. Methods for work planning and work organization, such as the REFA methods[8] or methods of operations research[9], represent the theoretical basis of traditional project management.

---

7) Cp. *Steinle* et al., (Instrument moderner Dienstleistung) 1995, p. 354
8) Cp. *Camra, J.J.*, (REFA-Lexikon) 1976
9) Cp. *Hillier, F.S.*, (Operations Research) 2001

For decades, project management was understood as the use of project scheduling methods, such as CPM and PERT, for scheduling projects as well as supporting resource and cost planning. Because of the CPM-based risks tied to unique tasks traditional project management uses methods for risk management as well as for controlling the project progress, project schedule, project resources and project costs.

Only through the definition of non-technical projects, such as marketing and organizational development projects, and the consideration of additional disciplines (such as organization, marketing, controlling), methods were introduced which were easy to use and to communicate, such as the work breakdown structure.

Organizationally, it appears that the most important element in traditional project management is the division of formal authorities between the project manager, the immediate supervisor of the project team member, and the team member. As possible solutions to this the pure project organization, the matrix project organization and the influence project organization are offered as standards.[10]

Chapter

B

The project management tasks are defined as the planning, the controlling and the organizing of projects. In traditional project management the objects of consideration of project management are the scope, the schedule and the costs. The relationships between these objects of consideration are depicted as the "magic triangle".

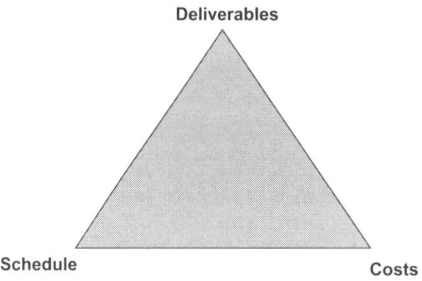

Fig. B2.1: Traditional objects of consideration of project management   ("magic triangle")

B 2.2   ROLAND GAREIS Project and Programme Management®

The influences of organization theory

The perception of projects as temporary organizations promotes the awareness that every project requires a specific organizational design which goes beyond the definition of the formal authority of the project manager. In addition to project planning, a situational design of the project organization should contribute to the success of the project.

The organizational design of projects includes the definition of project-specific roles, the development of project organizational charts, the definition of project-specific com-

---

10) Cp. *Reschke, H.*, (Aufbauorganisation) 1989, p. 874 ff

munication structures and the agreement on project-specific rules. Through the tempo-rary character of projects the designing of the project start and the project close-down obtain a special importance.

Relatively new management approaches, such as customer orientation, "empower-ment", flat organizational structures, team work, organizational learning, process orien-tation and networking, can be implemented in projects to contribute to project success. The management approaches "learning organization", "lean management", "process management" and "total quality management" are therefore to be seen as an additional, new theoretical basis for project management.

The perception of projects as temporary organizations also promotes the development of a project-specific culture. Such project management methods are, for example, the choice of the project name, and the formulation of the project mission statement and of project-specific slogans.

The influences of social systems theory

The perception of projects as social systems enables the use of views and models of so-cial systems theory for project management. A "systemic" project management builds not on traditional project management, but puts its methods into a new framework, in-terprets them and promotes the development of new project management methods.

Because of the need to manage the boundaries and the context as well as the complexity and the dynamics of projects, new potential and challenges arise for project manage-ment. A new understanding of the project management tasks to be fulfilled is enabled: Instead of planning, controlling and organizing the project the tasks of constructing the project boundaries and the project context, building up and reducing the project com-plexity and managing the dynamics of the project become relevant.

- **Construction of the project boundaries and the project context**

The construction of the project boundaries and the project context ensures a holistic view of the project. The definition of the project boundaries should enable an integrated con-sideration of technical, organizational, personnel and marketing objectives in the project.

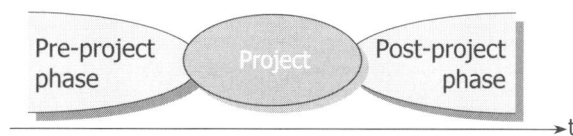

Fig. B2.2: Project boundaries and project context (time dimension)

For a detailed management of the project boundaries the following project management methods are available: Project objectives plan, objects of consideration plan and work breakdown structure, project schedule and project resource plan, project costs and project income plan, project organization, etc.

For an analysis of the project context and the design of project context relationships the project environment analysis, the analysis of the pre- and post-project phases, the business case analysis, as well as the analysis of the relationship of the project to other projects and to the company strategies can be used.

- **Building up and reducing project complexity**

Projects require a certain amount of complexity in order to be able to relate to the (infinitely) complex environment. The building up and reducing of complexity is a project management task.

A holistic project view, creativity in the project and the acceptance of project-related decisions can be ensured through adequate communication structures. The performance of project workshops at the project start process, at milestones and at the close-down process of the project, as well as the performance of project team and project owner meetings, promote the building up of complexity in a project. The differentiation of project roles, the definition of the relationships between the roles as well as the inclusion of different specialist disciplines and hierarchical levels in the project team are further organizational possibilities for the building up of complexity in a project.

By using different project management methods different perspectives for designing a project are chosen. Only the linking up of these different views in a "multi-methods approach" enables appropriate consideration of the project complexity.

To ensure continuity in a project, redundant structures should be created. A reducing of project complexity is achieved by the agreement on the project objectives within the project team. Furthermore, the use of project management standards, the establishment of project-specific rules and norms, the development of project plans as well as the performance of integrative project team meetings gives repeated orientation for the project work.

- **Management of the project dynamics**

The dynamics of a project result from the interventions of relevant environments as well as through the self-reference of the project. Examples of interventions from relevant project environments are new legal requirements from public authorities, a change in scope by the customer, cancellations from suppliers, an unexpected media response, a demotivated project team, etc.

The formal communication structures of a project enable its self-reference. Project management methods, such as the work breakdown structure, the milestone plan and the project environment analysis can support the communication in the project. The possibility of change in a project is dependent upon its relationship to relevant environments. Only when the functionality of the (relative) project autonomy is recognized and, therefore, the interventions of the permanent organization of the project-oriented organization are limited, is there a possibility of self-reference.

In order to promote change in a project, reflections and meta-communications, i.e. communications about communications, are necessary. Time, space and the corresponding know-how are all necessary for reflection. In a cyclical process the structures necessary for the performance of a project are formed, questioned and possibly adapted according to the new requirements.

CHAPTER
B

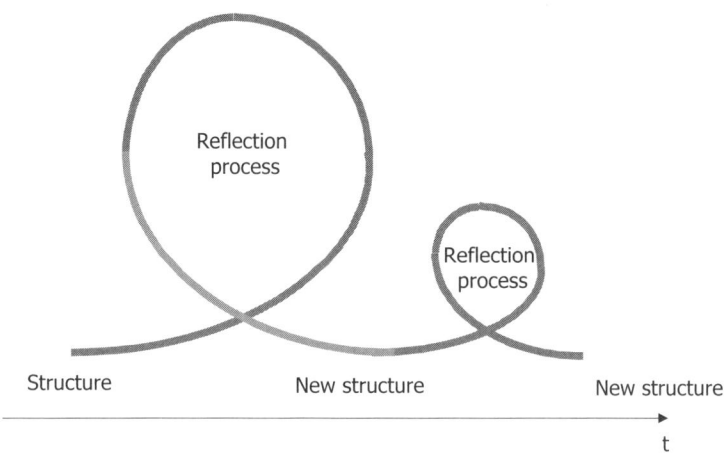

Fig. B2.3: Management of the dynamics in projects

Self-referencing processes in a project, or interventions from project environments, can lead to continuous or discontinuous changes in a project. Continuous changes in projects are considered in project controlling. Continuous changes in projects take the form of adaptations in the project structures, such as new project slogans, new formations of re-lationships to relevant environments, new definitions of project roles, new demands on the project team members, new planning of the scope and the project schedule, etc.

A discontinuous development in a project comes about when a change in the project identity takes place. This can result from a substantial deviation from the project objec-tives. A project discontinuity can take the form of a project crisis, a project chance or a structurally determined change in the project identity.

Process-oriented project management

A method-oriented project management focuses on the project management methods. The use of methods for planning and controlling project scope, project schedule, project resources and project costs, etc., is understood as project management. The success of project management is assessed on the basis method application. Competence for the application of the project management methods is achieved by training. There is a sup-position that good knowledge of methods ensures good project management.

ROLAND GAREIS Project and Programme Management®, as a process-oriented project management approach, defines project management as a business process of the project-oriented organization and focuses on its sub-processes. The project management per-sonnel requires competencies for managing the sub-processes project start, continuous project coordination, project controlling, project close-down and possibly resolving a project discontinuity. The success of project management is assessed on the basis of the professional performance of these processes, not on the basis of a project handbook that meets all formal demands. In doing so, the relationships between the sub-processes must also be considered and optimized.

For the performance of the individual project management sub-processes the corresponding project management methods are used. Their meaning does not get lost. The definition of the sub-processes of project management adds an integration level for ensuring the professional application of project management methods. Producing an optimal project schedule cannot be an objective in itself, but it must be an overall integrative objective to start the project in an optimal way.

The management of project objectives, the management of the project schedule, the management of the project cost planning, etc. cannot be accepted as project management processes[11], since only an integrated consideration of all methods of project management can lead to optimal results. The management of project plans as "processes" cannot ensure a holistic management.

CHAPTER

B

B 2.3   Definition: Project management

Project management can be defined in functional and in institutional terms.

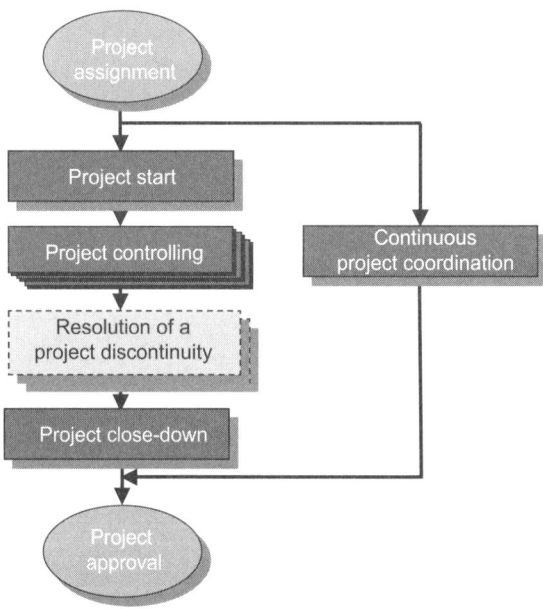

Fig. B2.4: The business process "project management"

---

11) Cp. *Duncan, W.R.*, (Body of Knowledge) 2000, p. 47 ff

In functional terms, project management is a business process of the project-oriented organization which contains the sub-processes project start, continuous project coordination, project controlling and project close-down. Project management may also contain the resolution of a project discontinuity (project crisis, project chance and structurally determined project identity change).

The project management process is to be differentiated form the business processes for the fulfillment of project deliverables. Therefore, project content-related processes such as procurement, the engineering of components and the testing of software, for example, are not management tasks.

The objects of consideration of project management as defined by traditional project management are to be added to the perception of projects as temporary organizations and social systems. The objects of consideration of project management are the project objectives, the project scope, the project schedule, the project resources, the project costs and project income, the project risks as well as the project organization, the project culture and the project context.

Fig. B2.5: Objects of consideration of project management

Roles representing project management institutionally are the project owner, the project manager, and the project team.

B 2.4   International project management approaches

Project-oriented companies develop company-internal guidelines for project and programme management, cooperate with other companies and with suppliers of training and consulting services, and in doing so apply (more or less consciously) different project management approaches. For the efficient formation of these cooperations it should be clarified which approach, on what theoretical basis, will be applied. A differing project and project management understanding can lead to grave misunderstandings and conflicts in a cooperation.

A project management approach applied can be original, i.e. an organization's own development, or it can be taken over from another source. Many companies manage projects on the basis of well known project management approaches, such as

- PM BoK® (Project Management Body of Knowledge) by PMI – Project Management Institute,
- ICB International Competence Baseline by the IPMA – International Project Management Association,
- PRINCE2® by OGC – the UK Office of Government Commerce or
- PROPS® by Ericsson.

User groups have been established which acquire the rights to use a particular project management approach; networks of project management trainers have been established which use the same approach and exchange experiences in relation to it.

In order to recognize a methodology as an approach several requirements must be fulfilled. Most importantly, the approach must

- be original and therefore clearly differentiated from other approaches,
- have a name ("label") in order to be communicated,
- be protected by a trademark,
- be documented in order to ensure its traceability,
- be communicated in the "community" and
- be accepted in the "community" as an approach.

When these requirements are not fulfilled it is a case of practiced methodology which contains elements from one or more approaches, but is not an approach in its own right.

ROLAND GAREIS Project and Programme Management® is a generic project management approach, i.e. it can be used for all types of projects in all branches. A central element of the originality of ROLAND GAREIS Project and Programme Management® is its process orientation. Project management and programme management are understood as business processes of project-oriented companies, whose outputs and quality are measurable.

In Figure B2.6 the systemic and process-oriented approach *ROLAND GAREIS Project and Programme Management*® is compared with the traditional, method-oriented PM Bok® approach by PMI.

Criterion	ROLAND GAREIS Project and Programme Management®	Project Management Body of Knowledge®
project definition	differentiation between small projects, projects, programmes	everything is a project
perception of a project	a temporary organization, a social system	a relatively unique, clearly defined task
project management understanding	project management as a business process	project management as a set of methods
differentiation between project and programme management	differentiation made	differentiation not made
use of project management methods	for supporting the project communication	for project control
objects of consideration of project management	objectives, objects, scope, schedule, costs, income, resources, roles, organization, culture, context, business case	scope, schedule, costs
specifics	"empowered" and "integrated" project organization; project culture; project context	contract administration, procurement
project success	optimization of the business case of an investment initialized by a project	meeting the objectives for scope, schedule and costs
management of the project-oriented organization	differentiation to project and programme management	no differentiation to project and programme management

Fig. B2.6: ROLAND GAREIS Project and Programme Management® compared to PM BoK®

CHAPTER

B

B 3   Managing different types of projects

Project types can be differentiated by branch, location, content, investment phase, degree of repetition, duration and relationship to business processes.

The differentiation of projects into different types makes it possible to analyze specific challenges and potentials for project management and to develop standard project plans.

Criterion for differentiation	Project type
branch	• construction, engineering, IT, pharmaceutical, NPO, etc.
location	• national, international
content	• customer relations (contracts), products and markets, infrastructure, personnel, organization
investment phase	• study, conception, realization, re-launch or maintenance
degree of repetition	• unique, repetitive
customer	• internal customer, external customer
duration	• short-, medium-, long-term
relation to business processes	• primary, secondary, tertiary processes

Fig. B3.1: Differentiation of project types

When differentiating projects by **branch** it is possible to distinguish between construction, engineering, IT or pharmaceutical projects, for example. For projects in different branches the project team members require branch-specific technology and market knowledge. A development of branch-specific project management careers (such as IT-project manager) is possible.

For the **location** of the project performance it is possible to distinguish between national and international projects. For international projects, specific personnel and organizational requirements are to be developed. The mobility of the project team members and their knowledge of foreign languages must be ensured, as well as the development of project documentation in a foreign language. Furthermore, different national cultures in the project team as well as potentially differing time zones need to be considered.

In a differentiation by **project content** projects can be differentiated into customer relations-related projects, product and market-related projects, infrastructure-related projects, as well as personnel and organization-related projects. For projects with different contents not only different competencies of the team members may be necessary, but different project cultures as well. Thus, contracting projects, for example, are characterized by a higher commitment than organizational development projects. Furthermore, project phases specific to each project type can be defined. For example, there exist procedures and project type-specific project management approaches for software and product development projects, in which all project phase milestones and detailed work package specifications are documented.

As regards the **phase of an investment process** for which a project is performed it is possible to distinguish between conception and realization projects. In order to reduce the risk of bad investments and to optimize the quality of investment decisions, it is advisable to create a concept, e.g. before the construction of a new industrial plant, or the realization of a new EDP system. The process of creating a concept can already be so complex and strategically important that it is advisable to do so in the form of a project. Conception projects are characterized by a high demand for openness with regard to content, for creativity and for project marketing. In case the decision makers decide on continuing the work at the end of a conception project, a chain of projects with a conception project and a realization project is created.

As regards the **degree of repetition** in the business processes to be performed by the project it is possible to distinguish between unique and repetitive projects. Attaining a quality certificate according to ISO 9000 is, for example, a unique project for each company. The performance of a contracting project by a construction company is a repetitive project since all contracting projects have fundamentally the same processes (design, procurement, preparing the construction site, construction, etc.). It is nevertheless recommended that (socially) complex contracts (new customers, partially new suppliers or partners, etc.) be performed in the form of a project.

For repetitive projects, as opposed to unique projects, it is possible to standardize some project management methods (such as work breakdown structure, milestone plan, cost plan). Less creativity is required for repetitive projects than for unique projects. The use of creativity techniques and multi-disciplinary teams for innovative problem solving are, therefore, primarily necessary for unique projects. In repetitive projects there is a threat that the project performance takes on a routine character, that standard project plans will not be questioned and adapted and that no project autonomy is possible.

A special form of unique project is the pilot project. Since it is assumed that after the performance of the pilot project repetitive projects with similar objectives will be performed, organizational and individual learning are explicit objectives of the pilot project. The organization of learning in relation to the technology employed, the market conditions, etc. are, therefore, part of the project content.

The differentiation between internal and external projects is made on the basis of different **customers**. In external projects (contracting) a customer external to the company assigns the company to perform a service in return for payment. The objective of an internal project, however, is the solution of a company-internal problem for an internal customer, without payment.

Only complex contracts are to be performed as projects. For less complex contracts which are small in scope, such as the delivery of components, technical planning etc., the use of project management is not necessary.

Fundamentally, all projects are internal projects except for the contracting projects. In external projects usually the objectives are more clearly defined than internal projects

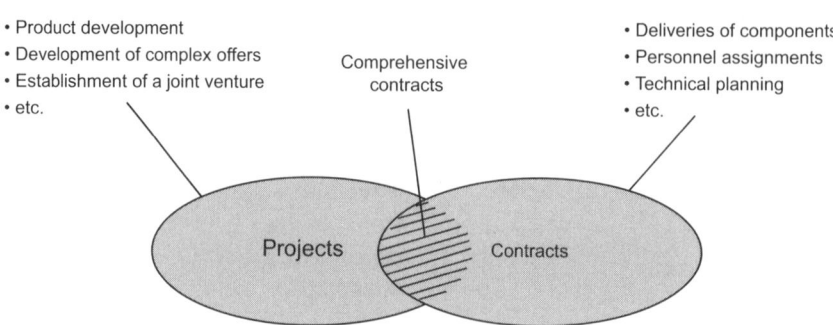

- Product development
- Development of complex offers
- Establishment of a joint venture
- etc.

Comprehensive
contracts

- Deliveries of components
- Personnel assignments
- Technical planning
- etc.

Projects   Contracts

Fig. B3.2: Relationship between contracts and projects

due to the extensive preparations during developing an offer. This difference should be less important in the future due to the frequent performance of conception projects. As for the setting of priorities, external projects are usually given precedence. This is surprising inasmuch as internal projects usually have a higher strategic importance than external projects.

As regards the **project duration** it is possible to distinguish between short-term, medium-term and long-term projects. Since no project should be shorter than six weeks or longer than twelve months, projects with a duration of 6-12 weeks are defined as short-term, 3-6 months as medium-term and 7-12 months as long-term. Projects for the realization of infrastructure (buildings, industrial plants, etc.) can last longer than twelve months. The project start and the project close-down of short- and medium-term projects are to be performed as quickly as possible, the frequency of formal project controlling is to be kept low.

In long-term infrastructure projects one should be aware of possible fluctuations in the project team. Changes in the project due to possible technological developments and new laws should be provided for.

··⫶> CASE STUDY B1: Identification of the project type

The establishment of a publishing house to publish a monthly business magazine was perceived by the managing directors – who, in their understanding, were editors – as a product development and product marketing project[a] The development of the magazine's content and layout as well as its marketing were, therefore, given special attention. In these areas, the magazine was a success.

Nevertheless, the project led the publishing house to insolvency. The mid-term financing was not ensured, there was no provision for the possibility of one of the managing directors leaving, not enough attention was given to the development of the corresponding legal framework.

> Would these actions have been taken, had the project been identified as a company start-up project?! This more extensive project understanding would have resulted in the definition of much broader project objectives and objects of consideration.

[a] Cp. *Doujak, A.*, (Krise ist wie Krieg) 1994, p. 342 ff

If projects are placed in relation to the **business processes** to be fulfilled in companies, it is possible to differentiate projects into the performance of primary, secondary and tertiary processes.

Projects for the performance of primary processes are offer projects and contracting projects. Projects for the performance of secondary processes can be, for example, product development or advertising campaigns. Projects for the performance of tertiary processes can be, for example, reorganizations or the introduction of an EDP system.

In the past, projects in different branches (such as construction, engineering, IT) were defined mostly with regard to the performance of primary processes, i.e. for the performance of contracts. Only in recent years a general project orientation can be observed and with it the use of projects for the performance of secondary and tertiary processes.

CHAPTER

B

References

Camra, J. J., (REFA-Lexikon) REFA-Lexikon, 2. Auflage, Beuth, Berlin, 1976

Doujak, A., Doujak, G., (Krise ist wie Krieg) Krise ist wie Krieg, in: *Gareis, Roland* (Ed.), Erfolgsfaktor Krise, Signum Verlag, Wien, 1994

Project Management Institute, (Body of Knowledge) A Guide to Project Management Body of Knowledge, Upper Darby, Pa. USA, 1996

Eschenbach, R./Kunesch, H., Prozessmanagement: Instrumente des Prozessmanagements, Wien, 1993

Fischer, T. M., (Sicherung unternehmerischer Wettbewerbsvorteile) Sicherung unternehmerischer Wettbewerbsvorteile durch Prozess- und Schnittstellen-Management, in: Zeitschrift für Organisation 5/1993

Gaitanides, M., (Prozessmanagement) Prozeßmanagement – Konzepte, Umsetzungen und Erfahrungen des Reengineering, Hanser, München, Wien, 1994

Hill, W., Fehlbaum, R., Ulrich, P., (Organisationslehre 1) Organisationslehre 1: Ziele, Instrumente und Bedingungen der Organisation sozialer Systeme, 5. Auflage Stuttgart, Bern: Verlag Paul Haupt (UTB), 1994

Hillier, F. S., Lieberman, G. J., (Operations Research) Introduction to operations research, McGraw-Hill, Boston, 2001

Kasper, H., (Organisierte Sozialsysteme) Die Handhabung des Neuen in organisierten Sozialsystemen, Springer, Wien 1990

Klimecki, R., Probst, G.; Eberl, P., (Systementwicklung als Managementproblem) Systementwicklung als Managementproblem, in: *Staehle, W.: Sydow, J.* (Ed.), Managementforschung, Band 1. – Berlin 1991

Kloock, J., (Flexible) Flexible Prozeßkostenrechnung und Deckungsbeitragsrechnung, in: Krp, 2/1993

Krüger, W., (Organisation der Unternehmung), Organisation der Unternehmung, Kohlhammer, Stuttgart, 1993

Luhmann, N., (Komplexität) Komplexität, In: *Grochla, E.* (Ed.), Handwörterbuch der Organisation, 2. Auflage, Poeschel Verlag, Stuttgart, 1980

Luhmann, N., (Soziale Systeme) Soziale Systeme: Grundriss einer allgemeinen Theorie, Suhrkamp Verlag, Frankfurt am Main, 1984

Luhmann, N., (Funktionen) Funktionen und Folgen formaler Organisation, Duncker und Humblot, Berlin, 1964

Motzel, E., (Projektmanagement Kanon) Projektmanagement Kanon – der deutsche Zugang zum Project Management Body of Knowledge, TÜV Verlag, Köln, 1998

Paslack, R., (Urgeschichte der Selbstorganisation) Urgeschichte der Selbstorganisation, Vieweg, Braunschweig, 1991

Reschke, H., (Aufbauorganisation) Formen der Aufbauorganisation in Projekten, in: *Reschke, H., Schelle, H., Schnopp, R.* (Ed.), Handbuch Projektmanagement, Band 2, TÜV Rheinland, Köln, 1989

Steinle, H., Bruch, H., Lawa, D. (Ed.), (Instrument moderner Dienstleistung) Projektmanagement: Instrument moderner Dienstleistung, Edition Blickbuch Wirtschaft, Frankfurt, 1995

Witt, F. J., (Aktivitätscontrolling) Aktivitätscontrolling und Prozesskostenmanagement, Poeschel, Stuttgart, 1991

CHAPTER

B

C

Project organization models and project roles

The adequate design of the project organization is the basis for a professional project performance. On the one hand, the organizational structure of the project is to be designed by the definition of project roles and the development of a project organization chart, and on the other hand, project communication structures are to be designed appropriate to the situation.

An "empowerment" of the project and the involvement of representatives of relevant project environments into the project organization enables an efficient project performance.

C Project organization models and project roles

Contents

C 1   Overview: Project organization

A prerequisite to understanding new developments in the design of project organizations is knowledge of the traditional organizational models, namely the influence project organization, the pure project organization and the matrix project organization.

By perceiving projects as temporary organizations an emphasis is placed on the design of the project organization.

The objects of consideration in the organizational design of projects are the organizational structure and the business processes. Elements of the organizational structure are the project roles and the relationships between the project roles, which can be depicted in a project organization chart. New organizational models for projects are described in this chapter.

Further methods for the design of the communication structures of projects are described in Chapter F1.7.

The business processes of projects are not described in this chapter.

It is possible to differentiate between business processes relating to the project content and the project management business process. Since the content business processes are dependent upon the project objectives and therefore vary for each project, they will not be dealt with here.

C 2   Traditional project organization models

Traditional project organization models, the influence project organization, the pure project organization and the matrix project organization, are differentiated by how the authorities are distributed between the project manager and the line manager of the department from which a project team member is recruited.

For the fulfillment of a work package by a project team member (or by a project contributor) decisions and directions from the project manager or the line manager are necessary. These are based on the formal authorities of these roles.

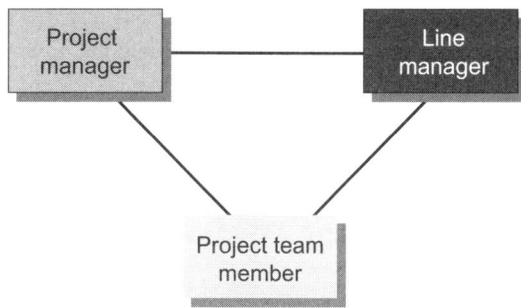

Fig. C2.1: Roles involved in the fulfillment of a work package

The formal authorities relevant for the formation of the project organization can be described with the help of the "6 questions":

What?	Authority to assign work packages and to control the progress of work packages.
How well?	Authority to control the quality of work packages.
Who?	Authority to select the person to perform to a work package.
How?	Authority to decide which methods, tools, etc. to apply to perform a work package.
For how much?	Authority to agree on project resources and costs for the performance of a work package and to control resources and costs.
When?	Authority to agree on the schedule for a work package and to control it.

Fig. C2.2: The 6 questions to define formal authorities in projects

The distribution of formal authorities between the project manager and the line manager can be regulated in different ways. These various distributions result in the different basic forms of the project organization, namely the influence project organization, the pure project organization and the matrix project organization.

C 2.1   The influence project organization

In the influence project organization the project manager has a staff function without formal authority. All formal authority lies in the departments of the permanent organization.

Since the project owner team expects the project manager to take on coordinating functions in order to ensure the project success, even without having formal authority, the project manager must use his/her informal competencies to do so. Informal competencies are to be understood as those which result from specialist knowledge, experience, personal relationships and friendships, informational advantage, the proximity to power and personal charisma. Informal competencies depend on the person, and not on the role. The distribution of formal authorities in the influence project organization can be depicted in a project organization chart. From this chart it is obvious that there are no lines to symbolize a formal authority between the project manager and the project team members..

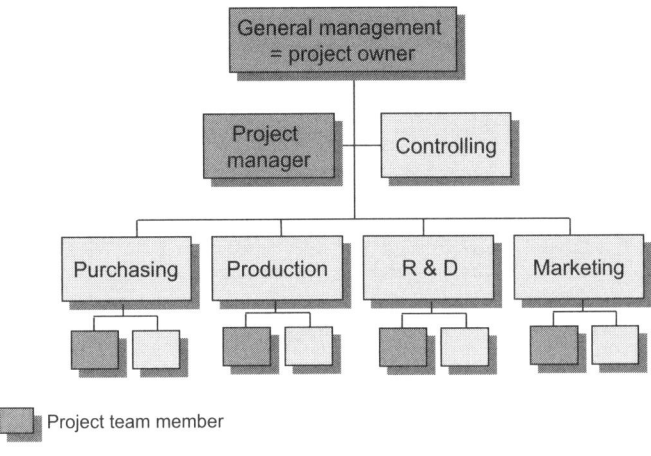

Fig. C2.3: Influence project organization

Advantages	Disadvantages
• lower organizational expense • the project team members formally report to their regular line managers in the permanent organization • the project team members work in their departments in a group of peers • no problems with recruiting and re-integrating staff • know-how is ensured in the departments of the permanent organization	• no formal authority of the project manager • dominance of the department's interests • little concentration on the project • slow decision-making

Fig. C2.4: Advantages and disadvantages of the influence project organization

C 2.2   Pure project organization

In a pure project organization the project manager has all formal authority over the project team members. Organizational independence of the project is mostly supported by the fact that the project team members are taken out of their various functional departments and moved together into a common workspace.

The graphic depiction of the pure project organization depicts an organization parallel to the permanent organization in which all project team members are under the guidance of the project manager in the form of a line organization.

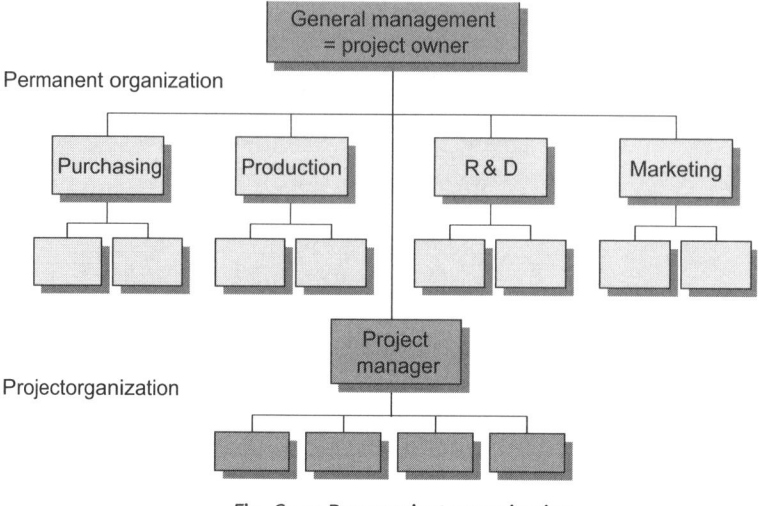

Fig. C2.5: Pure project organization

Advantages	Disadvantages
• all formal authority lies with the project manager • full concentration on the project • quick decision-making on the basis of short communications paths • strong identification with the project	• recruitment of project team members • full utilization of the members of the project organization not guaranteed • reintegration of the project team members into the departments of the permanent organization difficult • continuing the department-related tasks in the permanent organization difficult • ensuring know-how for the project-oriented organization difficult

Fig. C2.6: Advantages and disadvantages of the pure project organization

C 2.3   Matrix project organization

The matrix project organization is marked by a distribution of the formal authorities between the project manager (What? When? How much?) and the line managers (Who? How? How well?).

The distribution of the formal authorities in the matrix project organization can be presented in a project organization chart. Figure C2.7 depicts the double assignment of the project team members.

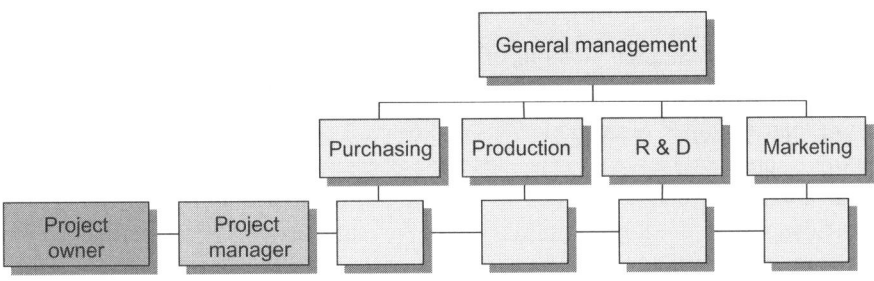

Fig. C2.7: : Matrix project organization

Advantages	Disadvantages
• distribution of formal authorities between the project manager and the line manager • coordination of the experts of the functional departments by the project manager • flexible use of personnel (no recruitment or reintegration problems) • experts work in groups of peers, possibility of exchange of ideas for ensuring know-how • intentional "constructive conflict" caused by double assignment of the project team members	• no full concentration on the project • limited identification with the project • high demand for organizational understanding of the members of the project organization • requirement for solving the intentional, "constructive" conflicts • high management effort for consensus

Fig. C2.8: Advantages and disadvantages of the matrix project organization

Matrix project organizations of "various strengths" are developed through various distributions of authorities between the project manager and the line managers. For instance, the project manager can also be given the authority to select the project team members.

The basic intention of the matrix project organization is direct communication between the project manager and the project team member. When the communication with the project team member is not direct, but only indirect via the line manager, the project manager's possibilities for coordinating communication are considerably limited.

C 2.4   Traditional project organization models in practice

In practice, companies will use the above described traditional project organization models according to their respective appropriateness. The matrix project organization has proven to be especially popular. Ensuring an organizational home for the project team members in the departments of the permanent organization and the inclusion of the department managers in project-related management tasks are considered the main advantages.

However, there are adapted versions of the traditional project organization models.

Instead of pure project organization large American engineering firms, such as Bechtel, also use matrix project organization instead of pure project organization for complex, "repetitive" projects. Here the project team members are still utilized in the project "full-time" and also have common office space. The line managers in the permanent organization, however, obtain formal authority as in the matrix project organization and "wander around" from project to project to check on the project team members.

For small, possibly secret projects the pure project organization is used instead of the expected matrix project organization. Because of the nature of the work the requirement for full-time utilization and joint office space for the project team members is given up, but full formal authority is given to the project manager.

Combinations of project organization models can also be found in practice. This is done in order to combine the strengths of different organization models.

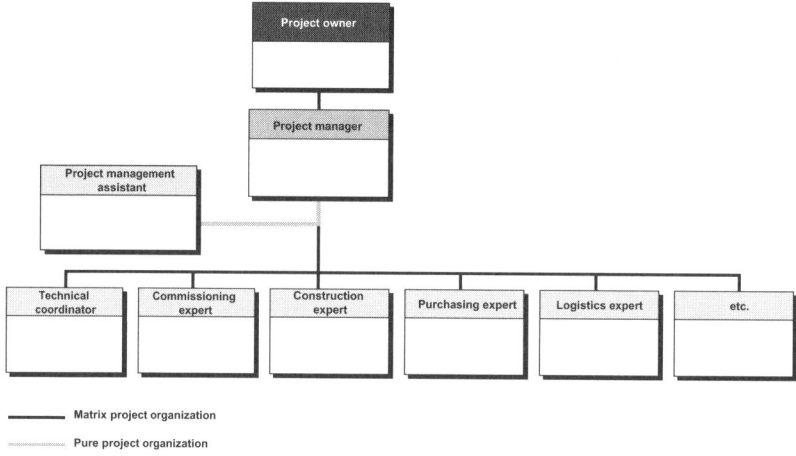

Fig. C2.9: Combination of matrix and pure project organization

In small, innovative projects the project organizations become flatter. It is attempted to ensure the creativity required for innovation through team orientation and flexible communications structures. A reduction of the hierarchical distance is the result.

New developments in the organizational design of projects are given in Chapter C4: New project organization models.

C 3 Project roles and project organization charts

It is possible to differentiate between project roles which are performed by individuals and project roles which are performed by teams. Project roles can be described in role descriptions but require project-specific adaptations.

The relationships between project roles can be visualized in a project organization chart.

C 3.1 Individual roles and team roles in projects

In addition to the roles of the permanent organization, such as managing director and department manager, in the project-oriented organization there also exist project roles, such as project owner team, project manager and project team member. Through the definition of project roles the organizational differentiation of the company is increased.

A "role" is a set of expectations which are bound to the fulfillment of the role. Project roles can be described through the objectives, the organizational position, the tasks to be fulfilled, the formal authority and the relationships to relevant project environments. Project roles are to be described in a relational manner, i.e. taking into account the relationships between roles. When, for example, a project manager has the authority to make purchasing decisions up to a value of €10,000, then the authority of the project owner team can only be for purchases over €10,000. The description of project roles serves to clarify the cooperation in the project and the (social) boundaries of the project.

The expectations regarding a particular role exist before a person or a team takes the role on. Roles are independent of individuals; but individuals have many possibilities for fulfilling a role.

ROLAND GAREIS Project and Programme Management® differentiates between roles performed by individuals and roles performed by teams. The project-related individual roles are, for example, project owner (as individual), project manager, project team member and project contributor. Project-related team roles are, for example, project owner team, project team and sub-team.

Basic descriptions of the project roles are given below. Although there are basic expectations which remain unchanged regarding each project role, the role descriptions for different projects can be adapted for each specific project when required.

The profiles of the project roles are described in Chapters C3.2-C3.5. The role descriptions used in practice, the impact of the role on project success, the objectives and the non-objectives of the role, the requirements of the person performing the role in the project and the recruiting possibilities are depicted. From this, it becomes obvious that project management is an task not only of the project manager, but of every member of the project organization.

C 3.2   Role: Project owner team

The significance of the project owner team

The success of a project is dependent upon the attention which the project receives from the management of the project-oriented organization. Management attention can primarily be provided by the appropriate constitution of the project owner team.

The organizational integration of a project in the project-oriented organization is mainly achieved by the project owner team. When the project owner team is not adequately staffed, its members will either not give the project enough support or they will show little interest in the project.

The tasks and responsibilities of the project owner team

The success of a project depends essentially on the professional fulfillment of the tasks of the project owner team.

The project owner team assigns a project to a project team in the project start process to realize the project objectives, and releases the project team with the project approval in the project close-down process. The project owner team provides the project team with context information, makes strategic project decisions and gives the project team feedback on the (intermediary) results which have been achieved.

An essential task of the project owner team is leading the project manager. The project manager has a right to leadership. For the continual contact to the project manager a speaker can be nominated from the project owner team. The project owner team attends to the application of the company's guidelines for project and programme management in the project. The project owner team has, therefore, an essential responsibility in relation to the assurance of management quality in the project.

The project owner team performs essential tasks in project marketing. Communication of the objectives and of the strategic importance of a project to the company performing the project as well as to the relevant project environments is something that members of the project owner team can perform with particular authenticity.

The role of the project owner team is an active one. Depending on the complexity of the project, the job of the speaker of the project owner team may require one half day to one day per week. The project owner team also works in meetings which should take place every two to four weeks, depending on the complexity and the duration of the project.

Appointing members of the project owner team

An appointment to the role of project owner is dependent upon the degree of complexity of the project. The more complex the project, the more areas of the company will participate in the project performance. The members of the project owner team should, therefore, be recruited from several (2-4 areas), though surely not from all the affected areas. Basically, the project owner should come from the lowest possible management level in the hierarchy in order to expand the circle of leadership in the project-oriented organization.

Competencies required of the project owner team

Certain competencies are required to fulfill the role of a member of the project owner team. Besides knowledge about the company performing the project and specialist knowledge regarding the project content, strategic orientation, social competencies and project management competencies are required. Members of the project owner team need not be able to develop work breakdown structures or project risk plans, for example, but should be able to assess their completeness and formal correctness, and to use the project plans as communication tools in the project.

Characteristics of the role: Project owner team	
Names	Project owner, project steering committee, project supervisor, project supervisory board, project sponsor, etc.
Importance for project success	Very high; though in practice usually not perceived as such
Objectives	Realization of project-related company interests; strategic project management; provision of context information
Number of persons	1 (for small projects) or 2 to maximum 4 (for projects); from the same or different levels in the hierarchy
Non-objectives	Performance of the tasks of the project manager, arbitrator for the project team
Competencies	Branch and company competencies, project management competence, strategic orientation, social competencies and decision-making abilities
Recruiting	Managers from the areas of the company affected by the project results

Fig. C3.1: Characteristics of the role: Project owner team

Role description: Project owner team

The role of the project owner team is generally described as follows:

Role: Project owner team
Objectives
• realize project-related company interests • coordinate project and company interests • assign the project to the project team • lead the project manager • support the project team
Position in the organization
• reports to the Project Portfolio Group • part of the project organization • project manager reports to the project owner team

Tasks
Tasks in the project start
• select the project manager and essential project team members • agree on objectives with the project team • contribute to the design of the project context • ensure the availability of resources • ensure the use of the company's project management standards • participate in the project start workshop • contribute to the initial project marketing
Tasks in project coordination
• communicate with representatives of relevant environments • continuous project marketing
Tasks in project controlling
• make strategic project decisions • strategic project controlling • continual information about the project context • continually ensure the availability of resources • contribute to conflict solving • hold project owner meetings • occasionally participate in project meetings and project workshops • contribute to project marketing • promote change in the project
Tasks in resolving a project discontinuity
• define a project discontinuity • collaborate in the resolution of a project discontinuity • collaborate in the development of and decisions about immediate measures • collaborate in the cause analysis • decide about alternative strategies • collaborate on the performance of corrective measures and checks on their success • end a project discontinuity
Tasks in project close-down
• performance evaluation • exchange feedback with project team • participate in project close-down workshop • ensure know-how transfer into the permanent organization • formal project approval • ensure adequate structures for the post-project phase
Environment relationships
• project Portfolio Group • project team, project manager • customers, suppliers

CHAPTER
C

Formal authority
• selecting the project manager • assigning the project team with the project performance • changing the project objectives • purchasing decisions over EUR ... • definition of a project discontinuity • project stopping • project approval

Fig. C3.2: Role description: Project owner team

C 3.3   Role: Project manager

The significance of the project manager

The project manager is the central integration role in the project. The project manager is the contact person for all members of the project organization and for representatives of the relevant project environments.

The project manager "drives" the project, is interested in the progress of the work and in the successful close-down of the project. He/she is that member of the project team which is primarily responsible for project management.

Tasks and responsibilities of the project manager

The task of the project manager is the design of the project management process. The project is to be started, continually coordinated, controlled and closed down. Possibly a project discontinuity is to be resolved. In addition, adequate communication forms are to be selected, adequate project management methods are to be used and adequate IT and telecommunications tools are to be applied. The project manager is responsible for professional project management and for adequate project marketing.

The fulfillment of the project management tasks by the project manager is a service to the project, not exercising a power position.

In fulfilling the project management tasks the project manager cooperates with other members of the project team and with the project owner team. Not the project manager alone is responsible for the project success, but also the project team members and the project owner team.

The project manager has a right to be led by the project owner team.

Appointing a project manager

In the project-oriented organization "project manager" is a profession in its own right. Project managers should, therefore, belong to an Expert Pool of "project management personnel", which they can be recruited from.

A combination of roles, combining project manager with other project roles, such as project team member or project contributor, is not to be recommended and requires a double qualification of the employee in question.

A project manager should not manage more than 2, maximum 3, projects at the same time. The focus on each project is important in order to ensure progress in the project.

Competencies required of the project manager

The core competence for fulfilling the role of project manager is project management competence. In addition, a project manager requires also company knowledge, knowledge related to the project content and – in international projects – also language knowledge and intercultural competence.

Characteristics of the role: Project manager	
Names	Project manager, project leader, project coordinator, etc.
Importance for project success	Very high; in practice often left alone
Objectives	Realization of project interests, strategic and operative project management, ensuring project information
Non-objectives	Work on the project content; expert on the project content
Number of persons	Always one person; in practice sometimes two persons
Competencies	Project management competence, branch, company and product competencies
Recruiting	From a project management Expert Pool (in the project-oriented organization); from the external personnel market

Fig. C3.3: Characteristics of the role: Project manager

Role description: Project manager

The role of the project manager is generally described as follows:

Role: Project manager
Objectives
• realize project interests • ensure the realization of the project objectives • lead the project teams and the project contributors • represent the project toward representatives of the relevant environments • develop and adapt the project management documentation
Position in the organization
• reports to the project owner team • member of the project team • leads the project team members and the project contributors

Tasks
Tasks in the project start

- Design the project start process (possibly together with selected project team members)
- Transfer know-how from the pre-project phase into the project (together with project team members and project owner team)
- Agree on the project objectives (together with project team members)
- Develop adequate project plans (together with project team members)
- Design an adequate project organization, team-building (together with project team members)
- Develop the project culture, establish the project as a social system (together with project team members)
- Risk management, avoidance and/or promotion of and provision for project discontinuities (together with project team members)
- Design project context relationships (together with project team members)
- Perform the initial project marketing (together with project team members)
- Develop project management documentation "Project start"

Tasks in project coordination

- Controlling the (intermediary) results of the work packages
- Disposition of project resources for work packages (together with project team members)
- Accept work packages
- Participate in sub-team meetings (periodically)
- Communicate with representatives of relevant environments
- Continuous project marketing

Tasks in project controlling

- Design the project controlling process (possibly together with selected project team members)
- Determine the project status (together with project team members)
- Agree on or perform controlling measures (together with project team members)
- Further develop the project organization and/or project culture (together with project team members)
- Adapt project objectives (together with project team members)
- Develop project progress reports (together with project team members)
- Redesign the project context relationships (together with project team members)
- Perform project marketing tasks (together with project team members)

Tasks in resolving a project discontinuity

- Propose the definition of a project discontinuity to the project owner team
- Design the processes for resolving the project discontinuity (together with project owner team)
- Work out immediate measures (together with project team)
- Perform cause analysis (together with project team)
- Work out alternative strategies (together with project team)
- Perform measures to resolve the discontinuity and check for success (together with project team)
- End the project discontinuity (together with project owner team)

Tasks in project close-down
• Design the project close-down process (possibly together with selected project team members)
• Plan the post-project phase
• Transfer know-how into the permanent organization (together with project team members and representatives of the permanent organization)
• Develop the project close-down report
• Perform closing project marketing (together with project team members)
• Perform emotional close-down of the project (together with project team members)

Environment relationships
• Project owner team
• Project team members
• Project contributors
• Cooperation partners, suppliers, media

Formal authority
• Holding project owner team meetings and project team meetings
• Purchasing decisions up to EUR ...
• Coordination of the project team members and the project contributors
• Selection of the project team (together with project owner team and line managers of the project team members)

Fig. C3.4: Role description: Project manager

C 3.4   Roles: Project team member and project contributor

The significance of the project team member and the project contributor

The differentiation between the roles "project team member" and "project contributor" serves to indicate the difference in intensity of their contribution to project management. The project team members, through their membership in the project team, must contribute to the optimization of the project success.

The project team member and the project contributor are of great importance for the project success, since it is not possible to achieve good results in the project without a relevant expert competence for the fulfillment of the work packages.

Tasks and responsibilities of the project team member and the project contributor

A project contributor fulfills content-related work packages in a project. A project team member fulfills content-related work packages in addition to project management and project marketing tasks. The project team member participates in project team meetings and communicates the project objectives and their results to relevant project environments.

The project team member and the project contributor are responsible for the quality of the results of their work packages.

Appointing a project team member and a project contributor

Project team members and project contributors are recruited either from company-internal Expert Pools (or departments) or from the external personnel market.

Competencies of a project team member and a project contributor

Project contributors require most of all expert competencies to fulfill the respective work packages. Project team members additionally require project management competencies.

Characteristics of the role: Project team member	
Names	Project team member, project expert, project engineer, etc.
Importance for project success	High; without content-related expert competence there can be no project results
Objectives	Fulfill work packages; possibly lead a sub-team; participate in project team meetings; contribute in project management, project marketing
Non-objectives	Only tasks as an expert
Number of persons	One
Competencies	Expert competence and project management competence
Recruiting	From an Expert Pool (of the project-oriented organization)

Characteristics of the role: Project contributor	
Names	Project contributor; expert, etc.
Importance for project success	High; without content-related expert competence there can be no project results
Objectives	Fulfill work packages; regular reports about the fulfillment of work packages
Non-objectives	Participate in project team meetings, project marketing
Number of persons	One
Competencies	Expert competence and little project management competence
Recruiting	From an Expert Pool (or a department) of the project-oriented organization

Fig. C3.5: Characteristics of the roles: Project team member and project contributor

Role description: Project team member

The role of the project team member is generally described as follows:

Role: Project team member
Objectives
• realize the project interests • contribute to the realization of the project objectives • possibly: Lead sub-team members/project contributors • represent the project internally and externally • fulfill work packages with the required quality and quantity • contribute to the fulfillment of project management tasks
Position in the organization
• reports to the project manager • part of the project organization • appointed by the manager of the Expert Pool • possibly: Sub-team members/project contributors report to the project team member
Tasks
Tasks in the project start
• contribute to the preparation, performance and follow-up of the project start workshop • agree on objectives with the project manager • contribute to the know-how transfer from the pre-project phase into the project • contribute to the development of the project plans • contribute to the design of the project organization • contribute to the project culture development • contribute to risk management, to the avoidance of and/or promotion of and provision for a project discontinuity • contribute to the design of the project context relationships • contribute to the performance of the initial project marketing
Tasks in project coordination
• contribute to the controlling of the intermediary results of work packages • contribute to the disposition of project resources for work packages • contribute to the approval of work packages • communicate with representatives of relevant environments in coordination with the project manager • contribute to the continuous project marketing
Tasks in project controlling
• contribute to the preparation, performance and follow-up of project controlling meetings and workshops • develop regular reports about the progress of the project to the project manager • contribute to the periodical project progress reports • give feedback to the project manager, to other project team members (content, social) • contribute to the design of the project-related environment relationships • contribute to project marketing • contribute to the agreement on and develop controlling measures

CHAPTER

C

• contribute to the further development of the project organization and the project culture
• review the project objectives together with the project manager and project owner team

Tasks in resolving a project discontinuity

• contribute to the development of immediate measures
• contribute to cause analysis
• contribute to the development of alternative strategies
• contribute to the resolution of a project discontinuity

Tasks in project close-down

• contribute to the preparation, performance and follow-up of the project close-down work-shop
• exchange feedback with the project manager and other project team members
• contribute to know-how transfer into the permanent organization
• contribute to the development of adequate structures for the post-project phase
• contribute to the project close-down report

Environment relationships

• project owner team
• project team members
• project contributor
• cooperation partners, suppliers, the media

Formal authority

• decisions about the way in which project-related work packages are to be fulfilled
• quality assurance for all project-related work packages
• possibly: Assign the performance of work packages to sub-team members/project contributors
• coordination of project contributors in the sub-team

Fig. C3.6: Role description: Project team member

Role description: Project contributor

The role of the project contributor is generally described as follows:

Role: Project contributor

Objectives

• contribute to the realization of the project objectives
• fulfill work packages with the required quality and quantity

Position in the organization

• appointed by the project manager and reports to him/her (or to a coordinating project team member)
• part of the project organization

Tasks

Tasks in the project start

• agree on objectives with the project manager or possibly with a coordinating project team member

Tasks in project controlling
• possibly participate in sub-team meetings • occasionally participate in project team meetings • report regularly to the project manager or a coordinating project team member about work in progress
Tasks in resolving a project discontinuity
• occasionally contribute to the resolution of a project discontinuity
Tasks in project close-down
• exchange feedback with the project manager and project team members • contribute to know-how transfer into the permanent organization
Environment relationships
• project owner team • project team members • project contributor • cooperation partners, suppliers, the media
Formal authority
• decisions about the way in which project-related work packages are to be fulfilled • quality assurance for all project-related work packages

Fig. C3.7: Role description: Project contributor

C 3.5   Roles: Project team and sub-team

The significance of the project team and sub-team

The project team and sub-teams are also roles in projects on which specific expectations are placed. The expectations regarding the project team are fundamentally different from those regarding the individual project team members, and the expectations regarding a sub-team are also different from those regarding the individual sub-team members. Only through this differentiation between team and team member does the meaning of team work and team meetings in projects become clear, as well as the necessity of developing competencies for efficient team work.

The significance of the project team lies in the development of added value in the project team work. The significance of the sub-teams lies also in the development of high quality, integrated work package results through team work. Sub-teams are required when the complexities of individual areas are too high for individual project team members to deal with and the work of several persons together is necessary. Examples of sub-teams in a complex engineering project are a sub-team "procurement", a sub-team "electronics", etc. These sub-teams are not relatively independent "sub-projects", but an integrated part of the respective project.

Tasks and responsibilities of the project team and sub-team

The project team has project management tasks to fulfill. The responsibility of the project team lies in the development of high quality project management solutions through their development, review and decision within the team.

The project team does not fulfill any content-related work, such as programming software, but may decide who is to be called in, in case of a problem in the programming of the software.

The tasks of the sub-team are the fulfillment of content-related project work and the co-ordination of this work within the respective sub-teams.

Appointing members of the project team and sub-team

The project team is made up of the project team members. Project teams can consist of a minimum of 3 persons, maximum 15 persons.

A sub-team of a project is made up of several project contributors and one project team member who represents the sub-team in the project team.

Competencies of the project team and sub-team

Not only the individual project team and sub-team members, but the respective project team and/or sub-team as a whole requires competencies for team work. The development of these team competencies is a project management task.

Characteristics of the role: Project team	
Names	project team, project group, etc.
Importance for project success	very high, as the continual review of the content and the assurance of synergies enable project success
Objectives	ensure synergies; solve conflicts; ensure commitment, develop the "big project picture"; organize learning in the project
Non-objectives	content-related project work (this is performed by the sub-teams or by the individual project team members and project contributors)
Number of persons	3 to 15
Competencies	team work competence, project management competence
Recruiting	project team members make up the project team

Characteristics of the role: Sub-team	
Names	sub-team, work group, etc.
Importance for project success	high, project results performed together in the sub-team
Objectives	joint fulfillment of work packages; regular reports about the progress of the work packages by a project team member coordinating the sub-team
Non-objectives	participation in project team meetings, project marketing
Number of persons	2 to 8
Competencies	expert competence and team competence
Recruiting	from a project team member and several project contributors

Chapter

C

Fig. C3.8: Characteristics of the project roles: Project team and sub-team

Role description: Project team

The role of the project team is generally described as follows:

Role: Project team
Objectives
• perform the project management processes • develop a common "big project picture" • ensure synergies in the project performance • develop commitment to the project • solve conflicts in the project • organize learning in the project
Position in the organization
• part of the project organization • members: Project team members and the project manager • assigned by the project owner team
Tasks
Tasks in the project start
• exchange information with the project team members • jointly decide on the design of the project organization and about project planning • jointly decide on the design of the project context relationships • agree on the project rules
Tasks in project controlling
• determine the project status • adapt project objectives, schedule, costs, etc. • adapt the project organization and the project context relationships • perform project marketing tasks

Tasks in resolving a project discontinuity
• suggest the definition of a project discontinuity to the project owner team • design the processes for resolving the project discontinuity • develop immediate measures • perform a cause analysis • develop alternative strategies • jointly perform measures for resolutions and checks for success • end the project discontinuity (together with project owner team)
Tasks in project close-down
• design the project close-down process • plan the post-project phase • transfer know-how into the permanent organization • perform closing project marketing • jointly perform emotional close-down to the project
Environment relationships
• project owner team • project team members • project contributor • cooperation partners, suppliers, the media
Formal authority
• agreements in the project team • agreements in the project owner team

Fig. C3.9: Role description: Project team

Role description: Sub-team

The role of the sub-team is generally described as follows:

Role: Sub-team
Objectives
• jointly fulfill work packages • ensure the quality of the work package results
Position in the organization
• reports to the project manager • part of the project organization • members: Sub-team members
Tasks
• fulfill the work packages • coordinate the work in the sub-teams
Environment relationships
• project owner team • project team members • project contributors • cooperation partners, suppliers, the media

Tasks
• fulfill the work packages
• coordinate the work in the sub-teams
Formal competencies
• necessary decisions for the fulfillment of the work packages

Fig. C3.10: Role description: Sub-team

C 3.6 Role conflicts and role potentials in the project-oriented organization

The high level of organizational differentiation in the project-oriented organization can lead to role conflicts and role potentials.

A conflict within a role (intra-role conflict) comes about when two or more parties have expectations regarding a role which are incompatible (such as the expectations of a customer that the project manager attends to the customer's interests, and the expectation of the project owner team that the project manager attends to the interests of the company).

Possible intra-role conflicts of the project owner team are shown in the following figure.

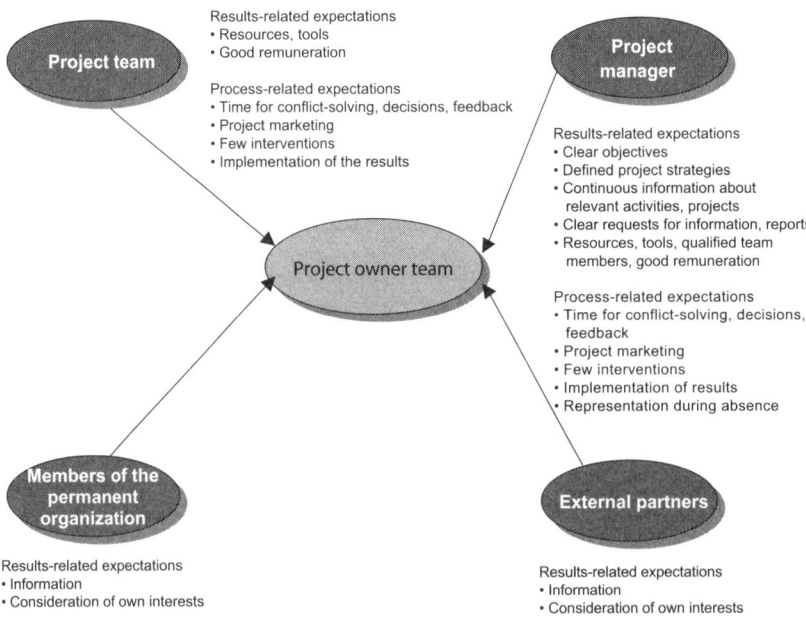

Fig. C3.11: Intra-role conflict of the project owner team

The simultaneous fulfillment of several roles in the project-oriented organization may give rise to inter-role potentials as well as to inter-role conflicts.

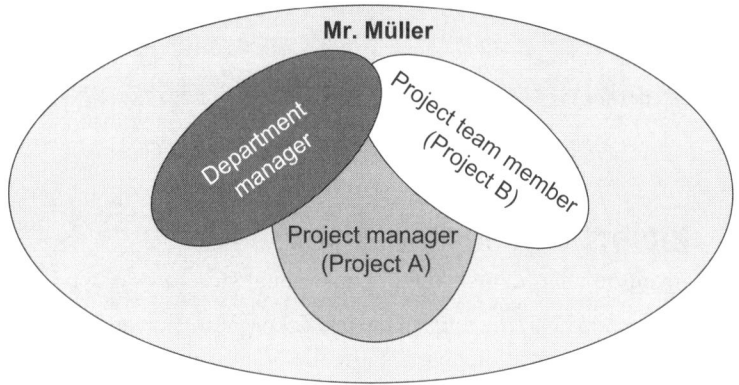

Fig. C3.12: Mr. Miller, performing several roles in a project-oriented organization

Caused by a possible incompatibility of expectations regarding the different roles of Mr. Miller, (inter-)role conflicts may occur. On the other hand, the simultaneous performance of several roles may have an important integration potential.

In small projects and in projects of medium complexity it is common for one person to take on both the role of the "project manager" and the role of "project team member".

Rules regarding one person taking on several roles in a project are shown in Figure C3.13.

Role combinations in a project	Rule
project owner and project manager	NO
project owner and project team member	NO
project owner and project contributor	YES
project manager and project team member	YES, but...
project manager and project contributor	YES

Fig. C3.13: Role combinations for a person in a project

The number of different roles a person is to take on in a project-oriented organization while still being able to ensure the proper amount of attention for the individual projects is limited. Figure C3.14 depicts a summary of related ratios. These depend on the size of the projects and are, therefore, to be understood as maximum values.

Rule	Number of project roles per person
project owner team member	4 – 6
project manager	2 – 3
project team member	4 – 5
project contributor	5 – 8

Fig. C3.14: Number of project roles per person

C 3.7   Project organization chart

A project organization chart depicts the organizational structures of a project.

In a project organization chart the roles in the project organization and their relationships are depicted. Project organization charts show the organizational structure of a project at a given point in time. The project organization chart is to be developed during the project start process. Since the project organization changes over time, the project organization chart is to be adapted during the project controlling process.

The project organization chart of *ROLAND GAREIS Project and Programme Management®* depicts the individual roles as well as the team roles. An ellipse around the project organization gives it a boundary and symbolizes the relative autonomy of the project as a social system; the line from the project owner team to the project team symbolizes the assignment of the project team (not only that of the project manager) by the project owner team. This visualization corresponds to the "empowered project organization" (see Chapter C4.1).

The systemic approach represented in this visualization of the project organization lies in the emphasis of the relationships between the project roles and in the depiction of the equal rank of the individual roles, in renunciation of the hierarchy and in the recognition of the different tasks of the roles in ensuring the project success.

In addition to the project roles and the relationships between the roles, the names of the persons performing the project roles can also be made visible in the organization chart. Should this lead to an "information overflow" this information can be presented in a separate list of the project roles. The department designations from the permanent organization should not be shown in the organization chart. A clear differentiation of the role titles of the project organization from those of the permanent organization is to be made. So, for example, the title of a project team member as "department manager, design" is not adequate. The formally correct title would be "project team member, design" since the department manager performs no line managerial tasks in the project, but rather those of a design expert.

For the design of the organizational chart not only different symbols can be used, such as boxes, circles, ellipses, arrows and lines, but also different colors, different sizes of the boxes and circles, different line widths, etc. The selected symbols coin a picture of the project and communicate the project culture. The following project organization

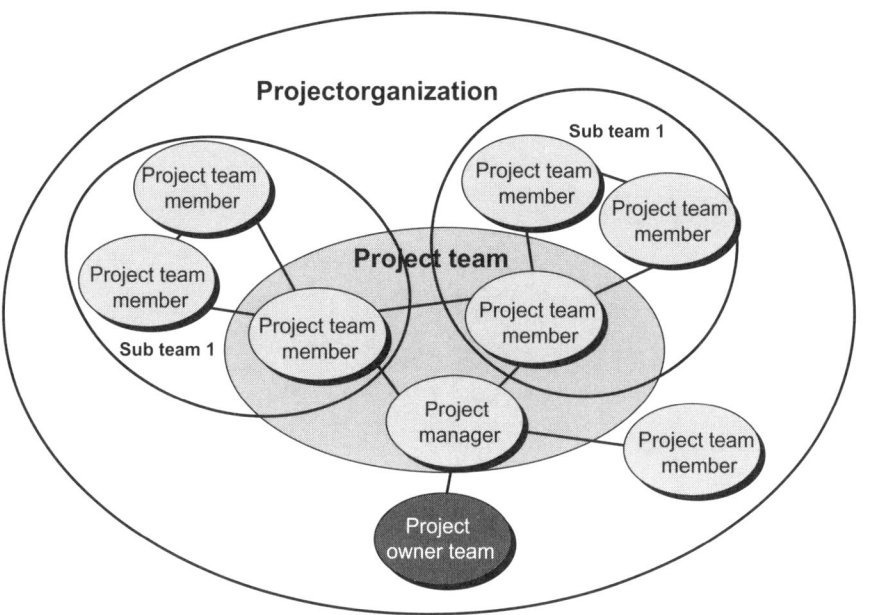

Fig. C3.15: Project organization chart

charts show two different examples for the performance of a customer contract, one by an Austrian engineering company, and the other of the reorganization of Swissair.

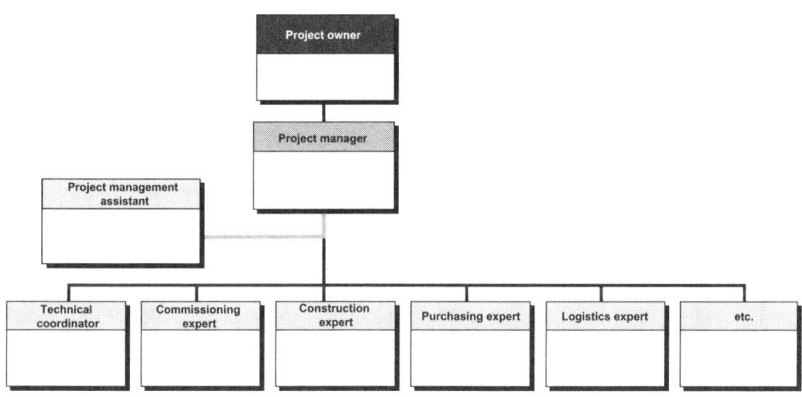

Fig. C3.16: Project organization chart: "Customer contract"

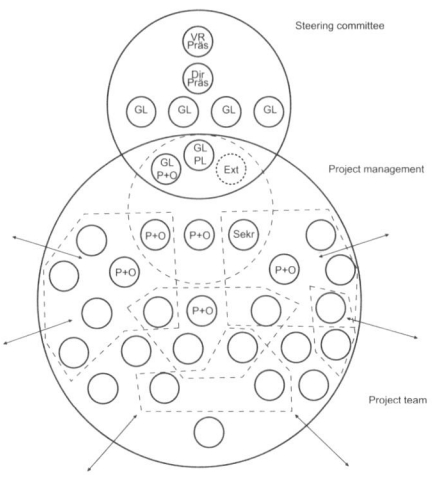

Fig. C3.17: Project organization chart of the reorganization of Swissair (1991)

The project cultures expressed in these organization charts have different degrees of attractiveness for potential members of the project organizations (see also Chapter D3.3: Methods of symbolic project management). In the project organization chart for the engineering project it is obvious that there is no project team defined, and that the manager of the project team members in the matrix project organization is not shown in the organization chart.

Different depictions of a project organization in different project organization charts provide different information. Therefore, a combination of several graphic solutions for the depiction of a project organization is possible.

C 4 New project organization models

The traditional project organization models originated in the 1960s. In the project organization model of ROLAND GAREIS Project and Programme Management® new managerial developments such as empowerment, integration of partners and virtual communication structures are considered.

C 4.1 Design element: Empowerment

Further development of the matrix project organization

The matrix project organization is a project organization model from the 1960s. The matrix project organization is a traditional, hierarchical organization approach, which separates the management from the performing tasks. The management tasks are divided between project manager and line manager of the project team members, the performing tasks belong to the project team member. The project team member receives his/her directions from these two managers (see Chapter C3.4).

The matrix project organization is

- not lean because of the large number of members in the project organization (for every project member there is a manager out of the line organization),
- too expensive and too slow because of the coordination requirements between the managers,
- not customer-oriented and therefore demotivating because of the lack of authority of project team members to make decisions.

New approaches to organizational theory, such as "lean management", promote the transfer of formal authorities to the group performing the task in order to speed up the decision-making process, to motivate the employee performing the task and to ensure a decided customer orientation in the fulfillment of the business processes.

The implementation of "empowerment" in the design of the project organization is a further development of the matrix project organization. In the "empowered project organization" not only the project team members are "empowered", but also the project team and the project as a whole.[1]

CHAPTER

C

"Empowerment" of the project team member

The empowerment of the project team member comes about through the transfer of the authorities of HOW? and HOW WELL? from the line manager to the project team member. The project team member can decide about the design of the fulfillment of an assigned work package and is responsible for the results of the work package. This means that the project team member can also make agreements with the project manager and the other project team members in project meetings and workshops.

In order to clarify the distribution of decisional authorities, the name of the role of the line managers should be changed from "department manager" to "Expert Pool manager". This Expert Pool manager is no longer the "super expert" who must take all the content-related decisions for the project team members, this is the manager of a pool of experts with different qualifications. The manager of the Expert Pool still has the WHO? authority and coordinates the disposition of personnel.

In case of a requirement for content-related work by an Expert Pool manager, which can often be the case in internal projects, he/she can take on any role, from member of the project owner team to project team member.

"Empowerment" of the project team

The empowerment of the project team comes about through the inclusion of the project team in project management decisions and through the transfer of the authorities of WHAT?, WHEN? and HOW MUCH? to the project team. The project team takes over the responsibility for the project success. The project manager, as a special role in the project team, is responsible for the professional project management.

---

1) The term "empowerment" means self-qualification, self-authorization, autonomy.

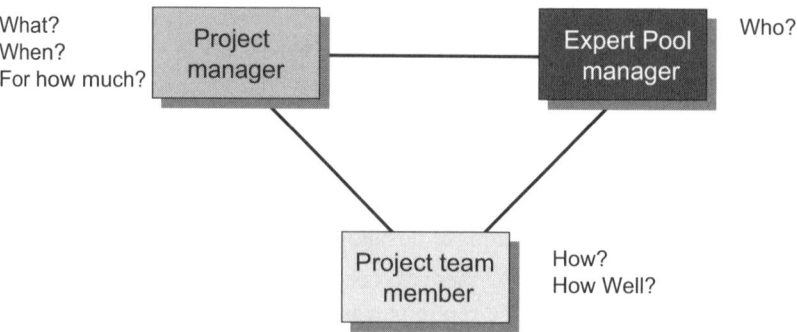

Fig. C4.1: "Empowerment" of the project team member

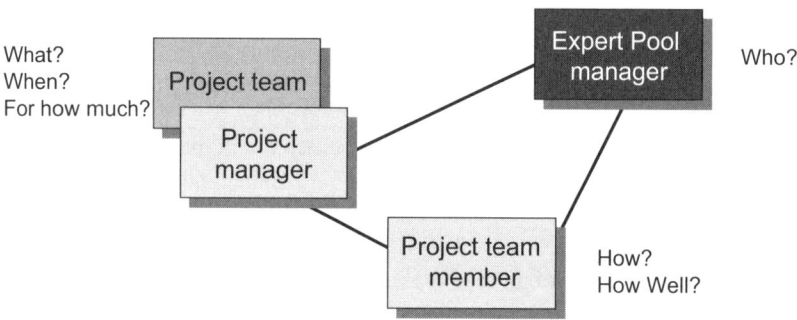

Fig. *C4.2: "Empowerment" of the project team*

"Empowerment" of the project

The empowerment of the project is effected by the definition of boundaries between the project and its relevant environment. It is most important to prevent frequent interventions from the permanent organization. A certain project autonomy is to be ensured in order to guarantee an efficient project performance.

The empowerment of the project is symbolically represented in the project organization chart by the boundaries which differentiate the social system from its relevant environment.

C 4.2 Design element: Integration

Hierarchies of project organizations and parallel project organizations

In practice, hierarchies of project organizations and several parallel project organizations are often used to perform a project.

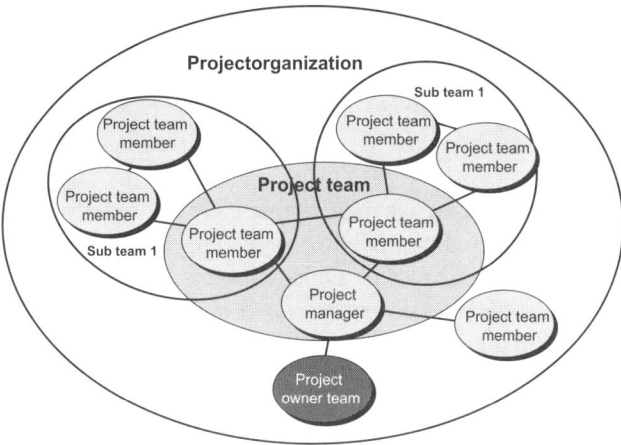

Fig. C4.3: "Empowerment" of the project

An investor establishes a project to implement an IT system, for example. A general contractor is appointed, who also defines a project for the fulfillment of the assignment. The general contractor, in turn, appoints one or more contractors, these appoint various subcontractors, and so on. Because of the scope of their works each participating company defines a project for the fulfillment of its part. The result is a hierarchy of project organizations according to the contract structure for the implementation of the IT system.

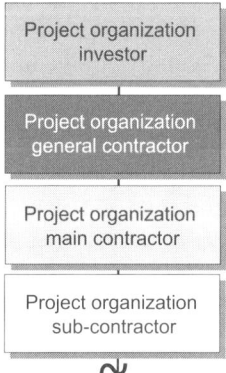

Fig. C4.4: Hierarchy of project organizations

Probably different project management approaches are practiced in the projects. Project organizations with different project managers as well as different project plans are developed. Thus, cultural and structural vagueness emerge. To perform the business process of the investor several project managers work with different project plans on the basis of different project management approaches and tools. The wishes of the investor, which should finally be realized by a subcontractor, will be misinterpreted due to the "grape-vine" phenomenon, which will lead to conflicts and to rework.

The objectives of "integration"

An "integrated project organization" is a project organization whose members are representatives of different organizations. The members of an integrated project organization can be representatives of a customer, his/her contractor and subcontractor or from a contractor and his/her subcontractor or from several partners of the same standing. Several, legally independent companies of a conglomerate can, for example, cooperate as partners in a research project, or as partners in a contracting project in an integrated project organization. Integrated project organizations can basically be divided into horizontal and vertical cooperations.

The project for which an integrated project organization is designed is to be defined from the viewpoint of the investor. The project organization is to be designed in such a way that the business case of the investor can be optimized. The objectives of the individual organizations working on the project are subordinate to this main objective. It is assumed that by optimizing the objectives of the investor the objectives of the cooperation partners can also be fulfilled. The project boundaries are therefore defined by the business case of the investors and not by the contract structures.

Cooperation in an integrated project organization requires a detachment of the organizational structures from the contractual structures. The multi-level relationships of customer and supplier are not taken into account in the design of the project organization: Instead a common, flattest possible project organization is developed, in which the representatives of the different organizations take on different roles. A representative of a main contractor could, for example, together with a representative of the customer, make up the project owner team, a representative of a contractor can be the project manager and a representative of a subcontractor can be a project team member.

This organizational design, even without contractual obligations, results in the application of a uniform project management approach, the use of a single project manager and the development of common, integrated project plans. All these tasks require a high level of openness and trust among the cooperating organizations.

The way the representatives of the different organizations perceive their roles must include the apprehension of a holistic project view, the assumption of the entire responsibility for the project and the fulfillment of project marketing tasks. It must be accepted that not every cooperation organization has its own project, its own project manager and its own project plans.

The objectives of the "integrated project organization" are:

- customer satisfaction with the project results,
- optimization in project performance through a holistic project view, a common "big project picture",
- avoidance of sub-optimization through elimination of the "grapevine" phenomenon, "cannibalization" of scarce resources,

- optimizations through a holistic project view, tailor-made organizational solutions,
- pursuit of common project objectives for all participants,
- development of win-win situations,
- the reduction of project costs through a single project manager, a single set of project management documentation, etc.
- development of common project plans, of "open books".

Dangers detected by potential partners in an "integrated project organization" are:

- loss of know-how through cooperation with eventual competitors,
- incompatible cultures among the cooperating companies,
- costly coordination processes,
- unclear responsibility and liability.

⤳ CASE STUDY C1: Parallel project organizations

Investor:
- Austrian telecommunications supplier; approx. 200 employees; products: Telephone, internet, ADSL, company networking, business solutions, terminals and internet service provider

Project objectives:
- building a regional network as a pilot infrastructure to provide telecommunications services for business users
- providing a billing system for the commercial aspect of the telecommunications services
- providing the organizational and the personnel prerequisites for the provision of telecommunications service

Contract structures:
- 2 main contracts to be awarded to a technical contractor to build the technical infrastructure and to a commercial contractor for the implementation of the billing system

1. Traditional organizational design according to the contract structures

- parallel project organizations from the investor, the technical and the commercial contractor
- informal communications between the persons performing the roles of the different projects
- no common project plans

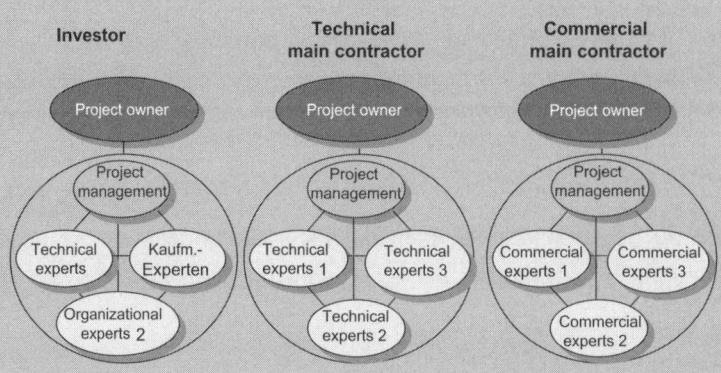

Fig. C4.5: Parallel project organizations

2. Integrated project organization:

- representatives of the investor and both contractors (possibly also selected sub-contractors) as members of a common project organization
- only one project manager
- common project plans

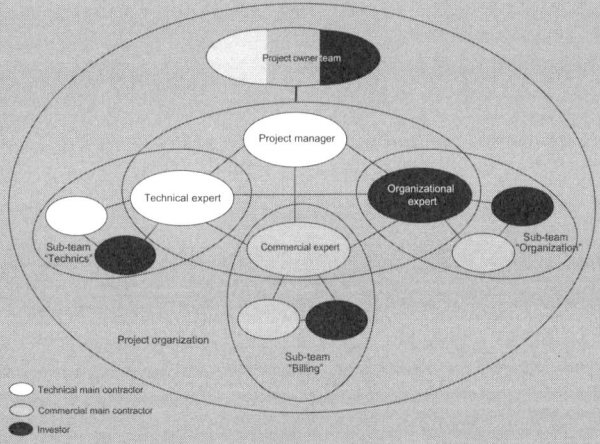

Fig. C4.6: Integrated project organizations

Organizational decision:

- decision for the traditional form of parallel project organizations

Interpretation of the organizational decision:

- fear of unclear liability questions in the case of the integrated project organization
- trust in the informal communication structures and in the power of the contract
- fear of the new, organization not recognized as a success factor
- dominance of the company's legal counsel

C 4.3   Design element: Partnering

"Partnering" or "strategic project alliances" goes further than designing an "integrated project organization", it supports the organizational integration with contractual provisions. A common incentive system for all the partners participating in the project supports the achievement of the common project objectives.

···⟩ CASE STUDY C2: Partnering in an engineering project[a]

Five Partners:
- Investor/Client: Ruhr Oel, Germany
- Engineering contractor: Fluor Daniel, The Netherlands
- Civil contractor: Strabag, Germany
- Mechanical contractor: Ponticelli, France
- Electrical contractor: Fabricom, Belgium

Project Objectives:
- Construction of a refining plant; engineering, procurement and construction (EPC) work for two refineries
- Start EPC work: January 2nd 2000; last plant ready for operations: August 1st 2001
- Cost should be fitting the low refining margins
- Safety: Meet Fluor Daniel standards

Selected Contract Form:
- Due to severe time constraints Ruhr Oel elected to execute the project in a for Germany new execution form, an alliance of 5 partners (partnering)
- One alliance contract (36 pages)
- All expenses paid at cost

- Sharing of over/under run of target price
- Bonus for timely completion
- No claims allowed against other partners

Project Execution:
- All partners had their own quality management system; the Fluor Daniel system formed the basis
- Common project execution plan
- Alignment meetings to share information and to get involvement and buy-in by project personnel
- An alliance board, which consisted of senior managers of each company and the project managers from Ruhr Oel and Fluor Daniel met monthly to monitor the project and check, that all parties remained aligned.

- Strong cooperation between and integration of all team members from all companies was supported.
- As no claims were possible according to their alliance contract, the only way to solve conflicts was through communication.

Key Project Results:
- The alliance managed to complete the EPC work for two refineries in 1 year and 7 months.
- In spite of 25 % increase in work due to scope development during the EPC phase the project was completed on time
- Cost: 9 % under the target price
- Plants operating as specified and meeting quality standards.
- All authority requirements fulfilled and positive publicity
- A very attractive way of working for all people involved
- Winner of the International Project Management Award in 2002.

Interpretation:
- Reduced competitiveness and the combination of core competences allowed to share risks and to create synergies
- This resulted in an improved quality and speed and allowed to meet the customer requirements, to deal with one entity

	Alliance	Normal
Basic Engineering (months)	6	8 - 10
Bidding and Evaluation EPC (months)	0	3
EPC phase	19	24
Delays due to changes	0	2
Change orders	open book	closed
Contract form	at cost with incentives	lump sum
Flexibility	high	limited
Cooperation between parties	high	limited
Team satisfaction	high	varying
Competitive bidding	very difficult	possible

Fig. C4.7: Benefits of partnering from Ruhr Oel's point of view

(a) Based on a presentation given at the IPMA World Congress in Berlin, 2002.

C 4.4   Design element: Virtuality

The perception of projects and programmes as virtual organizations

By definition, projects and programmes can be perceived as virtual organizations due to the cooperation of several organizations and due to members of the project organization working in different locations.

The perception of a project as a virtual organization promotes the fulfillment of integration functions through

- the adequate distribution of tasks between the cooperation partners which leads to the optimal use of core competencies,
- aconscious development of the project culture (see Chapter D3.1),
- the development of a common ICT infrastructure and
- the qualification of the members of the project organization for virtual cooperation.

New instruments of the ICT infrastructure are mainly project management groupware, project management portals and project-related telephone and video conferences.

The virtual characteristics of projects place new requirements on the members of the project organization. Corresponding ICT know-how, social competencies for team work with new cooperation partners and the willingness to apply core competencies without redundancies.

Team work in virtual projects is characterized by:

- team members in different locations without a common workspace,
- a lack of personal contact and informal contact between the team members, and
- a lack of a common history among the team members.

Undesirable developments are therefore more difficult to detect. Project team members can hardly be observed in their project work; attention to weak signals gains importance.

For this, adequate communication forms are to be combined (see Figure C4.8). A "big bang" project style is desirable in order to achieve the required project management quality, special rules for communication are to be established (see Figure C4.9) and on-site visits of the project manager to the project team members are to be performed..

		Time	
		Same	Different
Location	Same	Face-to-face communication • Workshops • Meetings • Informal contacts	Separate communication • Fax, e-mail • Documentation management • Discussion forum • Notice board
	Different	Communications at a distance • Fax, e-mail • Telephone conference • Video conference	Separate communication • Fax, e-mail • Documentation management • Discussion forum • Notice board

Fig. C4.8: Communication forms depending on time and place

Rules for communication in virtual project organizations
• E-mails instead of letters
• immediate confirmation of e-mails; answering e-mails within 24 hours
• 3 e-mails about the same problem are followed by a telephone call; then a visit
• the project manager visits the project team members every 2-3 weeks

Fig. C4.9: Rules for communication in virtual project organizations (example)

⋯⋗ **EXCURSUS C1: Virtual organization**

"Virtual" is something that exists, though not physically. "Virtual" means that at least one characteristic of the traditional object is missing. But supplementary characteristics create the additional benefit of the object.

A "virtual office" is, for example, factually in existence. Office functions are fulfilled but the traditional characteristics of an office, namely the common office space for the employees, face-to-face communication and continuous supervision of the employees, are dispensed with. With the additional characteristic "professional use of IT" the desired additional benefits, cost reduction and greater flexibility for the employees, are obtained.

A "virtual organization" is a goal-oriented cooperation between (legally independent) organizations. The organizations participate in the cooperation with their core competencies. To a third party they appear to be a single organization. The virtual organization has the potentials of a traditional organization, without having a comparable institutional framework (building, legal structures, uniform leadership, etc.).

Goldman, Nagel and *Preiss*[a] consider virtual organizations as a situational alliance of complementary core competencies which are distributed over a number of organizational units of a company or group of companies. *Bullinger*[b] defines virtual organizations as "...temporary horizontal and/or vertical location-independent cooperations of different companies". Cooperation in a virtual organization takes place along a value-added chain. Through the combination of the core competencies different companies can cross over organizational boundaries.

The characteristics of virtual organizations are:[c]

- temporary cooperation between independent organizations,
- integration of core competencies along a value-added chain,
- cooperation of organizations at different locations,
- ICT-supported communications,
- partnership-oriented organization, poor in hierarchy and with little formalism, and
- the existence of a culture of trust.

Figure C4.10 shows the visualization of a virtual organization.

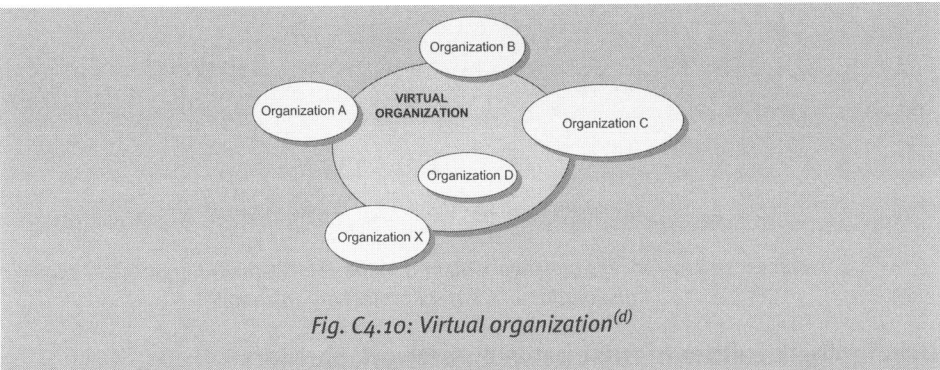

Fig. C4.10: Virtual organization[d]

Virtual organizations serve to accelerate business processes. The high speed of innovation being sought and the ability to react fast serve to increase the company's chances of survival. Efficient cooperations serve to reduce the costs of market entry. Creative, but undercapitalized companies can gain market influence by using virtual organizations.

CHAPTER

C

[a] Vgl. *Goldman, S. L.* et al, (Agil im Wettbewerb) 1996, p. 72.
[b] Vgl. *Bullinger, H. J.,* (Lernende Organisationen) 1996, p. 23.
[c] Vgl. *Gareis, R.,* (Virtuelle Organisationen) 1997, p. 2.
[d] Vgl. *Linde, F.,* (Virtualisierung von Unternehmen) 1997, p. 22.

The bases for virtual project organizations

Network Relationships

The prerequisites for virtual project organizations are network relationships between potential cooperation partners. Through networking the necessary basis is built up which makes the efficient and rapid establishment of virtual project organizations possible. The tasks of networks are, among others, the admission of new network partners, the development of a culture of trust between the network partners, the development of a common ICT infrastructure and the development of common "rules of play".

Conventional cooperations, such as strategic alliances or consortiums often require long preparations and because of that cannot make use of short-term market opportunities. In order to be able to design a situationally efficient virtual project organization the network relationships of the potential cooperation partners are necessary. "Networking" is therefore an important competence of project-oriented organizations.

The virtual project organization is very much dependent upon its context. The social network should serve as a basis for the temporary project organization and deliver a frame of reference which makes an efficient and rapid establishment, problem solving and following dissolution possible, without having to enter into time-consuming negotiations beforehand.

The process of networking and of the establishment, managing and dissolving of a virtual project organization is shown in Figure C4.11.

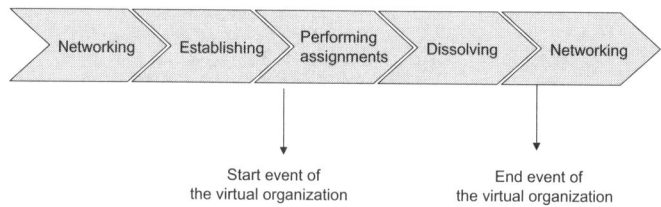

Fig. C4.11: The process of networking and of the establishment, managing and dissolution of a virtual project organization

Developing a culture of trust between network partners

The most important task of the social network is to promote trust between cooperation partners. "Partnership relationships can only flourish in a climate of mutual trust and open communications"[2]. To build up such a culture of trust integrative measures are of decisive importance. Among others, the practice of regular meetings of representatives of the project-oriented companies to deal with common themes has proved worthwhile. But social events, too, such as joint excursions or sports activities, promote informal communication and form a basis for good cooperation.

Common ICT infrastructure

A decisive factor for the success of virtual project organizations is the availability of a flexible ICT infrastructure with "plug-and-play" standardized interfaces which enable a quick adaptation of new products, processes or ICT systems of changing cooperation partners.[3] The ICT infrastructure should, on one hand, fulfill the prerequisite for decentralized access to information which is stored centrally and, on the other hand, support the rapid bringing together of expert knowledge which is distributed. In addition, the simultaneous distribution of information to all partners, also through telephone and fax, must be ensured. More than anything else, the simultaneous distribution of information contributes to the building up of trust.

Common "rules of play"

Networks require basic rules for cooperation. The processes of common product development and marketing as well as the acquisition and fulfillment of contracts are to be described, questions of liability for defects and damage are to be clarified, common conditions and practices are to be laid down, etc.

Common objectives, a common terminology and common standards, such as a common project and programme management approach, are to be agreed upon.

---

2) *Sydow, J.*, (Erfolg als Vertrauensorganisation) 1996, p. 16
3) Cp. *Mertens, P.*, Faisst, W. (Virtuelle Unternehmen) 1997, p. 65

Qualified contributors

The contributors to a virtual project organization must have certain competencies in order to do justice to the decentralized organization. Important competence characteristics are, for example, a distinct leadership competence, the willingness to take on responsibility, a high cultural sensibility, excellent communication and presentation abilities, a comprehensive willingness to pass on information undistorted and quickly, extensive ability in solving conflicts as well as a substantial technical knowledge.[4]

The above described ideal prerequisites for the design of a virtual project organization, i.e. the existence of networks of potential cooperation partners, are, however, in practice often not yet available. Nevertheless, the perception of the virtuality of projects should promote professional project management through the application of the models of the virtual organization.

C 4.5   Project organization model of *ROLAND GAREIS Project and Programme Managementt®*

The above described elements for designing project organizations offer various design possibilities:

*   "Empowerment" focuses on the transfer of decisional authorities and responsibilities to project team members, project teams and the project.
*   "Integration" promotes the design of a holistic project organization through the integration of representatives of customers, suppliers and sub-suppliers.
*   "Partnering" supports the integration of representatives of customers, suppliers and sub-suppliers through the use of a common incentive system.
*   "Virtual project communication" focuses on the design of an adequate, Internet-supported communication structure for the cooperation of project team members from different partners who are working in different locations.

These design elements complement each other. The project organization model of *ROLAND GAREIS Project and Programme Management*® provides for a combination of these design elements. In actual application this usually means an empowered and integrated project organization with virtual project communication structures. Supporting the integration of the partners through a common incentive system is a further design option. In the project organization chart both "empowerment" and the different partners integrated into the project organization can be visualized by suitable symbols.

---

4)   Cp *Krystek, U.*, et al., (Virtualität) 3/97, p. 15

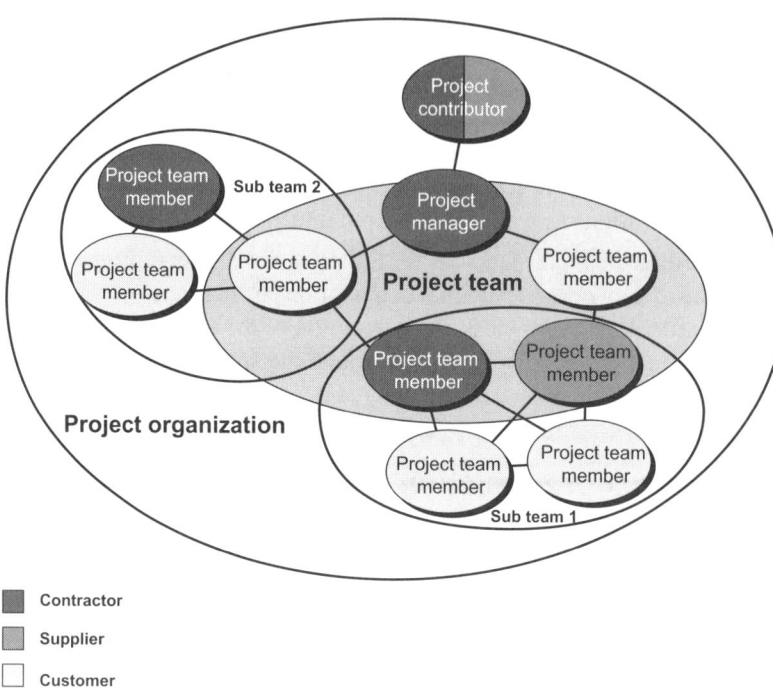

Fig. C4.12: Project organization chart of an "empowered" and "integrated" project organization

References

Bullinger, H. J. (Ed.), (Lernende Organisationen) Lernende Organisationen, Schäffer-Poeschel, Stuttgart, 1996

Gareis, R., (Virtuelle Organisationen) Projekte und Virtuelle Organisationen, Paper, PROJEKT-MANAGEMENT GROUP, Wirtschaftsuniversität Wien, 1997

Goldman, S. L., Nagel, R. N., Preiss, K., Warnecke, H. J., (Agil im Wettbewerb) Agil im Wettbewerb: Die Strategie der virtuellen Organisation zum Nutzen des Kunden, Springer, Berlin, Heidelberg 1996

Krystek, U., Redel, W., Reppegather, S., (Virtualität) Erfolgsfaktoren und Elemente der Virtualität, in: Gablers Magazin 3/97

Linde, F., (Virtualisierung von Unternehmen) Virtualisierung von Unternehmen, Dt. Univ. Verl., Wiesbaden, 1997

Mertens, P., Faisst, W., (Virtuelle Unternehmen) Virtuelle Unternehmen – eine Organisationsstruktur für die Zukunft? (Internet), 1997

Sydow, J., (Erfolg als Vertrauensorganisation) Erfolg als Vertrauensorganisation, Virtuelle Unternehmung, in: Business Review, 7 - 8, 1996

Van Wieren, H. D., (Alliance) Alliance, an Excellent Solution to Meet Project Execution Challenges, in: 16th IPMA World Congress on Project Management, *Making the Vision Work,* 4-6 June 2002, Berlin, 2002.

D

Teams, leadership and project culture

The conscious differentiation between groups and teams enables the application of adequate methods for group work or teamwork. In projects it is possible to differentiate between project owner teams, project teams and sub-teams.

"Leadership" is defined as a person-, group-, team- or organization-related intervention. Leadership in a project is a project management task which is to be performed by the project owner team, the project team, the project manager and the sub-team managers.

As a temporary organization a project has a specific culture. The culture of a project can be observed through the behavior of the members of the project organization as well as the methods and tools used in the project.

Symbolic project management supports the development of a project-specific culture. Through symbolic management the values and rules can be communicated to the members of the project organization. In contrast to the permanent organization, in which symbols are established and only seldom changed, symbols in projects (and programmes) are developed and used situationally.

D Teams, leadership in projects and project culture

Contents

D 1 Teams and teamwork in projects

(in cooperation with Michael Stummer)

D 1.1 Teams in projects

Teams versus groups

Both groups and teams are characterized by the participation of more than two people who, over a longer period of time, interact in a direct form and build up a specific norm and role structure. Groups and teams are established to develop and perform results which cannot be achieved by individuals alone.

The difference between groups and teams is mostly to be found in the quality of the underlying objectives, but also in the degree of responsibility for the achievement of the objectives. The group is characterized by being used primarily as a better means for achieving individual objectives. The team has a central common objective for which the team as a whole is responsible. A typical example of a group is the learning group, where the working form of the group serves to optimize the benefit for the individual group members. A football team is an example of a team where the achievement of the team objective ("winning") is the central point of the action. For the achievement of the team objective the team as a homogeneous whole is made responsible toward the relevant environments.

A typical characteristic of teams is that the abilities of the team members complement each other and the cohesion is higher than in groups. The work in teams requires a minimal amount of mutual trust and the building of "team spirit".

There are different types of teams. Each type of team has a specific leadership requirement (see Figure D1.1).

Types of Teams	
permanent teams	• set up for a long-term existence, continuity is to be ensured
temporary teams	• establishment, management and dissolution of team
working teams	• focus is on fulfilling the processes of the content work
management teams	• focus is on the performance the management tasks
large teams	• an appropriate size is up to a maximum of 12 persons
small teams	• up to 6 persons; the management and coordination tasks drop with the number of participating persons

Types of Teams	
heterogeneous teams	• different competencies of the team members
homogeneous teams	• similar competencies of the team members
local teams	• work together at the same location, have personal interaction
virtual teams	• separated from one another; use of various communication media

Fig. D1.1: Types of teams

In projects one can distinguish between project owner teams, project teams and sub-teams. Project teams and project owner teams are management teams, sub-teams are work teams. Projects have teams, but no groups.

An important feature of teams in projects is their temporary character. Teams in projects are usually earmarked by a relatively high time pressure. There is relatively little time in projects for team formation and teambuilding. In projects both large and small teams are used. A trend toward virtual teamwork can be observed.

CHAPTER

D

The team as social system

Teams can be perceived as social systems. Extending the typology of social systems developed by *Luhmann*, groups and teams are social systems to be ranked between interactions and organizations..

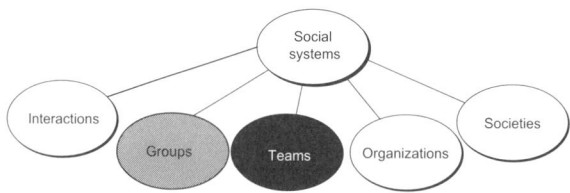

Fig. D1.2: Typology of social systems

A differentiation between groups and teams can be made on the basis of the level of formalization and the style of leadership. A high level of formalization exists when the actions are determined by norms and rules. Leadership can take place through emotions or through goal-oriented rationalism. In leadership by goal-oriented rationalism the orientation is mainly toward the achievement of goals.

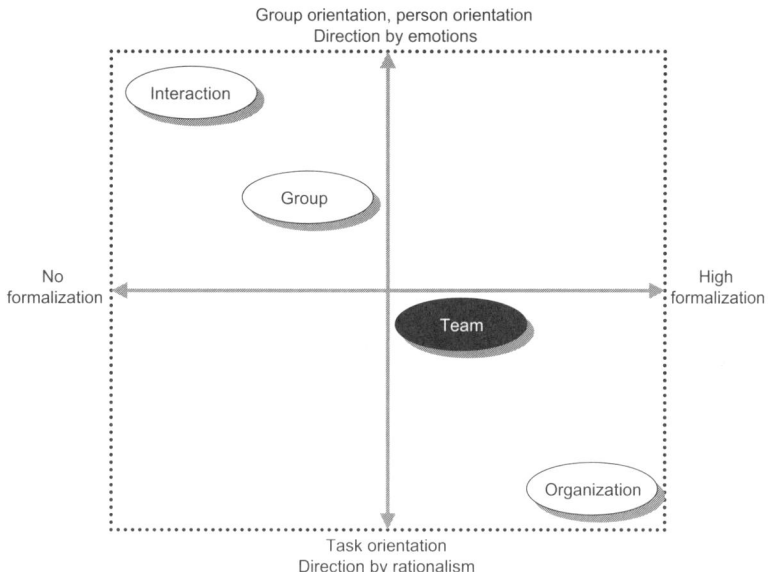

Group orientation, person orientation
Direction by emotions

Interaction

Group

No formalization

High formalization

Team

Organization

Task orientation
Direction by rationalism

Fig. D1.3: Classification of teams as social systems

As opposed to groups, teams are characterized by a higher degree of task-orientation and by a tendency toward a stronger formalism. Organizations, however, are characterized by an almost exclusive task-orientation. In addition, they demonstrate a higher level of formalism than teams and are characterized by the replaceability of individual persons.

D 1.2 Team lifecycles in projects

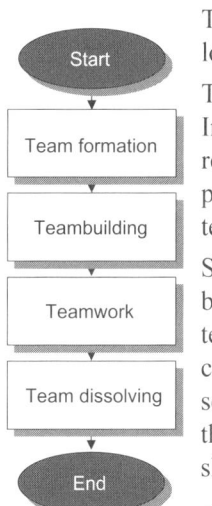

Start

Team formation

Teambuilding

Teamwork

Team dissolving

End

Teams in projects are characterized by their temporary nature. The following phases in the lifecycle of teams in projects can be identified.

The objective of team formation is the selection of the team members. Initial decisions about the formation of teams in projects will have already been made during the project assignment process. During the project start process the nomination of the members of the individual teams will take place.

Specialist competence and project management competencies are to be considered in the selection of the members of the project owner team, the project team and possibly the sub-teams. These competencies must be present to a different degree for each project role. The selection of the team members in projects is oriented not only toward the competencies of the team members, but also toward their relationships to other team members and to relevant project environments.

Fig. D1.4: Team lifecycle in projects

The selection of the members of the project owner team is made by the Project Portfolio Group (see Chapter I: Project portfolio management) of a project-oriented organization. The selection of the project manager is made by the project owner team. The selection of the remaining project team members is made by the project manager with the consent of the project owner team and the selection of the sub-team members by the responsible project team member with the consent of the project manager.

In order to reach the project objectives teams need team competencies. Team competencies are made up of the competencies of the individual project team members and the knowledge and the experiences of the team in regard to

- the common design of the project management process,
- learning in a team,
- the development of commitment in the team,
- the development of the "big project picture",
- the securing of synergies in the team and
- conflict-solving in the team.

The objective of teambuilding is the establishment of the team as a social system. The team competencies described above are to be developed during teambuilding. This happens through the promotion of a feeling of team membership and through the agreement to objectives, values, and rules for the team's work. The respective necessary communication process is supported by the use of project management methods, such as the work breakdown structure, project environment analysis, etc.

In addition to the formal structures which are to be agreed within the team, informal structures are also developed. Examples of the informal structures are informal communication paths, coalitions of team members and informal roles in the team (see Figure D1.5).

Informal role	Focus in the team
Performer	pushes through own opinion
Analyzer	develops solutions
Integrator	develops the relationships in the team
Controller	controls time and progress
Follower	avoids conflicts
Worker	performs the work

Fig. D1.5: Possible informal roles in teams

During the work of a team leadership tasks must be performed in order to enable the team objectives to be reached (see Chapter D2).

Once the team objectives have been reached the break-up of the team must be arranged. The social system "team" is to be dissolved, the existing relationships to persons and social systems are to be uncoupled and new relationships established. Methods for the dissolution of teams in projects are, for example, "social events", reflections in the team and feedback to the team members.

D 2   Leadership in projects

D 2.1   Leadership tasks and leadership roles in projects

Leadership in projects is a management task. The essential leadership tasks in projects are listed in Figure D2.1. Concrete examples of leadership tasks in projects are providing information about the project context through the project owner team or agreeing on objectives between the project owner team and the project manager or project team.

Leadership tasks in projects
• providing information
• agreeing on objectives and the distribution of tasks
• quality control and feedback
• making decisions
• contribution to solving conflicts
• creating the conditions for motivating the members of the project organization
• promoting learning and further development of both individuals and teams

Fig. D2.1: Leadership tasks in projects

Leadership tasks are to be fulfilled in all project management sub-processes. The context in which the leadership tasks are to be fulfilled is different in the individual project management sub-processes. For example, decisions involving the resolution of a project crisis are made under a greater time pressure than decisions in project controlling, and the provision of information in the project start process is more important than in the project close-down process.

Leadership tasks are to be differentiated from other management tasks, such as developing and updating project plans, writing protocols and reports, etc.

In projects, leadership tasks are performed not only by the project manager but also by the project owner team, by project team and sub-team managers. Leadership tasks in projects are to be fulfilled in regard to individuals as well as teams. The different leadership roles in projects and the respective "lead" are listed in Figure D2.2.

Leadership role in Projects	"The lead" in projects
project manager	project team member, project team, sub-team manager, sub-team and project contributor
project owner team	project manager and project team
sub-team manager	project contributor and sub-team

Fig. D2.2: Leadership roles and "the lead" in projects

D 2.2   Leadership styles in projects

The manner in which leadership tasks are fulfilled can be referred to as leadership style. Different classifications of leadership style can be found in reference works. One differentiation between a rather authoritarian and a cooperative leadership style is given, for example, by *Tannenbaum* and *Schmidt*[1]. According to *Tannenbaum* and *Schmidt* every leadership style can be classified on a continuum between complete decisional authority by the leader and complete decisional authority by the group..

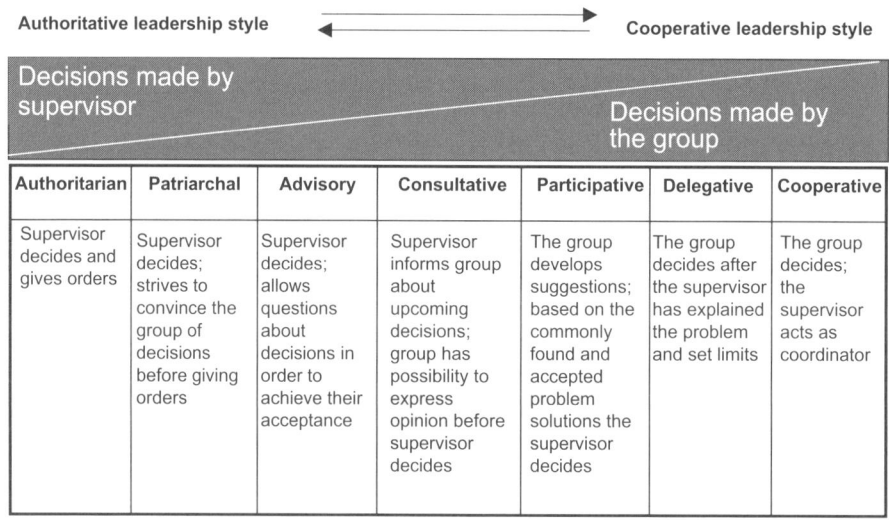

Fig. D2.3: Continuum theory of Tannenbaum and Schmidt

The leadership style of a person or a team can be adjusted to suit the respective context. In his contingency model, *Fiedler*[2] differentiates between the task-oriented and the person-oriented leadership style. Depending on the situation in which the leadership tasks are to be performed, he derives suitable leadership styles for different constellations.

---

1) Cp. *Weibler, J.*, (Personalführung) 2001, p. 300
2) Cp. *Steyrer, J.*, (Theorien der Führung) 2002, p. 202 ff

The context in which project leadership is to be fulfilled, is defined through the sub-processes of project management, the design of the project organization and the project culture.

D 2.3   Event-oriented leadership in projects

The motivation of the project team members and their productivity in the fulfillment of the objectives over the duration of the project cannot usually be kept at the same high level. But through the definition and the use of project events the "energy" in projects can be controlled.

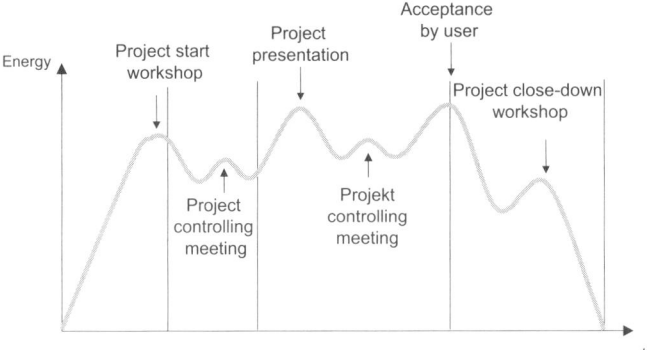

Fig. D2.4: Event-oriented leadership in projects

Possible ways of adding "energy" to a project are project meetings, project workshops and project presentations. One can refer to these as "project events". An objective of the periodic performance of such project events is the development of pressure within the project through the build-up of pressure from the outside. Pressure on a project from outside arises from the expectations of the participants in project events, regarding information about the status of the project and intermediary results being provided.

In order to give projects enough possibilities for regeneration and for continuing their work, it is important that project events do not take place very often. Decisions about the number of project events and their timing are to be made by the project manager and the project team together.

D 2.4   "Emotional" project management

Emotions in projects

In projects there are emotions, such as anger, fear, joy, sadness and surprise. Emotions are the intensive feelings of individuals, teams or organizations with a clear start and end. They relate to someone or something. The project team can, for instance, be happy about a successful project presentation or vexed by bad feedback.

Emotions in projects can be structurally caused or they can be specifically induced. It is a task of the management in projects to analyze the emotions that are expected (because they are structurally caused) and to plan and carry out strategies and measures for dealing with them. Conscious dealing with structurally caused emotions is a success factor in projects.

In the different sub-processes of project management different emotions exist. The uncertainty of the members of the project organization in the project start process is, for example, a structurally caused emotion. This uncertainty is present in every project during the start process.

Managing emotions in the project start process

Typical positive emotions of individuals to be expected in the project start process, are the joy about a new, interesting assignment, meeting new people (or a known person in a new role), of working in a new project team (with members of different cultural backgrounds). Typical negative emotions in the project start process are the fear of the new, of being overtaxed by the project work or by taking responsibility in the project (e.g. in the case of an "empowered" project organization).

Being overtaxed in a project leads to low motivation of the project team members as does underutilization (see Figure D2.5). High motivation of the project team can be ensured through the distribution of tasks according to the competencies of the team members.

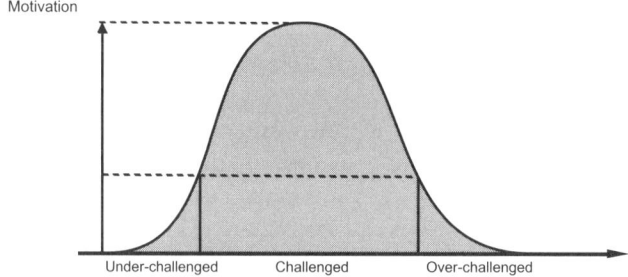

Fig. D2.5: Motivation of project team members according to overtaxation, underutilization and challenge

Measures for managing positive and negative emotions in the project start process are, for example:

- complete communication of the project objectives, the project organization, the project environment relationships, etc., for a conscious development of the project team's expectations. The communication should be supported by visual methods, such as a project organization chart and a project environment analysis. The desired project culture can also be projected through these symbols,

- communication of the project's contribution to the fulfillment of the strategic objectives of the project-performing organization to ensure the motivation of the project team members, to provide sense and meaning for their work,

- adequate allocation of assignments to individual project team members to assure the motivation of the project team members,

- clarification of the project roles to ensure the acceptance of the responsibilities carried by the project team members,

- development of an explicit process for teambuilding, for the integration of team members with different cultural backgrounds and for the agreement of organizational rules, and

- joint development of the work breakdown structure, the project environment analysis, etc. in the project team to ensure the acceptance of the project plans by the project team members.

Managing emotions in the project controlling process

Typical positive emotions to be expected in the project controlling process are the joy about reaching intermediary results, about good feedback and about creative problem solving. Typical negative emotions in the project controlling process are the fear of an unclear project status, of unjustified feedback, of the use of inadequate controlling methods and of too much competition within the project team.

Measures for managing the positive and negative emotions in the project controlling process are, for example:

- development of project controlling as an enjoyable and creative communication process, as a process to promote development in the project, and not as a mechanism to find mistakes,

- assurance of providing project information needed to communicate the "big project picture" for all members of the project organization. To support the communication the appropriate visual technology should also be used in project controlling (e.g. a project score card),

- use of adequate methods for project controlling to ensure understandable feedback. For example, the project results and the project score card assessments are to be interpreted (e.g. by the use of traffic light colors).

- interpretation of the benefit of the promotion of constructive conflicts in the project team for quality assurance in the project.

Managing emotions during the resolution of a project crisis

In the process of resolving a project crisis it is important to let negative emotions, i.e. insecurity and fear, develop in the project team as well as in the project owner team. A common project crisis reality must exist in order to create the basis for a common identity change in the project.

It may, therefore, be necessary for the project manager to create a certain shock in the project to prevent a delusion about the crisis situation. Such measures, for example a detailed analysis of the cause of the crisis and a demonstration of the potential damage, are unpopular. But they are functional because the necessary negative emotions are released.

Managing emotions in the project close-down process

During the project close-down process positive emotions should prevail. But the joy of having reached the project objectives can also stand face to face with sadness due to the pending separation from the other project team members, uncertainty about ones own career future and possibly also a burn-out syndrome when the pressure of the project work is gone.

The "burn-out syndrome" can be defined as a state of emotional exhaustion which leads to a reduction in personal productivity. A burn-out is associated with a feeling of emptiness. A burn-out is caused by distress such as time pressure, uncertainty of success, lack of support, problems with interactions, etc.

An example of burn-out syndrome of a project team is described in Case Study D1: Burn-out syndrome of the project team in the project close-down process of the pm days '02..

CHAPTER

D

···> **CASE STUDY D1: Burn-out syndrome of the project team in the project close-down process of the pm days '02**

Structures of the project pm days '02

- See Case Study E7: Project marketing of the project "pm days '02" in Chapter E.

pm days: General objectives and history

- perform a professional event, organized by the "project management experts" of ROLAND GAREIS CONSULTING (RGC) and the PROJEKT MANAGEMENT GROUP (PMG)

- use the event "pm days" as an instrument to control "energy" (motivation, commitment, management interest) in RGC and PMG (e.g. completion of marketing brochures to distribute at the pm days, caring for key accounts during the pm days, etc.)

- history of the pm days since 1983: "High-life" by means of interesting and innovative content, international participants, promotion of networking and of fun, good image and growth

Situation before and during the pm days '02

- Before the pm days '02:
  Missing project documentation, not enough project team meetings because of being a repetitive event (but partly with new team members); several parallel activities with access to the same resources (e.g. a start presentation of another project)

- During the pm days '02:
  Appearance of defects (such as projectors not functioning, wrong presentations on laptops); ad hoc measures, negative feedback from the project owner to the project manager, etc.

Emotions and emotional management during the pm days '02

- Emotions during the pm days '02:
  Surprise that not everything functioned correctly; insecurity on the part of the project team and the project owner; fear that the quality of the event would be bad; the project manager felt unjustly criticized; anger and frustration of the project owner whose resources were handled imprudently (additional stress, additional need for coordination), structural weaknesses were taken personally; a common sense of achievement for the project team did not come to pass

- Emotional management during the pm days '02:
  Only ad hoc measures were possible (settings for the projector were adjusted, acquisition of replacement equipment, etc.); integration measures for the project team in the framework of the pm days clubbing worked only partially; no time to work through the problems thoroughly

Emotions and emotional management after the pm days '02 events

- Emotions after the pm days '02: Insecurity about whether the weaknesses of the pm days '02 were apparent to the participants or not; break-out of "burn-out" in the project team during the reflection session on the evening after the practice conference (physical breakdown of several team members, common feeling of sadness and weakness, the immediate trigger was unjustified criticism from the project owner of a project team member during the reflection); little motivation for the performance of the expert seminar on the following day after the practice conference; "back in the office": All project team members and the project owner tired, distanced, slow, etc.; individual team members consider changing jobs!

- Emotional management after the pm days '02: Short-term closing of ranks to ensure a professional performance of the expert seminars; the situation was made subject of discussion both in individual discussions and in a team meeting in the following week; analysis of the problem; the Christmas holidays were used for rest and regeneration; planning of improvement measures in the future (planning of schedules, project management, etc.); introduction of a "mood barometer" at staff meetings

Emotionalizing in projects

Emotions are not only structurally caused but can also be specifically induced. "Emotionalizing" – the specific use of emotions in projects – is a leadership method to control energy in projects. Methods for "emotionalizing" in projects are, for example:

- telling of a secret by each project team member during the project start workshop to build trust within the team,

- performance of "project events" to develop periodic pressure in the project,
- discussing taboos in the project team in order to avoid fear and uncertainty,
- reflection of the cooperation process in the project team to reduce tension and solve conflicts,
- use of associative methods, such as pictures, parables, metaphors, etc., to create surprises and develop new perspectives in the project team,
- development of competitive situations in the project team in order to improve the quality of the project work.

The achievement of rational communication objectives in projects can be influenced by the development of the structural and emotional dimensions of communications. The model of the "three levels of communication" in Figure D2.6 differentiates between the rational, emotional and structural levels at which people (symbolized by a person with the emotional center in his/her belly) communicate.

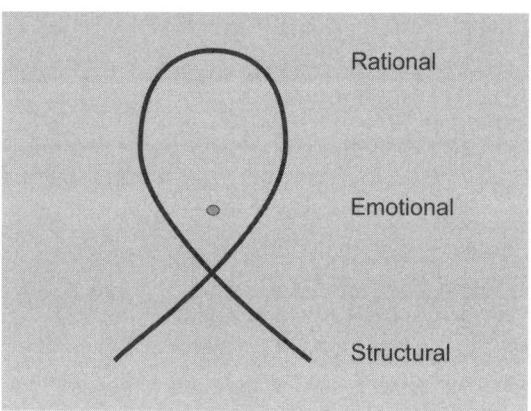

Fig. D2.6: Three levels of communication

The individual emotions of the participants in a meeting have a direct influence on the communication process. The structural conditions of a communication situation, such as the structure of the participants, the seating plan, the size of the room, the time available or the agenda of a meeting influence the possible results. The results are influenced directly and also indirectly by the emotions which are brought about by these structural conditions. The individual emotions of the meeting participants influence the communication process directly.

Emotional competencies in projects

To manage emotions in projects "emotional competence" is required. Emotional competence is especially required in projects by project managers, but also by the project owner team. It is a part of the social competencies and is, therefore, to be differentiated from other social competencies. Social competencies include the knowledge and experience needed to take action and make decisions in social situations, not only in emotional situations.

The "emotionally competent" project manager is aware of the meaning of emotions in the communication in projects. He/she is characterized by

- an emotional self-consciousness, i.e. a consciousness that emotions can influence the success of the project,
- an understanding of his/her own emotional position, i.e. an understanding for his/her own feelings and thoughts in a particular project situation,
- the ability to perceive the emotions of others and to analyze them, and
- the ability to plan and implement appropriate measures to manage emotions.

D 2.5 Leadership methods in projects

Descriptions of important leadership methods which should be used in projects are described below. A comprehensive description of the relevant leadership methods can be found in relevant reference works.[3]

CHAPTER

Feedback and reflection promote learning in projects, the "reflecting team" contributes to problem solving in projects and the moderation of project meetings and project workshops ensures an efficient provision of information and efficient decision-making.

Feedback and reflection for individual, collective and organizational learning in projects

The novelty of projects and the social demands made on teams make individual, collective and organizational learning a success factor in projects.

On the one hand, learning can be organized by feedback and, on the other hand, through a reflection of the respective social system.

Feedback is directed. The objective of giving feedback is to provide an outside point of view. The recipient of the feedback can be a person, a team or an organization. A feedback recipient can receive feedback from one or more feedback givers. When several feedback givers give feedback one speaks of 180° or 360° feedback (see Figure D2.7: 360° feedback to a project manager).

Since projects take place in a demanding social context, it is recommended to have as comprehensive an outside view as possible. The 360° feedback method promotes this objective and serves to offer the feedback recipient the broadest basis possible for further development.

---

3) Cp. *Stayrer, J.*, (Theorien der Führung) 2002, p 157 ff

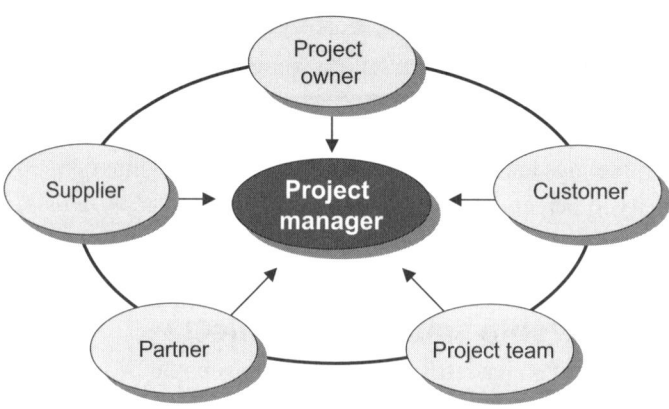

Fig. D2.7: 360° feedback to a project manager

For the professional arrangement of the feedback process the rules of giving and taking feedback are to be observed (see Figure D2.8).

Rules for giving and taking feedback	
Giving feedback	respecting others concrete tips acceptability positive suggestions maintain distance unmistakable; clear not in front of non-involved third parties relevant to current situation aware of consequences only when asked for
Taking feedback	ask if unclear no justifications accept what has been heard as subjective point of view thank for feedback

Fig. D2.8: Rules for giving and taking feedback

Projects and teams in projects are social systems whose control is possible through self-control. Social systems have the ability to observe themselves and to reflect. On the basis of this reflection the structures of the social system can be adapted.

Methods for reflection in projects are, for example, the flash light, the mood picture or the mood barometer. The objective of the reflection is to construct a common view of the current status in the social system and to recognize strengths and weaknesses. A common view makes it possible to agree on measures for the further development of the social system.

The flash light is performed with the help of a directed question (How are you?, How was it?, Are you satisfied with the intermediary results and with the working process?, etc.). The mood picture and the mood barometer use graphic tools to this end (see e.g. Figure D2.9).

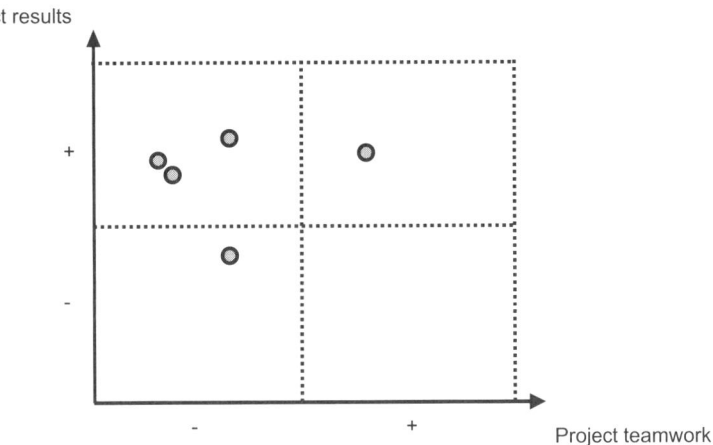

Fig. D2.9: Reflection with mood picture

"Reflecting team" for solving conflicts and problems

Looking at a problem from different perspectives is also important for solving problems. The "reflecting team" method makes efficient problem solving possible through clear structures and different roles.

The reflecting team method, which comes from family therapy[4], is based on the idea that not only the results of a work situation have meaning, but that the work process itself triggers changes. The special structures of the reflecting team method enable the observation of a problem from different perspectives.

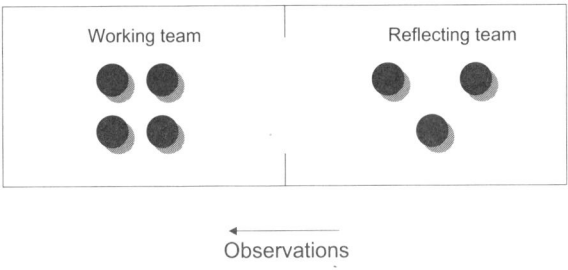

Fig. D2.10: "Reflecting team" situation

---

4) Cp. *Gester, P.,* (Systemische Gesprächs- und Interviewgestaltung) 1992, p. 153 f

In the problem-solving phase two teams are involved – the work team and the reflecting team – which sit clearly separated from one another. The reflecting team first takes on observer status and observes the work team as it works on the problem solution. Afterwards the reflecting team reflects its observations and adds its own views to the problem solution. This process of analyzing and working out a problem solution take place in moderated form with rigid time restrictions.

The clear distribution of roles and rules of play determine the working process. Through this work process, on the one hand, the complexity is reduced (through working on the problem in small groups) and, on the other hand, complexity is built up (through the additional point of view on the problem solution from the reflecting team).

Moderating project meetings and project workshops

A leadership task in projects is the moderation of meetings and workshops. Professional moderation has an important influence on the success of a meeting or workshop.

CHAPTER

D

The moderator is responsible for the development of the process of a meeting or workshop, content contributions are not expected. The nomination of a moderator is made by the project manager or project team. The role of the moderator can be fulfilled by the project manager or a project team member: They then fulfill two project roles simultaneously. In complex meetings or workshops it is recommended that an external moderator be used, in order to reduce the workload of the project manager. The project manager is then able to concentrate on the content of the meeting.

The expectations regarding a meeting or workshop can be shaped by the invitation to meetings and workshops, or by discussions with meeting and workshop participants beforehand. Shaping expectations is an important leadership task.

For the fulfillment of moderation tasks there are moderation methods which can be applied according to the situation.

Overview of moderation methods	
structuring/ overview	• metaplan • scenario • energy field • mood picture • mind map
innovation	• brainstorming • brainwriting
decision	• scoring • decision steps

D 3   Project culture

D 3.1   Development of a project-specific culture

As a temporary organization a project has specific values, norms and rules, i.e. a project-specific culture. The culture of a project can be observed by the behavior of the members of the project organization as well as from the methods and the communication forms used in the project.

The objective of the development of a project-specific culture is to develop a project identity which promotes the identification of the members of the project organization with the project and to give orientation within the project.

The identity of a project can be expressed in the project name and a project logo. Project-specific values provide a scale for measuring what is to be considered good, valuable and desirable in the project. It determines conscious and unconscious behavior of the members of the project organization. The project-specific values can be communicated internally and to relevant environments with the help of a "mission statement".

Project-specific rules should also provide orientation in the project. In addition to the existing organizational general rules in the project-oriented organization, project-specific rules can be made when needed. These can refer to signature authority, authority for financial decisions, guidelines for documentation and filing, rules for meetings, etc.

Time and space must be devoted to the development of the project culture during the project start process. A further development of the values and rules requires reflection in the project, which takes place mostly in the project controlling process.

⋯⋗ EXCURSUS D1: Company culture

Company culture is to be understood as "...a consistent whole made out of values, norms and symbols, which develops in a company as an answer to the requirements placed on the company as well as the needs of the people working there through the whole of the company's history. This is passed on consciously or unconsciously to new members of the company, particularly through the symbolic behaviors of roles models (dominant carriers of culture) and it affects the thinking and behavior of those working in the company in an unmistakable manner."[a]

Schein differentiates between the following three cultural levels:

- visible, but often not desipherable;

- greater level of awareness;

- taken for granted, invisible, preconscious.

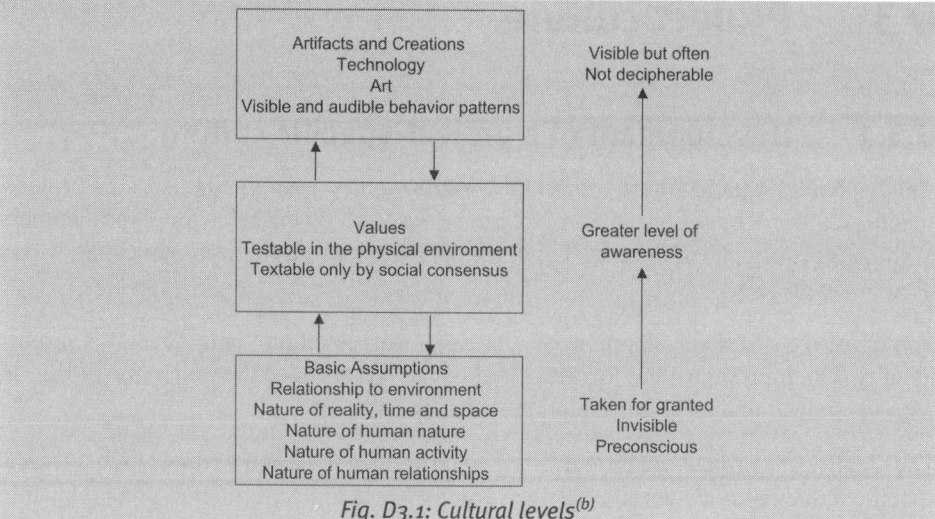

Fig. D3.1: Cultural levels[b]

Culture emanates from the human ability for socialization. Individuals bring those cultural standards, which they learned in their primary socialization as basic assumptions (Schein denotes this level in the organizational culture as "invisible" and "unconscious"[c]) into a company, in which these basic assumptions receive a virtual company-cultural specification via secondary socialization. For a scientific interpretation these society-, family-, etc. related cultural backgrounds are themselves only accessible in a limited way. They must be detected via the path of observable phenomena.

For this reason, a limitation on the mainly visible culture elements takes place. The internalized, socially anchored basis culture is taken as a prerequisite.

[a] *Hoffmann, F.,* (Unternehmenskulturen) 3/1989, p. 169 f.
[b] *Schein, E.,* (Organizational Culture) 1985, p. 6.
[c] Cp. *Schein, E.,* (Organizational Culture) 1985, p. 6 ff.

D 3.2 The objectives of symbolic project management

The use of symbols in project management serves to describe, to control the energy and to sustain the system.

Particularly in temporary organizations it is of great importance to describe the organization for representatives of relevant environments and also for new members of the project organization. With the help of symbols information can be passed on in a more complex manner. This way, values and norms can be passed on and orientation can be given.

Symbols can be used to motivate members of the project organization: Ceremonies (handing over the project assignment) and parties (to close down the project) positively influence the feelings of project team members toward the project.

Symbols "jog the memory" of feelings and experiences. For example, when a project team member looks at the project logo, he/she will perhaps recall a lively discussion which preceded the definition of the logo.

Symbols act to control energy in that they provide a possibility to vent negative feelings. If there are rivalries between project team members, these can be mellowed through the outlet of a contest (such as football, volleyball).

In projects in which project team members of several companies work together a professional symbolic management is necessary. Through the different pre-histories of the project team members there can be problems in understanding symbols. Symbols may be interpreted differently. This poses the challenge to provide a specific interpretation of the symbols.

The competencies for the adequate use of the methods of symbolic project management are of particular importance in the project-oriented organization. These competencies are organizational, for example, by providing tools and guidelines for project and programme management, as well as individual, and ensured by related training measures and coaching services for project managers.

CHAPTER

D

D 3.3　Methods of symbolic project management

Methods of symbolic project management are listed below, including the interpretation of the symbols, which surpasses the content of the artefact or the act.

Verbal symbols	Possible interpretation
project slogan	central objectives and values of the project
project-related anecdotes	central values and norms of the project
project language, jargon	belonging to the project
Interactional symbols	
availability of scarce resources	high strategic importance for the project
seating plan in project meetings	power in the project
social events	personal interest of the members of the project organization for each other
milestone party	start a new project phase
use of first names (in non-English-speaking culture the use of informal "you" as opposed to formal "you")	belonging to the project, setting boundaries to other organizations
burning old project plans	agreement of new project objectives
Objectified Symbols	
project organization chart	power in the project, meaning of individual roles, of relationships between roles
project logo, project name	challenges and objectives of the project

availability of a project room	high meaning for the project teamwork, home of the project
size and decoration of a work room	status in the project organization

Fig. D3.2: Methods of symbolic project management

These and other symbols and symbolic measures can be used in projects.

···ᐅ **EXCURSUS D2: SYMBOLIC MANAGEMENT**

Until the end of the 1970s symbols played a subordinate role in the research of business organizations and management.[a] At that time attention was focused on the development of efficient, rational and machine-like organizations. Since the beginning of the 1980s, however, importance was also given to the "soft facts" in organizations, which contribute significantly to the smooth operation of organizations. Apart from the company culture, "soft facts" can also be symbols, which can be viewed by themselves, but are also a part of the company culture.

A symbol is an action or an artefact which expresses meaning that goes beyond the actual content of the action or the object. A symbol enables interpretations of a larger contiguity. It is "a sign, which denotes something much greater than itself, and which calls for the association of certain conscious or unconscious ideas, in order for it to be endowed with its full meaning and significance."[b] "It is characteristic for symbols not to stand for a clear cut, reversible allocation of feature and meaning, but that this relationship is usually multi-valued or ambiguous and therefore leaves room for interpretation or for several possible connections."[c] Since both the coding and decoding process are mental, and therefore also cultural processes[d], different interpretations of symbols are possible. It is easy to understand that this happens most often between different companies. But also inside a company it may happen that employees live with symbol structures which they cannot interpret or understand.

Symbols can be divided into verbal, interactional and objectified symbols. Verbal symbols, such as anecdotes, slogans and speeches use speech as an instrument. Interactional symbols, such as parties, meetings and awards are symbolic actions. Objectified symbols, or symbolic objects, such as interior design, logo and organization chart, are artefacts of an organization.

A selection of these types of symbols is shown in Figure D3.3.

The need for symbolic management appears to be especially high in the following situations:
- when there is uncertainty about what is to be achieved,
- when acceptance should be increased,
- when individual, so far established views, meanings or objectives are to be changed,
- when employees can hardly be managed by their supervisors with regard to content,

- when organizations, should or must be given an altogether different identity, e.g. during crises,
- when loyalty, commitment and consensus are more important for success than expert knowledge.

The function of symbols in organizations can be divided into three categories: Descriptive, energy controlling and system sustaining.[e]

- Symbols are descriptive: A company can be described in different ways, for example by facts and numbers or by adjectives or with the help of symbols. One such symbol could be a story or anecdote. In a story the narrator can present information in a complex form, thereby clarifying the values of the organization to the listener.

- Symbols are energy controlling: Symbols can be used to motivate employees (through stories, ceremonies) and to recruit new employees (through slogans). Symbols make it easier to "jog the memory" about feelings and experiences. Symbols also work to control energy in that they provide a possibility to vent negative feelings about colleagues. This is, for example, the possibility presented by a sports contest.

- Symbols are system sustaining: Symbols are very important to protect a system, to stabilize it or to accompany a change in the system. These tasks can be fulfilled, for example, by a company party.

"Symbols help employees to interpret and understand the organization and their role in it by providing information about status, power, commitment, motivation, control, values, and norms."[f]

Symbolic management is not a stand-alone management theory; it should broaden it by a new facet. One cannot manage non-symbolically. The question is, whether one becomes conscious of it and whether one does it consciously.

Verbal symboles
stories, anecdotes, myths, parables, legends, sagas, fairy tales
beliefs, superstitions, rumours
slogans, mottoes, maxims, principles
phraseology, jargon, taboos
metaphors
speeches at parties, rhetoric
style and language of memos
nick-names for people and office equipment
humor, jokes
songs, hymns, poems

Interactional Symbols
rites, rituals, ceremonies, traditions, parties, feasts, jubilees
conventions
conferences, meetings
visits from the board
organizational development
selection and introduction of new employees, promotions
demotion, dismissal, voluntary dismissal, retirement, death
complaints
magic acts (employee selection, strategic planning, etc.)
taboos
games, relaxation
pranks
gestures
food-sharing

Objectified Symbols
status symbols
insignia, emblems, gifts, flags
logos
prizes, certificates, incentive travels
idols, totems, fetishes
clothing, outward appearance, dress codes, uniforms
architecture, design and decoration of the work space, office furniture Working conditions
quality and placement of office machines
posters, brochures, company newspapers, photos
systems which are fixed in writing (wage calculation, pay grades, promotions)
organization chart, handbooks
bulletin board (placement, content)

Fig. D3.3: Verbal, interactional and objectified symbols

(a) *Weibler, J.,* (Symbolische Führung) 1995, p. 2017.
(b) *Pondy, L. R.* et al, (Organizational Symbolism) 1983, p. 4 f.
(c) *Neuberger, O.,* (Führen und geführt werden) 1990, p. 245.
(d) Cp. *Jones, M. O.,* (Organizational Symbolism) 1996, p.9 ff.
(e) Cp. *Dandridge, T. C,* (Symbols) 1983, p. 71.
(f) *Daft, R. L,* (Symbols in Organizations) 1983, p. 198.

References

Daft, R. L., (Symbols in Organizations) Symbols in Organizations: A Dual-Content Framework for Analysis, in: *Pondy, L. R., Frost, P. J., Morgan G.* (Ed.): Organizational Symbolism, JAI Press Inc. Greenwich, Connecticut, 1983

Dandridge, T. C., (Symbols) Symbols´ Function and Use, in: *Pondy,* L. R. et al. (Ed.), Organizational Symbolism, Jai Press, Greenwich, Conneticut, 1983

Gester P., Warum der Rattenfänger von Hameln kein Systemiker war? Systemische Gesprächs- und Interviewgestaltung, in: *Schmitz, C./Gester P./Heitger B.*, (Ed.), Managerie – Systemisches Denken und Handeln im Management, 1. Jahrbuch, Carl Auer Verlag, Heidelberg, 1992

Hoffmann, F., (Unternehmenskulturen) Erfassung, Bewertung und Gestaltung von Unternehmenskulturen – Von der Kulturtheorie zu einem anwendungsorientierten Ansatz, in : zfo 3/1989

Jones, M. O., (Organizational Symbolism) Studying Organizational Symbolism, Sage Publications, Thousand Oaks, 1996

Neuberger, O., (Führen und geführt werden) Führen und geführt werden, Ferdinand Enke Verlag, Stuttgart, 1990

Pondy, L. R., Frost, P., Morgan, G. (Ed.), (Organizational Symbolism) Organizational Symbolism, JAI Press Inc., Greenwich, Connecticut, 1983

Schein, E., (Organzational Culture) Organzational Culture and Leadership, Jossey Bass, San Francisco/Washington/London, 1985

Steyrer, J., Theorien der Führung, in: *Kasper, H., Mayrhofer, W.,* Personalmanagement, Führung, Organisation Linde Verlag, Wien, 2002

Weibler, J., (Symbolische Führung) Symbolische Führung, in : *Kieser, A., Reber G., Wunderer, R.* (Ed.), Handwörterbuch der Führung, 2. Auflage, Schäffer-Poeschel Verlag, Stuttgart, 1995

Weibler, J., (Personalführung) Personalführung, Verlag Franz Vahlen, München 2001

CHAPTER

E

Project management

Project management is a business process of the project-oriented organization. It comprises the sub-processes of project start, project coordination, project controlling, (possibly) resolving of a project discontinuity and project close-down.

The objectives, processes, responsibilities and the quality of the results of the project management process and its sub-processes can be documented and measured.

E Teams, leadership in projects and project culture

Contents

E 1   Project management and its context

Project management is a business process of the project-oriented organization which is fulfilled in projects.

The methods of process management can be used to describe the project management process. At a macro level the project management process is to be given boundaries and differentiated from other processes. At the micro level the objectives, tasks, responsibilities and results of the project management process and its sub-processes are to be described.

The documentation of the project management process promotes communication about the objectives, tasks, responsibilities and results of project management. The definition of the results of the project management sub-processes makes it possible to measure and evaluate the performance of project management and the quality of project management.

The description of the project management process also provides the basis for a targeted further development of the individual and the organizational project management competencies in the project-oriented organization.

E 1.1   Overview: Project management

The objective of the business process of project management is the professional management of projects. A prerequisite for the realization of the project content objectives is the professional fulfillment of the sub-processes of project start, project coordination, project controlling, (possibly) resolving of a project discontinuity and project close-down.

In project management the project objectives, project scope, project schedule, project resources, project costs, project earnings, project risks, project organization, project culture and project context are considered. The dimensions of the project context are the pre- and post-project phases, relevant project environments, other projects, the company strategies and the business case for the investment which is initiated by a project.

The process objectives, the process boundaries and a rough process structuring can be defined for the project management process as a whole. The objectives, tasks, responsibilities and results can be presented in detail in the descriptions of the project management sub-processes.

Objectives of the project management process
- developing the structural prerequisites for the realization of the project objectives
- efficient performance of the project start, project controlling, project close-down and the continuous project coordination
- possibly: Efficient resolution of a project discontinuity
- management of the social-, time- and content-related project boundaries
- management of the relationships of the project to the project context
- building up and reducing of project complexity

- management of the project dynamics
- non-objective: Realization of the content work of the project (Note: This is an objective of the project and not of project management)

Time boundaries of the project management process
- Start: project assigned
- End: project approved

Structure of the project management process

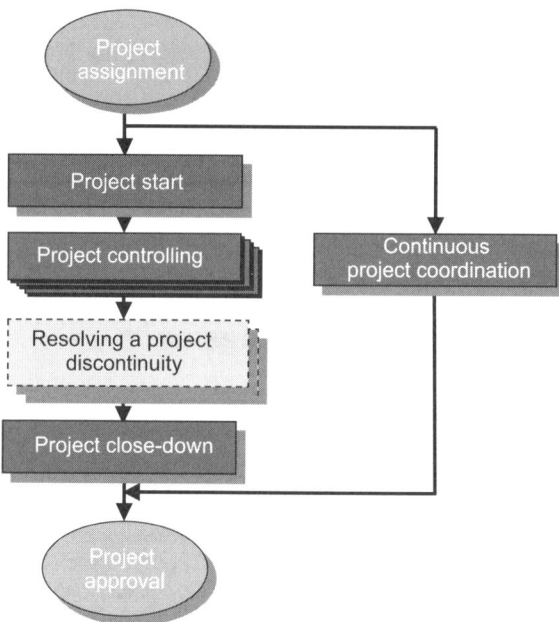

Fig. E1.1: Project management process

From a systemic point of view, it is the objective of the project start to establish the project as a social system. The objective of project controlling is to promote the evolution of the project and the objective of the project close-down is to dissolve the project as a social system. The objective of resolving a project discontinuity is to develop a new project identity in order to resolve the discontinuity. The objective of continuous project coordination is to ensure the project progress.

The project coordination process is performed continuously. The performances of the other project management sub-processes are limited in time.

By definition, the project start and the project close-down are each performed only once. Project controlling is performed several times in a project and takes place either periodically or at project milestones. The necessity of resolving a project discontinuity depends on the situation.

The benefit of a common view of the project management processes lies, on the one hand, in ensuring the uniformity of the project management approach used and, on the

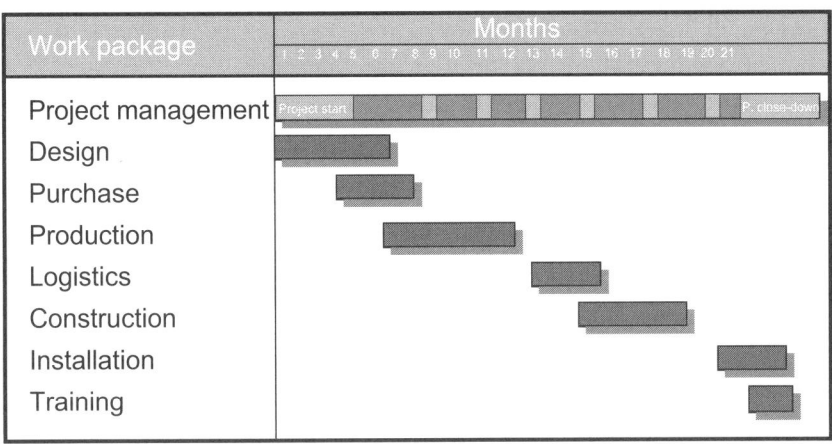

Fig. E1.2: Project management sub-processes in an engineering project

other hand, in considering the relationships between the sub-processes of project management.

The application of a uniform project management approach ensures that uniform terminology and methods are used in all sub-processes. A professional project management considers the relationships between the sub-processes in order to optimize the project management results. The following relationships between the project management sub-processes exist. For example:

• at the project start the structures for project controlling and project close-down are planned,

• the criteria for evaluating the project success at project close-down are determined at the project start by defining the project objectives,

• at the project start the working methods to be applied during project controlling and during project close-down are established (e.g. project meetings, project workshops, developing minutes, reflections),

• through the application of the scenario technique and the development of alternative plans at the project start, potential measures for the resolution of a project discontinuity are provided,

• the management of any structurally determined change of identity of the project is planned at project start,

• in project controlling the project plans developed at the project start will be controlled and possibly adapted,

• when managing a project discontinuity the alternative plans developed at the project start and/or the current project plans from the latest project controlling can be used,

- at project close-down the plans which were developed at the project start and adapted during project controlling form the basis for the evaluating the project success and for ensuring organizational learning, and
- project marketing is performed in all sub-processes of project management based on a uniform project marketing strategy.[1]

E 1.2   The context of the project management process

The context of the project management process in terms of time are the project assignment and the investment controlling processes. In terms of content these are the content-related business processes.

The content-related business processes depend on the project type. For an ICT-conception project, the following content-related business processes have to be performed, according to the project phase structure: Gathering information, analysis of the current situation, definition and description of alternative solutions, implementation plan for each alternative, decision-making. For a contracting project the engineering, the procurement, the production and the logistics, the construction and the installation as well as the training and the commissioning make up the content-related business processes.

To increase the efficiency of projects in the project-oriented organization business processes for the performance of the project contents are to be documented and standardized. Standards can be adapted for each project by considering individual project requirements.

Content-related business processes are performed during a project parallel with the project management process. The relationship between the project management process and the business processes for the performance of the project contents is immediate, as it is an objective of the project management process to develop appropriate structures for the fulfillment of the project contents.

The project assignment process is performed before the project start, while that of the investment controlling process is performed during the project and after the project close-down. The project assignment process is especially important for project management because in it the basic structures for the project are determined. The definition of the project objectives, the planning of the project organization and the drafts of the project plans are roughly worked out.

Project management is important for the investment controlling process since the project documentation constitutes and an essential foundation for the controlling.[2]

E 1.3   Description of the project start process

The project start process can be described with regard to its objectives, time boundaries, tasks and responsibilities as well as the tools to be used.

---

1) Project marketing as an integrative project management task is described in Chapter D1.8.
2) The project assignment process and the investment controlling process are described in Chapters I2 and I5.

Objectives of the project start process

- information transfer from the pre-project phase into the project,
- definition of expectations regarding the post-project phase,
- development of adequate project plans for managing the project objectives, scope, schedule, resources, costs, income and risks,
- design of the project organization, adequate integration of the project into the permanent organizations,
- development of the project culture,
- establishment of communication relationships between the project and other projects and relevant project environments, initial project marketing,
- communicating the "big project picture" to all members of the project organization,
- planning of measures for discontinuity management,
- definition of the structures for the following project management sub-processes,
- developing the documentation "Project start" and
- efficient design of the project start process

Time boundaries of the project start process

- Start: project assigned
- End: documentation "Project start" filed
- Duration 2-3 weeks

Demand for a professional project start

Due to the time pressure of projects once they are assigned it is tempting to start the content-related business processes immediately, without having performed the corresponding project start process. This lack of willingness to perform the project planning and the design of the project organization together with the project team often results in:

- unrealistic project objectives and unclear definitions of roles,
- project plans which are inadequate and not binding,
- unclear agreements regarding the design of project environment relationships and missing organizational rules.

A professional project start is to be performed in order to ensure adequate project management quality.

Tasks and responsibilities in the project start process

Tasks \ Responsibilities	Project owner team	Project manager	Project team	Project team members	Project management consultant	Expert Pool manager	Representatives of relevant project environments	Documents
Planning the project start								
• Checking the project assignment and the results of the pre-project phase		P						
• Selecting start communication form		P						
• Selecting project team members (and a PM consultant)		C				P		
• Selecting PM methods and PM templates to be used		P						
• Agreeing with project owner	C	P						1)
Preparing the project start communications								
• Hiring of a project coach (possible)		P			(C)			
• Preparing start communications I, II, etc.		P			(C)			
• Inviting participants		P					C	2)
• Documenting the results of the pre-project phase		P		C	(C)		C	
• Developing drafts for planning, organizing and marketing the project		P		C	(C)		C	
• Developing information material for start communication		P		C	(C)			3)
Performing the project start communications								
• Distributing information material to participants		P						
• Performing start communication I	C			P	(C)		C	
• Developing draft of PM documentation "Project start"		P			(C)			
• Performing start communication II, etc.	C			P	(C)		C	
Follow-up to the project start communications								
• Completing draft of PM documentation "Project start"		P			(C)			
• Agreeing with project owner	C	P						4)
• Project marketing: Initial information	C			P	(C)		C	
• Distributing PM documentation "Project start"		P				I		
• Filing of PM documentation "Project start"	C			P			C	
Performing first work packages (parallel)				P			P	

Legend:
P ... Performance
C ... Contribution
I ... Information

Documents:
1) List of project management methods to be used
2) Invitation of participants to the project start workshop
3) Information material for the project start workshop
4) Project management documentation "Project start"

Fig. E1.3: Description of the project start process

Phases of the project start process

The project start process as well as the other project management sub-processes can be structured into the phases planning, preparation, performance and follow-up.

The planning of the project start process includes a brief situation analysis based on ex-isting project documents (e. g. project assignment, initial project plans), minutes of meet-ings and interviews with members of the project organization and representatives of rel-evant environments. The design of the project management process is to be developed, i. e. the use of project management methods and of standard project plans, the selection of communication forms, the use of ICT, the use of a project management consultant and/or project management coach have to be decided on.

To prepare the project start communications the results of the pre-project phase are to be documented, the initial project plans are to be developed and the project marketing is to be planned. Project crisis-avoiding and/or project chance-promoting measures are to be planned and provisional measures for the management of project discontinuities are to be taken. A structurally determined change of a project identity may possibly have to be prepared.

The participants in the individual communication situations are to be selected and in-vited. The performance of the project start communications is usually a combination of individual meetings with project team members, a kick-off meeting and a project start workshop. For projects of high complexity several workshops and also joint social events may be necessary.

The follow-up work to the project start communications includes the development of the documentation "Project start" and its distribution to the members of the project or-ganization, as well as the performance of initial project marketing.

Organization of the project start process

The project owner team, the project manager, the project team and the individual project team members are responsible for the performance of the project start process.

The project manager may be supported in the documentation work by a project man-agement assistant or a project controller. In a socially complex project start process a project management consultant or a project management coach can be called in. When a project-external consultant takes over the moderation and documentation tasks the work load of the project manager and the project team members are reduced.

In small projects the preparation of the project start communications can be developed by the project manager alone. In projects of high complexity the project manager can perform the task in cooperation with 2-3 selected team members.

The project owner team, and possibly representatives of relevant environments, can be presented with the results of the project start workshop in authentic form with the help of developed flip-charts, graphics, etc. at the end of the workshop.

The resources (personnel, equipment, ICT, external consultants, etc.) and costs required for the project start process should be determined and assigned to the work package "project start" in the work breakdown structure.

Project start without a formal project assignment process

For the performance of a professional project start the results of the project assignment process, i.e. a formal project assignment document, a business case analysis and the initial project management documentation, are required. In practice, however, these ideal prerequisites are not always available.

When no formal project assignment process has been performed before the project start, the definition of the project objectives, the development of a business case analysis, the selection of a project owner team and the development of the initial project management documentation, is to be made up during the project start in a cyclical process of concretization.

In the case of starting a project without having performed a formal project assignment process there is a high risk that the project will not be performed at all. By developing a business case analysis it may become apparent that the investment to be initialized by the project is not profitable.

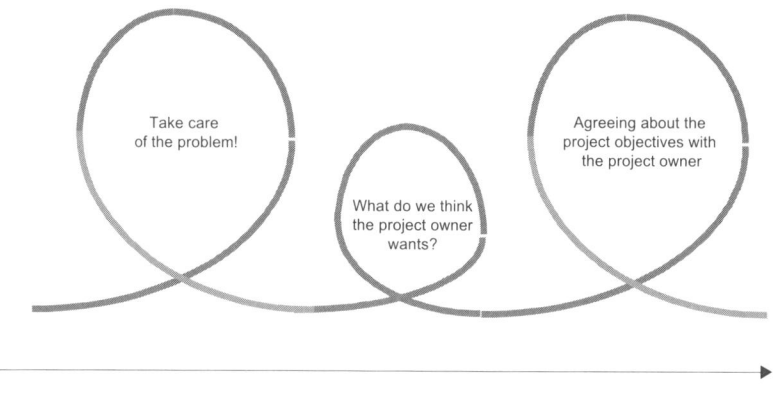

Fig. E1.4: Concretizing the project assignment

Quality of the project start process

The quality of the project start process can be measured by an evaluation of the results and the duration and the costs of the project start process. The results of the project start process are, above all, the developed project plans and the established project as a social system. The project plans can be evaluated in terms of their existence and their quality (completeness, level of detail, fulfillment of formal criteria, etc.). The project start process is documented in the meeting and workshop minutes.

It must be an objective of the project to maintain the high level of project management quality achieved by a professional project start process. Therefore, analogous to the project start regular project controlling meetings and/or workshops are to be performed.

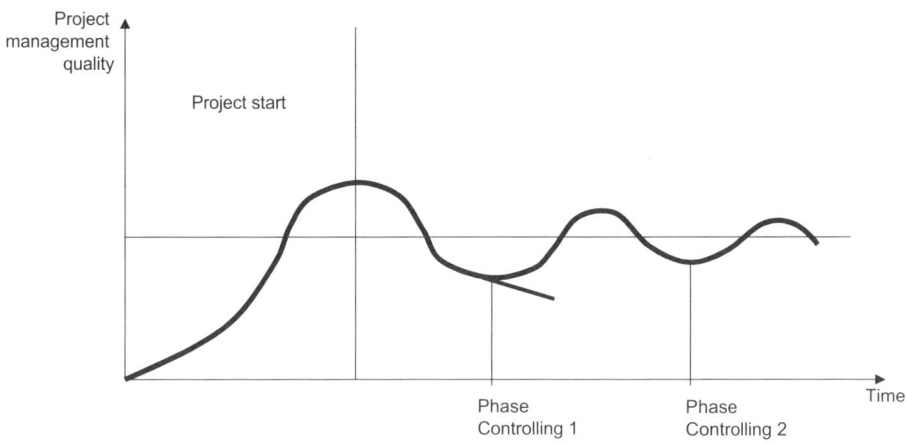

Fig. E1.5: Project controlling for maintaining the project management quality

Different project starts for different types of projects

Different types of projects lead to different demands and potentials for the project start:

- A conception project and a realization project are an example for a chain of projects. In a conception project drafts of project plans are developed for the realization project which might follow. These constitute an essential basis for the project start of the realization project.

- For repetitive projects project plans can be standardized. For unique projects, however, new solutions are needed. Therefore, specific working forms and creative techniques are to be used in the project start. Unique projects usually attract a great deal of attention at the project start. The project start process for repetitive projects can be upgraded, for instance, by inviting external participants to the project start workshop.

- In contracting projects representatives of the customer are to be actively involved in the project start process. They may become members of the project owner team and/or members of the project team.

⋯⋗ **CASE STUDY E1: Start process for the project "Restructuring the personnel management system"**

Company performing the project:
- an Austrian service company; approximately 500 employees

Project objectives:
- restructuring the personnel administration and the personnel development systems for all employees
- adaption of the database structures and improving the IT support

Project duration:

- 6 months

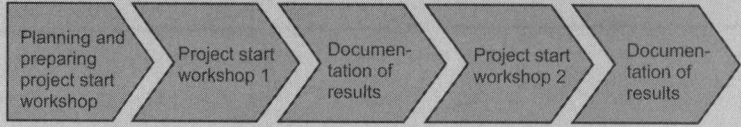

Planning and preparing the project start process:

- planning and preparing by the project manager and 2 project team members)
- restructuring concept as the basis
- developing initial project plans
- invitation to 2 one-day workshops for the project team, project owner and guests (union representatives)

Design of project start workshop 1:

- definition of the project boundaries
- information about the project objectives, objects of consideration, context (relationship to company strategies, with other projects, with relevant environments), milestones
- expectations of the project team members, roles in the project
- further development of the initial objects of consideration plan and the work breakdown structure
- definition of central values, selection of the project name
- presentation of intermediary results to the project owner team

Documentation of the results of project start workshop 1:

- development of the initial documentation "Project start" by the project manager
- IT documentation of the work breakdown structure, etc.

Design of project start workshop 2: Project planning

- discussing the initial documentation "Project start"
- description of selected work packages
- project scheduling, project cost planning
- adaptation of the descriptions of the project roles, planning of the project communication forms for project controlling, project close-down
- rough project risks analysis
- presentation of the results to the project owner team

Documentation of the results of project start workshop 2:

- completion of the documentation "Project start" by the project manager
- sending the documentation to all members of the project organization
- project briefing with representatives of the relevant project environments, initial project marketing

E 1.4   Description of the project coordination process

The project coordination process is depicted in Figure E1.6 with regard to its objectives, time boundaries, tasks and responsibilities.

Objectives of the project coordination process

- continuous information of the members of the project organization and of representatives of relevant project environments,
- ensuring the project progress and the quality of the work package results by controlling of the work package progress and acceptance of work packages results,
- continuous coordination of the project resources,
- control of the work package progress and
- acceptance of work package results.

Time boundaries of the project coordination proce

- Start: project assigned
- End: project approved

Tasks and responsibilities in the project coordination process

Tasks \ Responsibilities	Project owner	Project manager	Project team members	Sub-team	Expert Pool manager	Representatives of relevant project environments	Documents
Communicating with project owner team	C	P	C				1)
Communicating with project team members and project contributors					I		2)
Communicating with representatives of relevant environments	C	P	C			C	1)
Continous project marketing	C	P	C				
Participating in sub-team meetings			C		P		3)

Legend:
P ... Performance
C ... Contribution
I ... Information

Results/documents:
1) To-do lists
2) Acceptance certificates
3) Meeting minutes

Fig. E1.6: Description of the project coordination process

Demand for a professional project coordination

In contrast to periodically performed project controlling project coordination is performed continuously. Project coordination is the ongoing task of the project manager.

Beside continuous information of the members of the project organization and of representatives of relevant project environments, objectives of project coordination are continuous project marketing, coordination of the project resources, ensuring the project progress and the work package quality. The project progress is ensured by the project manager controlling the progress of individual work packages, coordinating the relationships between the work packages and accepting the results of individual work packages.

Performing the project coordination process

The project manager communicates in meetings with individual members of the project team, via telephone, via e-mail or fax or in video conferences. The participation of the project manager in sub-team meetings represents a further communication form in the project coordination process.

Since the project coordination process is ongoing, it must not be explicitly planned, prepared, performed and followed-up. The individual communications of the project manager, however, require this structuring – on a micro-level.

Organization of the project coordination process

The project manager performs the project coordination process by communicating with the project owner team, project team members and project contributors as well as representatives of relevant project environments. The project manager can be assisted by a project management assistant or a project controller.

Quality of the project coordination process

The quality of the project coordination process depends on the quality of the communication of the project manager with the other members of the project organization and representatives of relevant project environments. On the one hand, this requires appropriate social competence and, on the other hand, appropriate communication tools are to be used. These are, above all, the project management documentations developed in the project start process, such as the project objectives plan, the work breakdown structure or the project environment analysis, but also specific tools of project coordination, i.e. to-do lists, meeting minutes and work package acceptance certificates. The commitment established in the communications determines the quality of the project coordination process.

E 1.5   Description of the project controlling process

The project controlling process is depicted in Figure E1.7 with regard to its objectives, time boundaries, tasks and responsibilities.

Objectives of the project controlling process

- determining the project status, constructing a common project reality,
- agreeing on directive measures,
- adapting the project objectives, further developing the project culture and the project organization,
- updating the project plans,
- developing project controlling reports (project progress reports, project score cards),
- organizing organizational learning of the project,
- efficient design of the project controlling process.

Time boundaries of the project controlling process

- Start: initialization of the project controlling
- End: project controlling report filed
- Duration: 1-2 weeks

Tasks and responsibilities in the project controlling process

Demand for professional project controlling

Since changes occur in a project, such as changes in the objectives, changes in the availability of resources, etc., and the level of information improves during the performance of the project, it is necessary to periodically perform a project controlling.

The periodical project controlling is to be planned subject to the project duration. In a product development project of 6 months duration it is recommended to perform a formal project controlling every 2-3 weeks, in an engineering construction contracting project with a duration of, for example, 24 months, formal project controlling meetings with the project owner team will be necessary every 2 months, and short project progress reports once a month. It is recommended to perform project controlling upon reaching project milestones.

The evolution of the project, which results from the dynamics of the project itself and from the dynamics in the project environment relationships, must be followed. Possible deviations of the actual data from the planned data are to be identified and directive measures for utilizing new potentials and/or for correcting undesirable deviations are to be set. Chances for organizational learning of the project are to be used.

The project controlling refers to all objects of consideration of the project management, not only the project scope, the project schedule and the project costs. Within the framework of "social" controlling, above all, the project organization, the project culture and the relationships to the relevant project environments are to be controlled.

In the project controlling changes are to be promoted, deviations are to be viewed as learning chances. Project controlling requires rather an approximate approach than a detailed one, a thinking in alternatives rather than adhering to wishful thinking.

Tasks / Responsibilities	Project owner team	Project manager	Project team member	Project team	Expert Pool manager	Representatives of relevant project environments	Documents
Planning the project controlling							
• Adapting established project controlling structures		P					
• Reviewing project controlling structures	C	P					1)
Preparing the project controlling communications							
• Acutal data collection and planned versus actual analysis		P	C				
• Deviation analyses, planning directive measures		P	C		C		
• Developing drafts of adapted project plans and project controlling reports		P	C				
• Preparing project controlling communications							2)
Performing the project controlling communications							
• Distributing information material to participants	I	P		I	I		
• Performing project controlling communication I	C	C1			P	C	
• Performing project controlling communication II	C	C1			P	C	
Follow-up to the project controlling communications							
• Completing adapted PM documentation, project controlling report		P					3,4)
• Initializing updating of project portfolio database		P					
• Project marketing	C	C1	P			C	
• Distributing project controlling report, project score card		P		I	I	I	
Performing work packages (parallel)			P			P	

Legend:
P... Performance
C... Contribution
I ... Information

Documents:
1) Adapted project controlling structures
2) Invitation of the participants to the project team meeting
3) Adapted project management documentation
4) Project controlling reports

Fig. E1.7: Description of the project controlling process

Phases of the project controlling process

The project controlling structures must already be planned in the project start process (frequency of controlling, content and form of the reports, communication forms, etc.). This planning possibly needs to be adapted when starting the project controlling. The decision about the minimum requirements project controlling has to meet, must be taken by the project owner team. This decision depends on the type of project. For different types of projects different controlling standards are functional.

For preparing the project controlling communications the following tasks must be fulfilled by the project manager and the project team members:

- project control: Determining actual data, performing planned versus actual analyses, performing deviation analyses,
- project direction: Planning directive measures,
- adaptation of the project plans: Updating the project plans,
- development of the project controlling reports: Developing project progress reports, project score cards, deviation trend analyses.

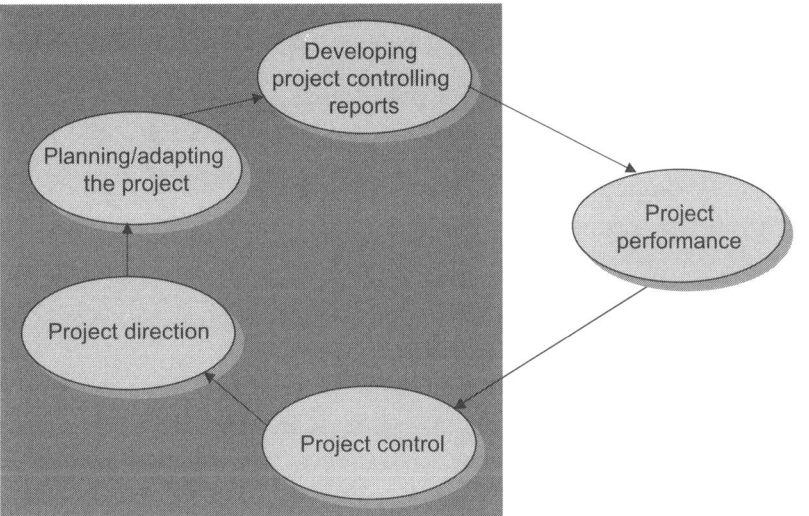

Fig. E1.8: Project controlling cycle

The performance of the project controlling communications usually includes a project team meeting and a project owner team meeting. In these meetings a common project reality should be constructed, which forms the basis for agreeing on directive measures. The prepared project controlling reports are amended and corrected in the meetings.

The follow-up to the project controlling communications includes completing the project controlling reports, possibly initializing adaptations to the project portfolio database, a project marketing based on new intermediary results of the project and distributing the project controlling reports.

Organization of the project controlling process

Project controlling is a sub-process of project management. Therefore, it is performed by the role performers of the project, i.e. the project manager, the project owner team and the project team, and not by project-external roles, such as a division controller. In projects of high complexity the project manager may delegate some project controlling tasks to a project controller or a project management assistant.

The division controlling and the company controlling in the project-oriented company as well as the project portfolio management thus represent the context of project controlling.

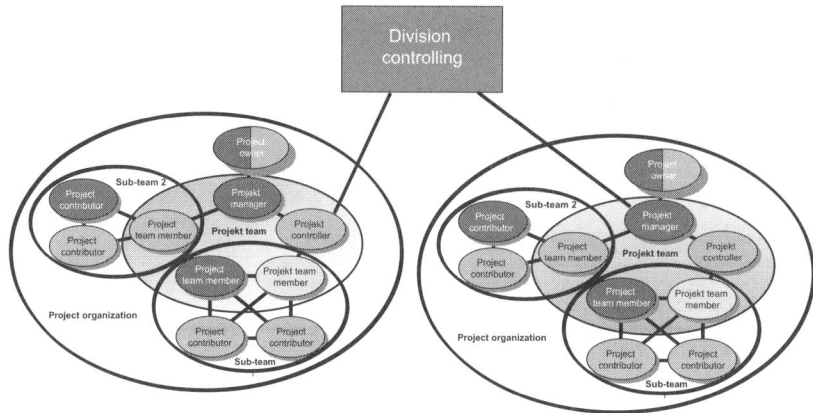

Fig. E1.9: Relationship between division controlling and project controlling in the project-oriented company

Quality of the project controlling process

In practice, a formal project controlling is often performed only in financially risky projects. For such projects the necessary effort of data collection, analysis and reporting seems justified. In the project start process often a project planning is carried out, yet, a formal controlling during the project performance does not take place.

An adequate project controlling quality is only ensured by applying project management methods and communication forms in the project start process as well as in the project controlling process. Those project plans which were developed in the project start, are to be controlled in the project controlling and to be adapted, if required. Thereby the necessary effort in the project start process is justified.

···❯ CASE STUDY E2: Controlling the ABS (antibiotics strategy) project

Company performing the project:
- Austrian Ministry for Social Security and Generations

Project objectives:
- Contributing to the further development of the antibiotics "culture" in Austrian hospitals for optimizing the treatment of patients with antibiotics, for increasing the efficiency in the use of antibiotics, for reducing resistances and for reducing costs.

Project content:
- Establishing an antibiotics "culture" in Austrian hospitals,
- developing guidelines for the further development of the antibiotics culture,
- communicating the guidelines in an ABS road show,
- consulting 5 hospitals during implementation of the antibiotics guidelines.

Project duration:
- 18 months

Project organization:
- Project owner team (2 representatives of the Austrian Ministry for Social Security and Generations),
- project manager and project management assistant,
- 10 ABS experts (microbiologists, infectious disease experts, pharmacists, hygienists, etc.).

Project contents:
The contents of the ABS project is depicted in the work breakdown structure.

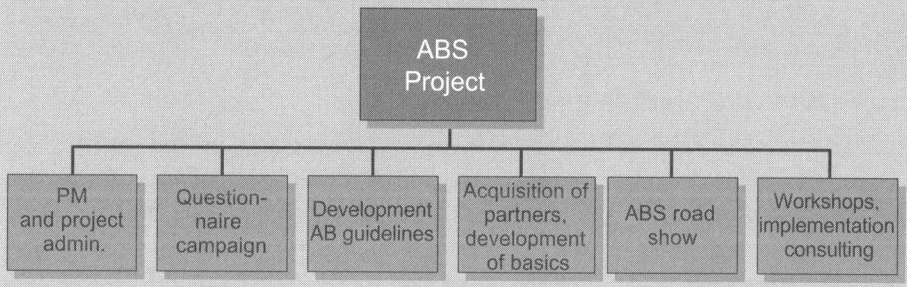

Fig. E1.10: Work breakdown structure of the ABS project

Project controlling process:

Planning and preparing the project controlling communications:
- Gathering information on the project status by the project management assistant and the project team members,
- draft of the project controlling reports and updating of the project plans by the project manager and the project management assistant.

Project controlling workshop with the project owner team:

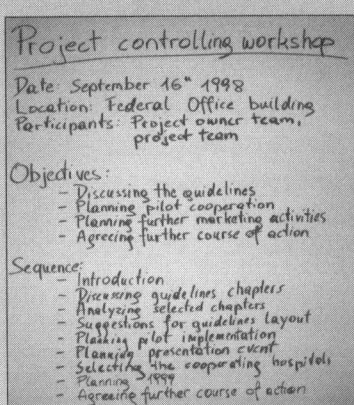

- Participants: Project team, project owner team, project manager, project management assistant
- Frequency: every 6-8 weeks
- Duration: 3-4 hours
- Roles: Moderation by the project manager, documentation by the project management assistant, contributions by the project owner team

Fig. E1.11: Design of the project controlling workshop with the project owner team

Preparing the project controlling workshop with the project team:
- Adapting the project controlling reports,
- inviting the project team members and sending them the project controlling reports.

Project controlling workshop with the project team:
- participants: Project team, project manager, project management assistant
- frequency: Every 6-8 weeks; a few days after the project owner team workshop
- duration: 3-4 hours
- roles: Moderation by project manager, documentation by project management assistant, content contributions by project team members

Follow-up to the project controlling communications:
- Final editing of the project controlling reports,
- distributing the project controlling reports to all members of the project organization.

Interpretation:
- What is special about this case is that the workshop with the project owner team took place before the workshop with the project team. This was due to the pronounced content-related commitment of the project owner team on the one hand, and the use of an external project manager, on the other hand. The use of an external project manager required close collaboration for coordination with the project owner team.

This case study is continued in Case Study E4.

E 1.6   Description of the process of resolving a project discontinuity

The process of resolving a project discontinuity is depicted in Figure E1.12 with regard to its objectives, time boundaries, tasks and responsibilities. The process of resolving a project discontinuity is depicted for the alternative that a redesign of the project is possible. Other alternatives are project stopping and project interruption. The processes of project stopping or of project interruption correspond with the project close-down process.

Objectives of the process of resolving a project discontinuity

- resolving a project crisis or a project chance,
- limiting the possible damage or optimizing the possible potentials for the project,
- creating the basis for a successful continuation of the project,
- efficient resolution of the project discontinuity.

Time boundaries of the process of resolving a project discontinuity

- Start: defining the project discontinuity
- End: ending of the project discontinuity communicated
- Duration: several weeks

The process of resolving a project chance is analogous to the process of resolving a project crisis. Both processes differ only in regard to their causes – existential threat on the one hand, new potentials on the other hand – and therefore also in their objectives.

The process of resolving a structurally caused change in identity differs fundamentally from the resolution of a crisis or a chance. As regards the objectives and the phases, it corresponds with the project start process and is therefore not described as a business process in its own right. A case study of a structurally caused change in the identity of a project is described later.

Tasks and responsibilities in the process of resolving a project discontinuity

Tasks / Responsibilities	Project owner team	Project manager	Project team	Project team member	Project management consultant	Expertenpool-Manager	Representatives of relevant project environments	Documents
Defining the project discontinuity								
• Definition of the project discontinuity	P	C	C			I		
• Communicating the project discontinuity to relevant project environments	C	C1	P		C			1)
Planning and performing immediate measures								
• Planning immediate measures	C	C1	P		C			
• Deciding on immediate measures	P	C	C			C		2)
• Performing immediate measures		C1		P	C		C	
• Controlling the success of immediate measures	P	C1	C					
• Communicating the results of immediate measures	C	C1		P	C	I		
Cause analysis, planning resolution strategies								
• Cause analysis	C	C1	P		C			
• Planning alternative resolution strategies		C1	P		C	C		
• Deciding on a resolution strategy	P	C						3)
• Communicating the decision	C	C1	P		C	I		
Alternative: Redesigning the project: Planning and performing additional measures								
• Planning additional measures	C	C1	P		C	C		4)
• Performing additional measures		C1		P	C		C	
• Controlling the success of additional measures	P	C1	C					
• Communicating the results of additional measures	C	C1		P	C	I		
Ending the project discontinuity								
• Evaluating the resolution of the project discontinuity	P	C	C					
• Defining the end of the discontinuity	P	C	C					
• Adapting project management documentation		C1	P		C			5)
• Defining learning points		C1	P					
• Communicating the ending of the project discontinuity	C			P	C	I		
Performing work packages				P			P	
Alternative: Stopping the project (see project close-down process)	C	C1		P	C			
Alternative: Interrupting the project (see project close-down process)	C	C1		P	C			

Legend:
P ... Performance
C ... Contribution
I ... Information

Documents:
1) Communication plan
2) Immediate measures plan
3) Documentation of the alternative resolution strategies
4) Additional measures plan
5) Adapted project management documentation

Fig. E1.12: Description of the process of resolving a project discontinuity

Tasks in managing project discontinuities

Tasks in managing discontinuities are on the one hand resolving project discontinuities and on the other hand avoiding project crises or promoting project chances, and providing for project discontinuities..

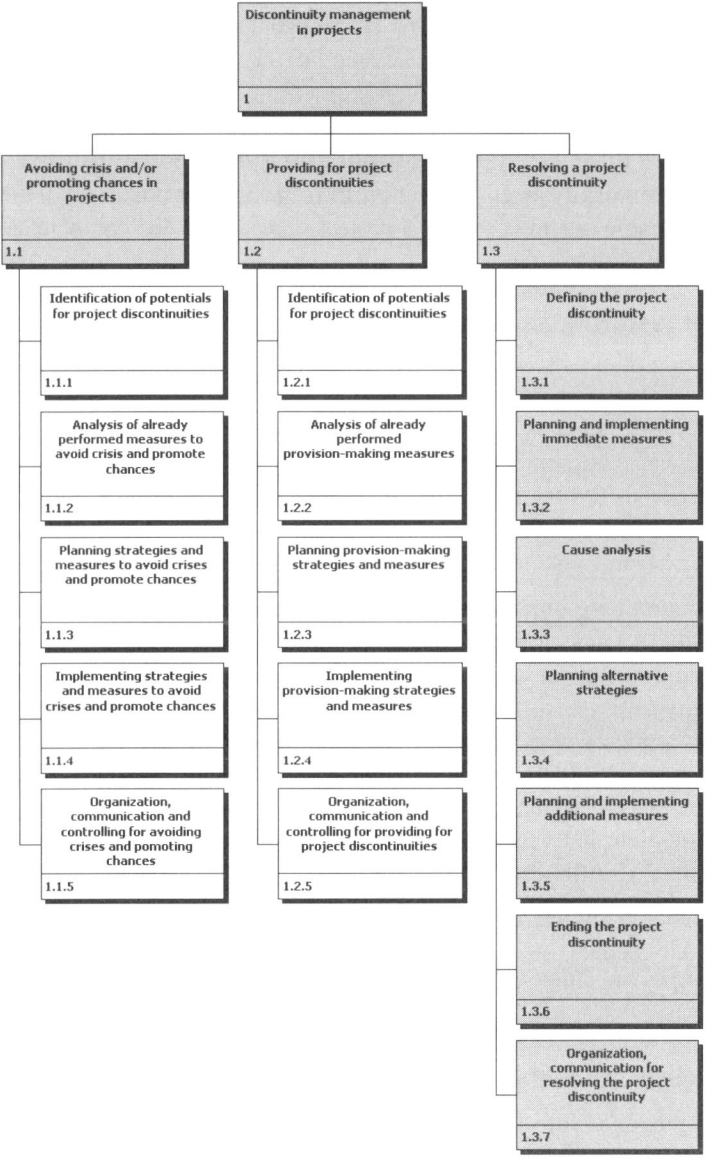

Fig. E1.13: Tasks in managing discontinuities in a project

It is the objective of avoiding project crises to plan and implement strategies and measures that avoid crises. It is the objective of promoting project chances to plan and implement strategies and measures that promote project chances. It is the objective of providing for a project discontinuity to develop strategies and measures for the case that a discontinuity occurs. This should enable an efficient resolution in the case of occurance..

Avoiding project crises and promoting project chances and providing for project discontinuities are not sub-processes of project management in their own right, but are tasks to be fulfilled in the project start process and in the project controlling process. Therefore, in the following, only the resolution of a project discontinuity will be described.

Demand for a professional resolution of a project discontinuity

Due to their complexity and dynamics projects have a high potential for discontinuities. In order to be able to still complete a project successfully the competence to resolve a project discontinuity professionally is needed in the project-oriented organization.

Phases of resolving a project discontinuity

The definition of the discontinuity is a central task in the process of resolving a project discontinuity. The existence of a project discontinuity cannot be measured by means of objective criteria, such as project ratios, but must be constructed in a communication process. A loss in a contracting project amounting to € 50,000 can lead to the definition of a project crisis or not. That depends on the size of the project and of the structures and cultures of the project-performing organization. If a crisis is defined this leads to the use of crisis management, otherwise "normal" project management is practiced.

The resolution of a project discontinuity thus requires a conscious construction of the discontinuity as a new project reality. Watzlawik assumes that there exists no objective reality but only a subjective construction of reality.[3] Only the conscious definition of a discontinuity gives the situation a specific sense and clarifies its social meaning. By defining a discontinuity the "crisis" or the "chance" are differentiated from normality. A difference is made "that makes a difference". On the one hand, the identification of a situation as crisis or as chance serves as a "label", which aims at securing special management attention, and on the other hand, it legitimizes the use of (radical) measures for resolving the discontinuity.

The process of resolving a project discontinuity comprises the phases of planning and performing immediate measures, cause analysis, planning alternative resolution strategies, as well as planning and performing additional measures.

General strategies for resolving a project discontinuity are:

- redesigning the project,
- stopping the project and
- interrupting the project.

---

3) *Watzlawick, P.,* (Wirklichkeit) 1976, p. 69.

Redesigning the project may lead to appointing a new project owner team, a new project manager or new individual project team members, may require redefining the project objectives and project content, may necessitate a new design of the project environment relationships and may include creating a new project culture. By redesigning the project a new project identity is created. This "revolution" in the project serves as the basis for a successful continuation of the project.

The resolution of a discontinuity is characterized by a high demand of creativity and discipline. Central weaknesses are to be identified and eliminated, strengths are to be preserved and expanded, important existing environment relationships are to be secured and new ones are to be developed. The necessary strategies and measures are to be operationalized to enable traceability and measurement of success.

Stopping a project poses a catastrophe in the development of the social system "project", its survival is no longer guaranteed. From the point of view of the project-performing company, however, the decision to stop a project may be reasonable. Interrupting a project is a further strategic alternative. It presupposes that the project can be successfully continued after the period of interruption.

The objectives and the processes of stopping and interrupting a project correspond to those of the project close-down. In project stopping the project environment relationships are to be dissolved, the achieved results are to be secured and the members of the project organization are to be given feedback and new orientation for their work. In the event of a project interruption the relevant environments are to be informed about the project interruption, the achieved results are to be secured, the members of the project organization are to be given feedback, and the new start of the project is to be planned. The availability of important members of the project organization at the new start of the project is to be ensured.

Like its definition, the ending of the project discontinuity is also an act of symbolic management. The ending of the discontinuity should take place as early as possible and as late as necessary for establishing a new project identity. In the course of the ending of the project discontinuity it is to be agreed which new project rules and project values apply after the discontinuity.

Organization of the resolving of a project discontinuity

IGenerally, in addition to the members of the project organization external experts are required for resolving a project discontinuity, to provide the know-how required at short notice. In case the members of the project owner team and the project manager are not themselves the cause of the project discontinuity, they should retain their roles in the resolution of the project discontinuity.

The extent and the intensity of the project communication increases during the resolution of the project discontinuity.

Quality of the process of resolving a project discontinuity

Crises and chances often occur in projects, but are rarely resolved professionally in practice. The process of resolving a project discontinuity is formalized only in few compa-

No-nos in resolving project crises
• delaying and concealment tactics • negating the situation • recriminations • delegation to superiors

Fig. E1.14: No-nos in resolving project crises

nies, organizational competencies concerning this process are not existent. The quality of the resolution of project discontinuities depends exclusively on the individual competencies of members of the project organization.

The effort of avoiding crises and providing for crises in projects is also rarely practiced.

⋯⟶ CASE STUDY E3: Resolving a crisis in the project "Conception: Smoke Signals"

Company performing the project:
- An Austrian health insurance company

Project objectives:
- Developing a concept for the protection of non-smokers at the workplace against smokers.

Context:
- introduction of the Health and Safety at Work Bill, 1994,
- introduction of project management in the health insurance company.

Project plans:
- The structures of the project are presented by the work breakdown structure, the project environment analysis, and the description of the project organization.

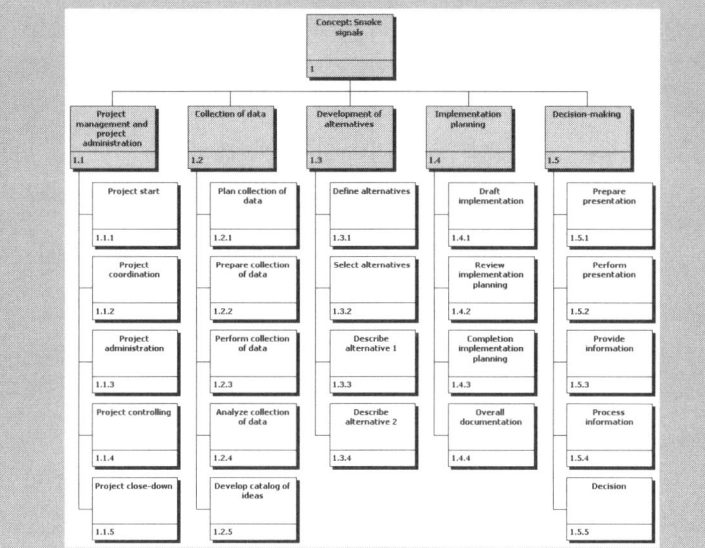

Fig. E1.15: Work breakdown structure of the project "Conception: Smoke Signals"

Project organization:

- multifaceted team (smokers, non-smokers, HR representatives, medical doctor, technical departments), approx. 10 people,
- project owner: One member of the board of the health insurance,
- use of a project consultant for consulting in the project start and the project controlling process,
- conflicts during project start workshop: Objectives unclear, different interests apparent.

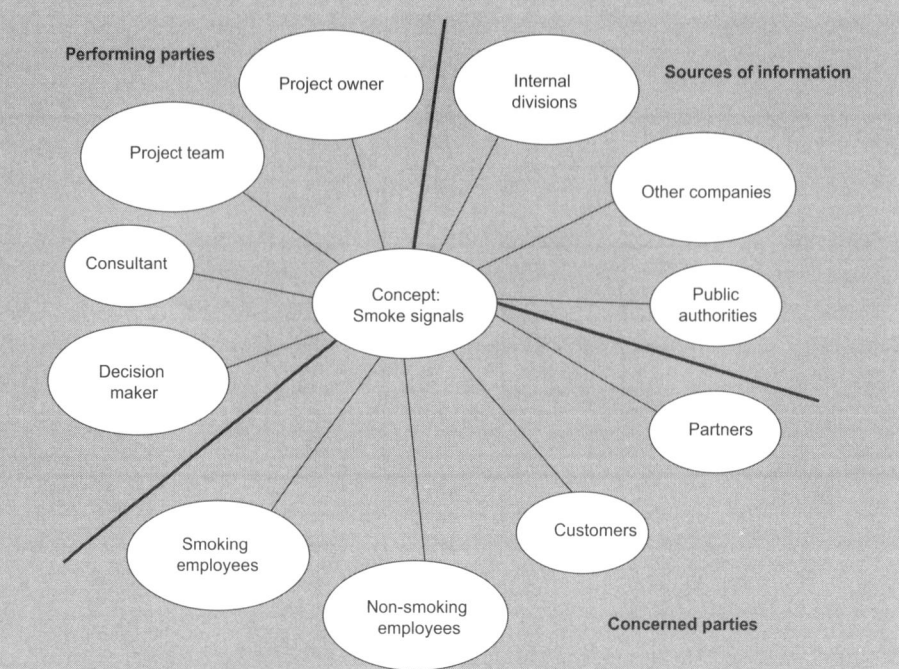

Fig. E1.16: Project environment analysis of the project "Conception: Smoke Signals"

Situation at the first controlling workshop:

Statements of project team members at the beginning of the controlling workshop:

- "Colleagues from the departments are making fun about us"
- "Departments don't provide information"
- "Project owner does not give enough support to the project"
- "Project manager is overtaxed"
- "Team has differing expectations regarding the project results"
- "Weariness within the team to continue work"
- "This is a project crisis!"

Project Progress:

- The work package "Analyze collected data" is being performed, "Define alternatives" is completed.
- Actual schedule and actual costs as planned, quality of the achieved intermediary results is assessed differently by different project team members.

Process of the resolution of the project crisis:

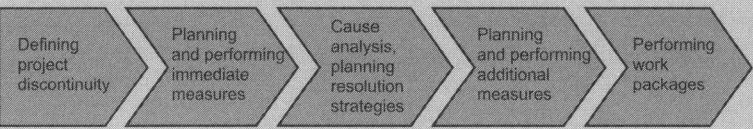

- The situation was defined as a project crisis, yet there were problems incon-
  structing this new, common reality.
- Measures performed for resolving the crisis:
  - Cause analysis was performed
  - Alternative resolution strategies were defined and described (see Figure
    E1.17)
  - Plans were presented to the project owner for decision making
  - Decisions made resolving the project crisis
    - realizing the strategy "Leap forward"
    - realizing planned measures
- no new approaches or other methods

Environment	Resolution strategies		
	"Continue to fiddle about"	Leap forward	Stopping the project
Department managers, employees	▪ Use informal contacts for information	▪ Presentation: Information about the relevance of cooperation in the project	▪ Information about stopping, about causes and consequences, available results
Project owner	▪ Inform him about problems in an informal manner	▪ Active project marketing ▪ Closer contact to the project team	▪ Discharge project team ▪ Information of the managing board
Project team	▪ Additional resources?	▪ No additional resources	▪ Protect image ▪ Clear documentation of the achieved results

Fig. E1.17: Description of alternative resolution strategies

⋯⋗ CASE STUDY E4: Resolution of a structurally caused change in identity of the ABS project

Company performing the project:
- Austrian Ministry for Social Security and Generations, see Case Study E2

Project plans:

- The project phases "Questionnaire campaign" and "Development AB guidelines" can be referred to as "Guidelines phase", the project phases "Acquisition of partners, development of basics", "ABS road show" and "Workshops and implementation consulting" as "Implementation phase".
- Each of these phases required a fundamentally different project identity, which was expressed, above all, in the project organization charts and in the different project cultures.

Project organization in the "Guidelines phase":

- In this phase the project organization consisted solely of a questionnaire team and a guidelines team, each of which comprised the project team members of the ABS experts.

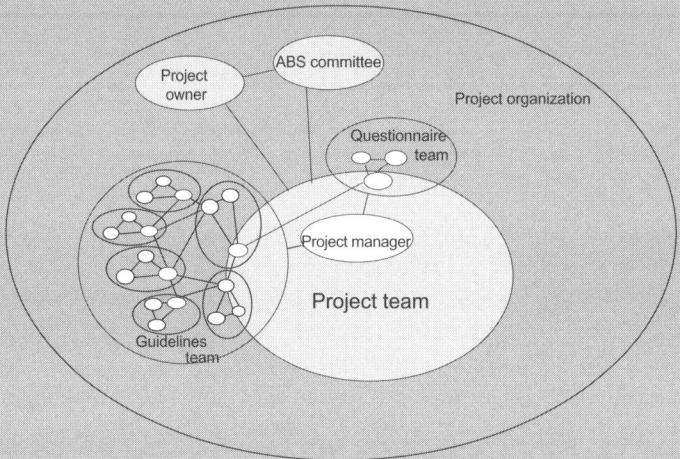

Fig. E1.18: Project organization chart of the ABS project in the "Guidelines phase"

Project organization in the "Implementation phase":

- In the "Implementation phase" the ABS experts cooperated with different external partners in sub-projects to organize the ABS road show, the ABS implementation workshops and the ABS implementation consulting. To organize the ABS road show a cooperation was undertaken with hospitals in 6 regions, with offices of state governments and with state sanitary authorities. Further approximately 50 employees from 5 hospitals were involved in the ABS implementation activities.

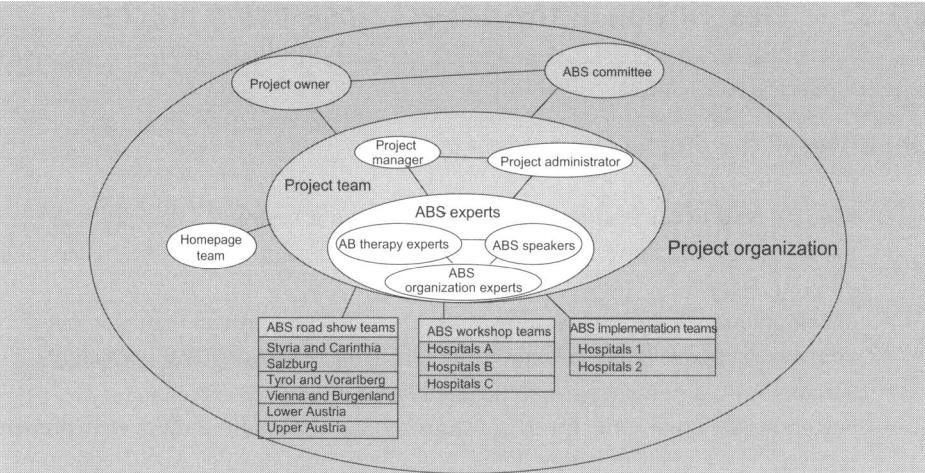

Fig. E1.19: Project organization chart of the ABS project in the "Implementation phase"

Project culture in the "Guidelines phase":

The following elements characterized the project culture in the "Guidelines phase":

- implementation of existing know-how, no broad surveys,
- helping people to help themselves,
- teamwork of the Austrian ABS experts,
- creativity for innovative, user-friendly solutions,
- exclusive financing by the Austrian Ministry for Social Security and Generations

Project culture in the "Implementation phase":

The following elements characterized the project culture in the "Guidelines phase":

- cooperation with local organizations in the ABS road show and with hospitals in ABS workshops, and ABS implementation consulting,
- performing services for money,
- flexible design of small "service teams",
- role of the project team: Integration, exchange of experience, overall direction, marketing, but no content work.

Management of the structurally caused change in identity:

- Already during the project start it was apparent that a change in identity of the project would be necessary.
- The project manager, the project management assistant and the project owner became conscious of it in good time and documented the changes to be expected in the project organization, the project culture and the project environment relationships.
- The change in identity was analyzed in the project team in the course of an extensive project controlling workshop at the end of the "Guidelines phase".

E 1.7 Description of the project close-down process

The project close-down process is depicted in Figure E1.20 with regard to its objectives, time boundaries, tasks and responsibilities.

Objectives of the project close-down process

- Completion of remaining work packages and planning of the post-project phase, handing over of the business case analysis to the investor,
- transfer of gained know-how into the project-performing organizations and into other projects,
- development of project close-down reports and actual project management documentation, possibly performance of an exchange of experience workshop,
- agreements regarding a possible investment evaluation,
- designing the environment relationships by dissolving the project environment relationships and establishing new environment relationships, final project marketing,
- evaluating the project success, assessing the contribution of the members of the project organization to the project success, dissolution of the project teams,
- project approval,
- efficient design of the project close-down process.

Time boundaries of the project close-down process

- Start: initialization of the project close-down
- End: project approved
- Duration: 2-3 weeks

Tasks and responsibilities in the project close-down process

Tasks / Responsibilities	Project owner team	Project manager	Project team member	Project team	Expert Pool manager	Representatives of relevant project environments	Documents
Planning the project close-down							
• Establishing the project close-down structures	C	P					1)
Preparing the project close-down communications							
• Documenting the remaining work to be fulfilled		P					
• Planning of tasks and responsibilities in the post-project phase		P	C				
• Preparing the evaluation of the project success		P	C			C	
• Preparing the dissolution and establishment of environment relationships		P	C				
• Developing drafts of the project close-down reports		P	C			C	
• Developing drafts of the actual PM documentations		P	C				
Performing the project close-down communications							
• Distributing information material to participants of the close-down communication	I	P			I	I	2)+3)
• Performing close-down communication I, II, etc.	C	C1			P	C	
• Performing exchange of experience workshop		P			C	C	
• Performing "social" end event	C	C1		P	C	C	
Follow-up to the close-down communications							
• Dissolution and establishment of environment relationships		P	C			C	
• Completing project close-down reports, special reports, actual documents		P	C			C	
• Initializing updating of project portfolio database		P					
• Distributing project close-down reports	I	P	I		I	I	4)
• Closing the project cost center	I	P					
• Final project marketing	I	P					
Performing remaining work packages (parallel)				P		P	

Legend:
P ... Performance
C ... Contribution
I ... Information

Documents:
1) Adapted project close-down structures
2) Invitation of the participants to the project close-down workshop
3) Information material for the project close-down workshop
4) Project close-down reports, actual project management documentation, special reports

Fig. E1.20: Description of the project close-down process

CHAPTER

E

Demand for a professional project close-down

When the objectives of a project are achieved, the project as a social system has no right to persist. This contradicts the general objective of social systems to secure their viability. Therefore, like the project start, the close-down of a project requires much effort.

The uncertainty and fear of customers (or users) during the project close-down process, of not being able to appropriately deal with the project results or of not having received qualitatively appropriate project results, may lead to a considerable prolongation of the project. From a project point of view, therefore, measures are to be taken which convey to the customers the necessary confidence and the competencies for utilizing the project results. Every close-down is at the same time a start for the customers (or users) of the project results. From a joint analysis of the project close-down situation and this start situation potentials for the design of the project close-down process can emerge.

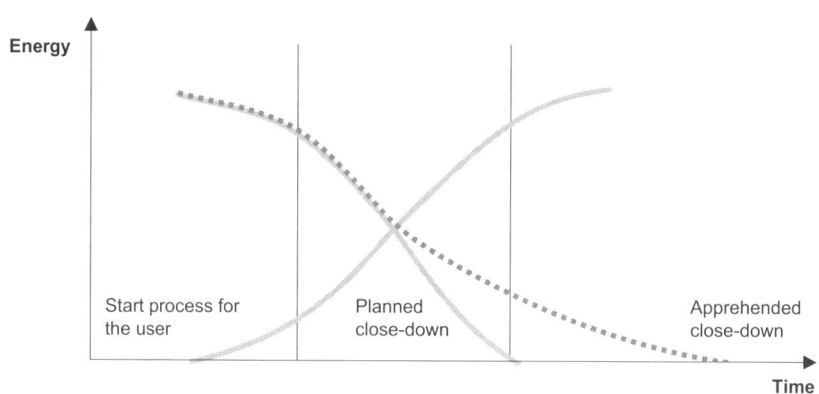

Fig. E1.21: Relationship between close-down and start

The project close-down process is characterized by the fact that often unattractive work packages remain to be performed, some project environments, e.g. the project owner team, the customer, individual project team members, may be interested in the persistence of the project, and some former project team members may already work in new projects.

In the project close-down process the content and time boundaries of the project are to be decided definitively. Objectives and contents not belonging to the project are dispensed with or are postponed to the post-project phase. A definite end event should be defined. The project end is to be communicated internally as well as externally.

The formal ending of the project serves to release resources and energy for new tasks. Know-how gained in the project should be transferred into the project-performing companies and into other projects by the project documentation and by exchange of experiences. A contribution to the knowledge management of the project-oriented companies is to be made.

Phases of the project close-down process

A draft for the planning of the project close-down process should be developed in the project start process, since the costs of the project close-down are to be considered in the project cost plan. This draft is to be concretized by the project manager after the formal project close-down process has been initialized by the project owner team.

The preparation of the project close-down communications comprises the documentation of the remaining work as well as drafts for planning the tasks and responsibilities in the post-project phase, for evaluating the project success, for dissolving and establishing environment relationships, and drafts of project close-down reports and the actual project management documentation.

In the project close-down process several closing communications are to be performed: A close-down workshop with the project team, a close-down meeting with the project owner team, a closing "social event" and possibly an exchange of experience workshop.

In order to properly close down a project in the project team, it is recommended to perform a project close-down workshop. The project close-down workshop is an instrument of symbolic management. It promotes the concluding utilization of teamwork potentials and of a holistic view of the project, and fulfills an important integration function. The objective of the team-internal close-down is to evaluate the project success, to assess the performance of the members of the project organization and to provide information on the disposition of personnel after the project end.

The close-down meeting with the project owner team deals with the assessment of the project by the project owner team, with the reflection of the fulfillment of the project owner team role and the formal project approval.

The follow-up to the project close-down communications comprises the completion of the project close-down documentation and its distribution, the final project marketing and the closing of the project cost center.

Organization of the project close-down process

In addition to the members of the project organization also representatives of the project-performing organizations are to be involved in the project close-down process to ensure organizational learning. The dissolution and establishment of environment relationships also involves representatives of relevant environments.

Quality of the project close-down process

In practice a formal project close-down is rarely performed as yet. The right of the members of the project organization to adequate feedback and an emotional close-down as well as the learning potentials for the project-oriented companies concerned, conflict with the dynamics of everyday business and (yet) a lack of professionalism in project management.

Similar to the project controlling process, the quality of the project close-down process is assured by the continuity of the project management methods used. For example, the project objectives plan, the business case analysis or the project environment analysis are central methods of the project close-down process. On the basis of the business case analysis and the project objectives agreed in the project start and adapted in the project controlling the project success can be evaluated in the project close-down. A "retrograde" sense-making by a redefinition of the project objectives in the project close-down process is often useful.

The project environment analysis enables the dissolution of existing project environment relationships in an appropriate quality, and the establishment of new relationships of the project-oriented company with relevant environments for the post-project phase. It also forms the basis for the final project marketing.

In order to enable a social project close-down – and not only a close-down on paper – adequate social competencies are required. Giving and taking feedback as an individual and as an organization, addressing positive and negative points, requires adequate competencies and experience.

Which mistakes can be made in the project close-down process?
• no definite project end event – "never ending stories"
• no planning of the post-project phase
• unconscious project close-down, direct changeover to new objectives
• no project close-down documentation – "No job is finished until the paperwork is done" (graffiti)
• lack of strategic orientation regarding the close-down of the project relationships with customers, suppliers. etc.
• no reflection of the learning experiences in the project
• no feedback to the members of the project organization; inhuman project management
• individual instead of joint leave-taking of the project team members; the project manager remains
• project cost center not closed

Fig. E1.22: No-nos in the project close-down process

Project close-down in different types of projects

The necessary reflection work, the evaluation of the project success and the assessment of the contribution of the members of the project organization for the project success is much easier in the close-down of successful projects than in unsuccessful projects. Dealing with flops creates a high degree of social complexity which requires a high degree of social competence of the project. Reasons for the loss of a contract as a result of an offer project, for example, are rarely analyzed.

The basis for the transfer of the gained know-how into realization projects is to be created in the project close-down process of conception projects. This can be achieved, on the one hand, by adequate project documentation and, on the other hand, by involving employees who will later work in the realization project.

In repetitive projects the learning for similar projects is to be organized in the project close-down. Knowledge management is of great importance in pilot projects.

····} CASE STUDY E5: Project close-down of the project "Bhumibol"

Company performing the project:
- Austrian construction engineering company

Type of project:
- rehabilitation project, contracting project for a Thai customer
- contract value: Approx. € 40 million
- project duration: 18 months
- a "pilot" as rehabilitation project for the company

Project objectives:
- iIncreasing the efficiency of a hydraulic plant by replacing individual parts of the plant (turbines, control equipment, etc.),
- first-time use of a digital control.

The close-down process:

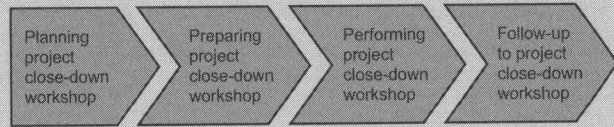

Planning and preparing a project close-down workshop:
- recruiting a project management consultant, informing the consultant about the project (documents, meetings),
- planning the workshop design, inviting the participants of the project close-down workshop,
- developing drafts of the project close-down report and of the special report "Digital control".

Performance of the project close-down workshop:
- Duration: 1 day, approx. 3 weeks before the definite project end (project approval by the project owner)
- Location: Seminar room of the project-performing company
- Participants: Project owner, project manager, project team members, line managers, business division manager, project management consultant (workshop moderation)
- Objectives: Organizational learning from the pilot project, professional project close-down
- Agenda:
  - introduction, objectives and scedule, information on the project close-down process

- presentation: Project status and project close-down reports (by project manager)
- reflection of the learning experiences: Specifics of the management of rehabilitation projects, use of the digital control (project manager and digital control expert)
- planning remaining work and planning the post-project phase (all)
- planning the dissolution of the project environment relationships and the establishment of new environment relationships (all)
- evaluation of the project success (all), see Figure E1.23
- planning further course of action: Assessment of performance in the project team (project team)

Interpretation of the results:

- Definite agreements for the successful project close-down were made.
- Some additional project objectives were redefined.

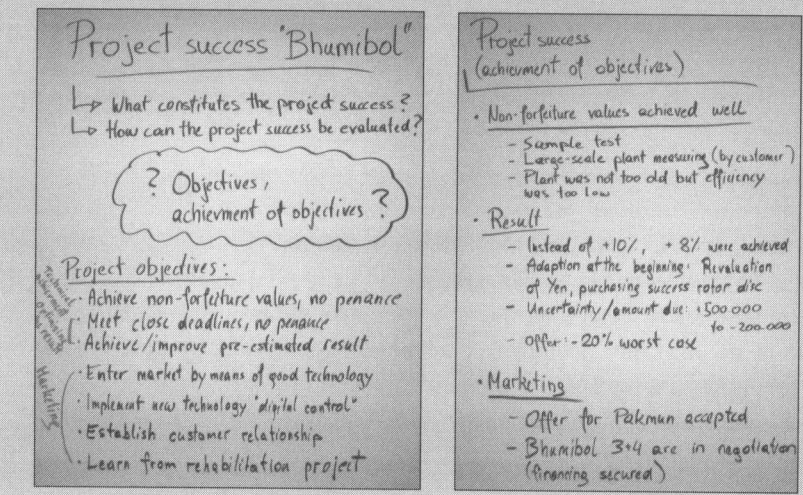

Fig. E1.23: Extract of the flipchart protocol of the close-down workshop

E 1.8   Project marketing

There is not enough project marketing!

Many projects are characterized by a high degree of content orientation but a low degree of marketing orientation. The members of the project organizations concentrate mainly on fulfilling the work packages. They do not realize that an appropriate communication of the objectives, of the content and of the organization of the projects to the relevant project environments is also necessary to ensure the project success.

When there is not enough project marketing there is a risk that the project is not getting enough management attention and is provided with non-adequate resources.

Not only is the quality of the content results of a project to be ensured by an adequate communication, but also their acceptance. The success (S) of a project can be defined as the product of the quality (Q) of the content results of a project and its acceptance (A).

$S = Q \times A$

If, in the extreme case, acceptance of the content project results is zero despite good quality, the project success is also zero.

Objectives of project marketing: Do well and talk about it!

Project marketing is a success factor in projects. It supports the management of the project environment relationships.

Objectives of project marketing are:

- ensuring appropriate management attention,
- ensuring adequate resources for the project,
- ensuring acceptance of the (intermediary) results of the project,
- minimizing conflicts in the project and
- promoting the identification of the members of the project organization with the project.

By means of adequate project information conflicts in the project and in the relationships to relevant environments are to be minimized and adequate expectations regarding the project results are to be developed. Feedback for the project is to be ensured and a dialog with the relevant environments is to be established.

Project information to the outside creates constructive pressure on the inside. The project organization is to satisfy the developed expectations. Thus, project marketing promotes identification of the members of the project organization with the project.

Project marketing also complies with the personal interests of self-marketing of the members of the project organization.

Assumptions for project marketing
• The competition for management attention and for scarce resources between projects increases in the project-oriented company. • The project success is dependent on the quality and acceptance of the project results. • The project environments influence the project success. • Members of the project organization have to sell their projects. • Project marketing is part of professional project management.

Fig. E1.24: Assumptions for project marketing

Project marketing: Definition

Project marketing can be defined as project-related communication with relevant project environments. Project marketing is a project management task, which is pointed out here because of its importance for the success of a project. Project marketing focuses on the communication of the project. Other functions of marketing, i.e. product policy, distribution policy, price and conditions policy, are covered by fulfilling the "other" project and project management tasks.

Relationships exist between the marketing of the project initializing an investment, and the investment marketing to be performed in the course of the whole investment process. Investment marketing relates to the investment initialized by a project. In the case of an investment in a product, investment marketing corresponds to product marketing, if the investment is in a new office building, than the benefits of this new building have to be communicated to the employees, customers, partners, etc..

Acceptance for the performance of an investment is to be ensured in the project assignment process (see Chapter I2). In the project assignment process the proposal team performs marketing activities to ensure a positive investment decision and to obtain the project assignment. The reasonability of the investment is also to be communicated during the project. The business case analysis can be used as an instrument in doing so.

Investment marketing relates to the object, product or service developed in the project. Marketing, especially product marketing, can already begin during the project, but will have to be performed mainly in the post-project phase. Often, it is not useful to dissociate product marketing from project marketing. E.g. in event projects, such as the pm days, both product and project marketing objectives are realized by sending out announcements.

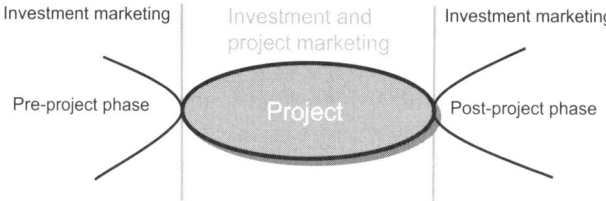

Fig. E1.25: Focal points of investment marketing and project marketing

Project marketing in the project management-sub-processes

Project marketing is a project management task to be fulfilled in all sub-processes of project management. Project marketing is therefore rarely considered a work package in its own right in projects.

Project marketing starts with the project assignment. The definition of a project for fulfilling a business process of medium to large scope represents an important marketing measure. Project marketing is especially important in the project start process. Initial information about the project and the project identity can be communicated.

Subsequently, project marketing measures should be sustained at a relatively even level of intensity over the whole project duration and according to specific requirements. During ongoing project coordination it is possible to perform informal project marketing at lunch, over a cup of coffee, in the lift, etc., with all those interested in the project. During project controlling the intermediary results of the project and changes in the project structures can be communicated.

In resolving a project discontinuity the communication of the project discontinuity is a work package in its own right, since communication is of special importance in the resolution of a project discontinuity.

In the project close-down not only the project results but also the process of the project work and the contribution of the members of the project organization are to be communicated.

Instruments for project marketing

The project environment analysis and the development of the project culture form the basis for project marketing. The project environment analysis enables the differentiation of project-specific marketing strategies and concrete measures by environments (target groups of marketing).[4]

In developing the project culture the project name, a project logo, project slogans and project-specific values are defined, which can be used in developing the instruments for project marketing. To ensure consistent use the font to be used, the font sizes, the project colors, etc., have to be clearly defined.

Instruments of project marketing are:

- print media: Project folder, project information sheet, project newsletter, project wall newspaper, project report (in the company newspaper)
- project-related events: Project vernissage, project presentation, project meeting, project-related competition, project start event, project end event, press conference
- give-aways: Project stickers, project T-shirts, etc.
- online: Project homepage, etc.
- visits: Visiting a construction site, visiting a project room, etc.
- project management documentation: Project handbook, project progress report, project score card, etc.

Rules for non-project marketing
• Keep your project secret!
• Your project should by no means be different from other projects!
• Do not consider the expectations of the project environments by any means!
• For the involvement of others in the project, follow the motto: The later, the better!
• Should anyone be interested in your project, this can only be a bad sign!

Fig. E1.26: Rules for non-project marketing

---

4) Objectives of the project environment analysis, however, are not only the planning of marketing measures, but also the planning of organizational, contract-designing and personnel measures.

Responsibilities in project marketing

All members of the project organization are responsible for project marketing. Not only the project manager but also the project owner team and the project team members are responsible for adequate communication of the project. Project contributors, who only work in the project selectively, do not bear responsibility for project marketing.

The members of the project organization require marketing competence. Project marketing is therefore not only an instrument but also a matter of attitude, it is part of the self-understanding, which is especially important for the project manager. This self-understanding requires adequate ethical standards, which rule out misinformation and manipulation in project marketing.

Budgeting project marketing

The costs of project marketing are to be considered in the project management budget. Personnel costs as well as material costs for various project marketing instruments (project folder, project newsletter, project vernissage, etc.) incur.

Project marketing for different types of projects

Project marketing is especially important in internal projects (e.g. in reorganization projects or in personnel development projects). A central challenge is to ensure acceptance of the project objectives striven for. In contracting projects the degree of social complexity is lower than in internal projects. Professional project marketing contributes to building up trust in the customers and in the partners. In reference and pilot projects project marketing is to be combined with product marketing.

Successful projects are certainly easier to communicate than unsuccessful projects. As regards project marketing, attractive projects, such as developing a new product, can be differentiated from unattractive projects, such as implementing rationalization strategies in the company.

An objective of conception projects is also to ensure acceptance of the implementation measures to follow.

Dangers in project marketing
• "hot air": A lot of project marketing without any content substance • misinformation due to "political interests" • too late or too early project information • too much project marketing in relatively secret product development projects" • project marketing that is not "compatible" with the company culture

Fig. E1.27: Dangers in project marketing

Examples of project marketing in a reorganization project and in an event organization project are depicted in Case Studies E6 and E7.

···⟶ **CASE STUDY E6: Project marketing in the project "ABS optimization"**

Project objectives:

- further development of individual and organizational competencies for the appropriate antibiotics (AB) use in approximately 30 cooperating Austrian hospitals,
- contribution to promoting the efficiency of AB use in the cooperating hospitals,
- creating bases for optimizing the treatment of patients, reducing AB resistances and reducing the costs of AB use in the cooperating hospitals,
- initializing a lasting change process, establishing an ABS expert network for further implementation of AB strategies.

Project organization:

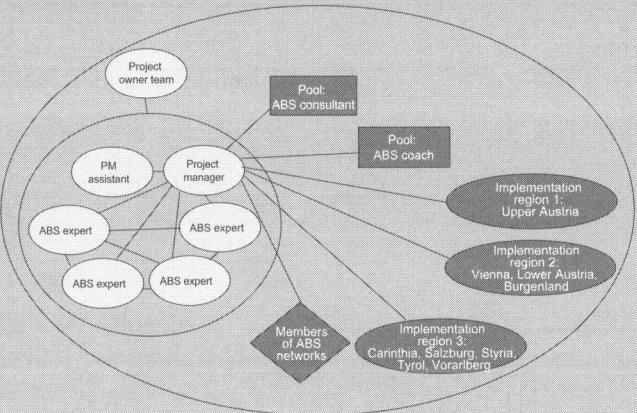

Fig. E1.28: Project organization

Project content:

Optimization ABS use in hospitals (1)

1.1 Project management and project administration
- 1.1.1 Project start
- 1.1.2 Project coordination
- 1.1.3 Project controlling 1-6
- 1.1.4 Project administration 1-3
- 1.1.5 Project close-down

1.2 ABS marketing
- 1.2.1 Start presentation for cooperation partners
- 1.2.2 ABS marketing 2002
- 1.2.3 ABS marketing 2003
- 1.2.4 ABS marketing 2004
- 1.2.5 Final presentation

1.3 Recruitment of cooperating hospitals
- 1.3.1 Analysis of existing ABS activities
- 1.3.2 Development tools for recruitment
- 1.3.3 Recruiting cooperating hospitals region 1
- 1.3.4 Recruiting cooperating hospitals region 2
- 1.3.5 Recruiting cooperating hospitals region 3

1.4 Preparation of implementation
- 1.4.1 Planning ABS networks
- 1.4.2 Selecting additional ABS experts
- 1.4.3 Planning performance of ABS training for ABS delegates
- 1.4.4 Planning performance of ABS training for doctors, etc.
- 1.4.5 Developing training standards
- 1.4.6 Developing implementation standards
- 1.4.7 Developing ABS maturity model

1.5 Pilot cooperation region 1
- 1.5.1 Planning implementation pilot cooperation region 1
- 1.5.2 Performing ABS training 1
- 1.5.3 Analyses 1
- 1.5.4 Planning implementation 1
- 1.5.5 Draft ABS tools 1
- 1.5.6 Completion ABS tools 1
- 1.5.7 Presentation 1
- 1.5.8 Evaluation cooperation 1

1.6 Pilot cooperation region 2
- 1.6.1 Planning implementation pilot cooperation region 2
- 1.6.2 Performing ABS training 2
- 1.6.3 Analyses 2
- 1.6.4 Planning implementation 2
- 1.6.5 Draft ABS tools 2
- 1.6.6 Completion ABS tools 2
- 1.6.7 Presentation 2
- 1.6.8 Evaluation cooperation 2
- 1.6.9 Developing network ABS delegates 1

1.7 Pilot cooperation region 3
- 1.7.1 Planning implementation pilot cooperation region 3
- 1.7.2 Performing ABS training 3
- 1.7.3 Analyses 3
- 1.7.4 Planning implementation 3
- 1.7.5 Draft ABS tools 3
- 1.7.6 Completion ABS tools 3
- 1.7.7 Presentation 3
- 1.7.8 Evaluation cooperation 3
- 1.7.9 Developing network ABS delegates 2

1.8 Documentation
- 1.8.1 Draft documentation
- 1.8.2 Completion documentation
- 1.8.3 Preparation evaluation (when ended)

Fig. E1.29: Work breakdown structure

Relevant project environments:

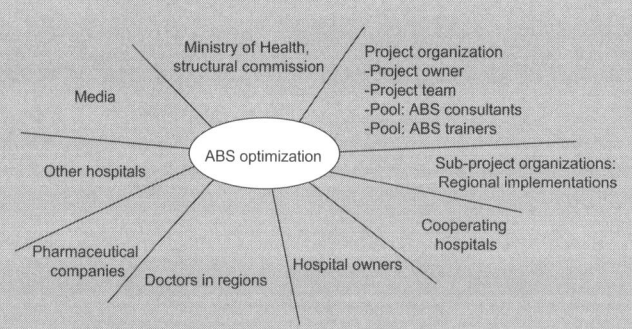

Fig. E1.30: Project environment analysis

Project marketing strategies:

- building on the corporate design of the previously performed ABS project (colors, logo, font, etc.),

- regular presentation of those ABS experts whose performance has an impact on the project success (by that personalizing the abstract topic),

- combination of project and product marketing (communication of the project objectives, the project content and the project organization partly together with antibiotics strategies (ABS) and ABS tools),

- ensuring appropriate number of cooperating Austrian hospitals (approx. 30) and appropriate quality (provision of the necessary resources); long-term interest in implementing the antibiotics strategies,

- no active competition with other projects funded by the ministry.

Project marketing measures:

- Project start presentation on April 10th 2003; development of a project folder; distribution of the "Guidelines for further development of the antibiotics culture in Austrian hospitals"; design of the ABS homepage (www.antibiotika-strategien.at),

- presentations and press conferences in the 3 implementation regions; project management documentation (e.g. project handbooks for the implementations in the 3 regions),
- periodical project owner meetings; project close-down presentation on September 11th 2003.

Fig. E1.31: Folder of the project "ABS optimization" and ABS homepage

Interpretation:

- marketing continuity with the previously performed ABS project and with the "product" ABS had to be ensured; small adaptations in appearance were possible

- ABS experts were foregrounded, which corresponded with the service-oriented nature of the project.

⋯⋗ CASE STUDY E7: Project marketing in the project "pm days '03"

Project
- Organization of the "pm days", the annual international project management event

Organizers
- ROLAND GAREIS CONSULTING (RGC) in cooperation with the
- PROJEKTMANAGEMENT GROUP (PMG), University of Economics and Business Administration, Vienna

Structures of the project "pm days '03"
- See Case Study D1 in Chapter D and Case Study F3 in Chapter F5.

Project objectives:
- organization and performance of the annual international event for project management and for the management of project-oriented companies in Austria,
- high content quality, innovations,
- workshops in addition to lectures, open space working form,
- 15 exhibitors,
- 120 participants at the research conference, 220 participants at the practice conference, 30 participants at expert seminars,
- financial contribution: € 20,000.

Additional objectives:
- Establishment of long-term relationships with cooperation partners,
- cooperation with universities of applied science and universities for the pm days '03 student award,
- enhanced marketing after the pm days (telemarketing, newsletter) for customer recruitment (RGC, PMG).

Project organization:

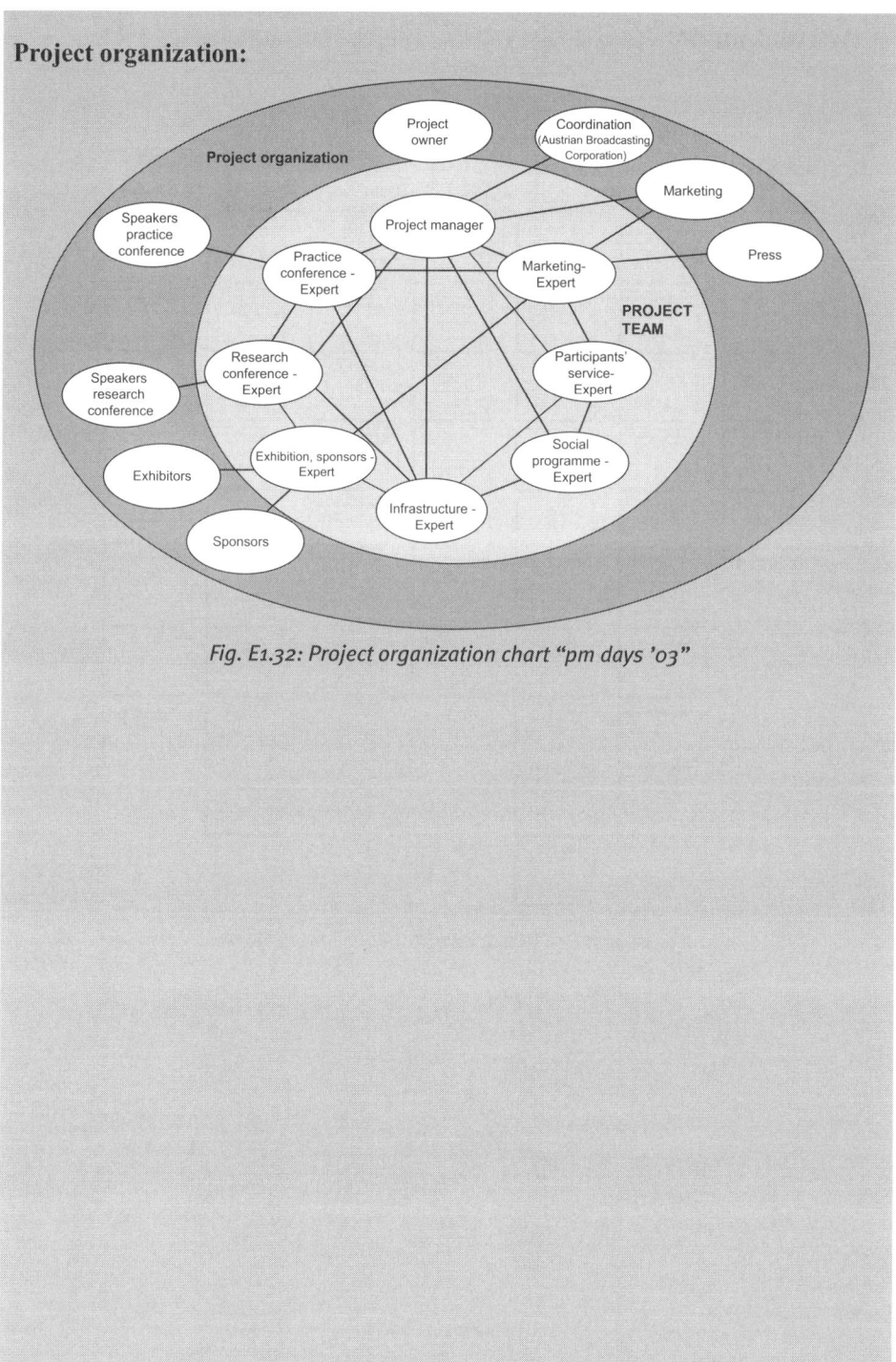

Fig. E1.32: Project organization chart "pm days '03"

Project content:

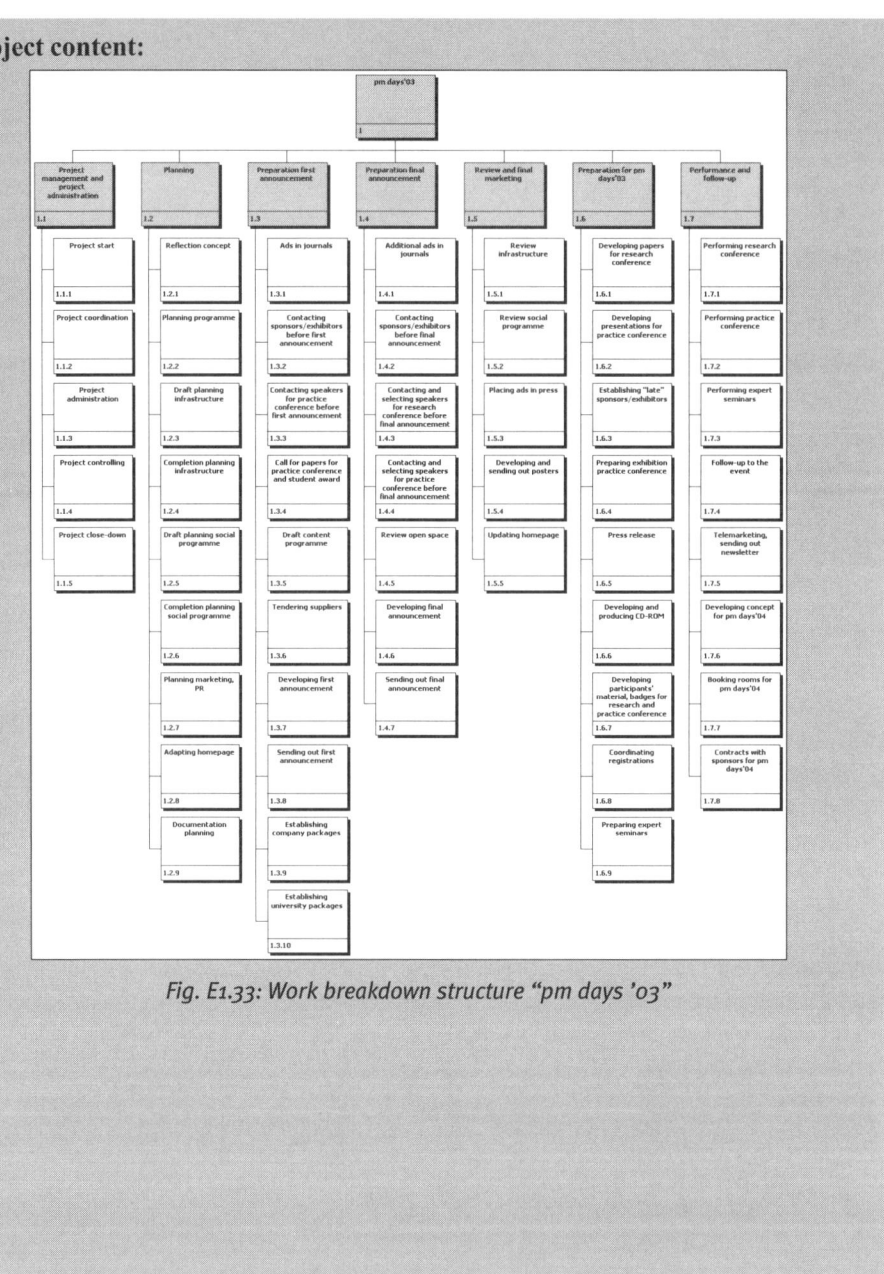

Fig. E1.33: Work breakdown structure "pm days '03"

Relevant project environments:

Project organization
•Project owner
•Project manager
•Project team
•Project team member

Organizers
•RGC
•PMG

Competitors

Media

Cooperation partners

Investors
•Congress office
•EU

University of Economics and Business Administration, Vienna

PM researchers
•PM research network
•Scientific committee
•Young researchers

pm days '03

University training course "International project management"

Speakers

Customers of the event
•Participants
•Cooperation partners, exhibitors, sponsors
•Companies
•Students

Suppliers
•Catering
•Interpreters
•Decoration
•Technics
•Graphics
•CD-ROM producer
•Printer's

PM associations
•IPMA
•PMI
•EPCI

Fig. E1.34: Project environment analysis "pm days '03"

Project marketing strategies:

- Perception of cooperation partners as members of the project organization; use of project management documentation to support communication,
- continuity in the appearance of the pm days as a repetitive, annual project management event, yet enhanced professionalism,
- foregrounding of the project team to promote identification with the event,
- clear differentiation of communication by target groups (participants, sponsors, exhibitors, marketing cooperation partners, student sponsoring); national and international target groups,
- combination of project and product marketing; project marketing conveys to cooperation partners the competence of the organizer in the project management of event projects.

Project marketing measures:

- pm days homepage,
- first and final announcement in English and in German,
- cooperation brochure for sponsors and exhibitors in English and German,
- project handbook: To all members of the project organization for information,
- project team meetings (motivation, ensure resources and management attention for the project "pm days '03"),
- presentation of the project teams at the event.

Fig. E1.35: Front page of the final announcement of the pm days '03

> **Interpretation:**
> - Externally, product marketing was foregrounded; yet, project marketing also conveyed the project management competence of the organizer to participants, exhibitors and sponsors.
> - Internally, it was the task of project marketing to ensure the availability of scarce resources.

Project marketing and programme marketing

In programmes not only the individual projects but, above all, the programme as a whole is to be marketed. Due to the high strategic importance of programmes programme marketing is highly important. The project marketing of the individual projects of programmes is to be performed in accordance with the strategies and measures of programme marketing. Examples of programme marketing are described in Chapter G3.7.

E 1.9 Project administration

Project administration: Definition

Project administration is the business process of the administrative handling of projects, comprising the administration of the project personnel, the administration of project-related customer contracts and supplier contracts and the administration of the project infrastructure as well as the filing of the project correspondence and of the project documentation.

Project administration is not a sub-task of project management, but an additional task. A clear differentiation of the project management tasks from the project administration tasks contributes to the clarification of project management and thus to its professionalization.

Objectives of project administration

Objectives of project administration are:

- ensuring traceability of personnel-, customer- and supplier-related documents and of correspondence, and
- ensuring immediate access to personnel, customer and supplier data.

Tasks in project administration

Project administration comprises the followings tasks:

- administration of the project personnel,
- administration of customer contracts (in external projects) and of supplier contracts,
- administration of the project infrastructure (project rooms, ICT infrastructure), and

- filing of the project correspondence and of the project documentation.

Project administration is a task in its own right in the project and, in contrast to project marketing, it is not part of the project management tasks. Project administration should, therefore, be depicted as a work package in the work breakdown structure.

Roles in project administration

The assignment of project management tasks and project administration tasks to different role performers is depicted in Figure E1.36. In projects of high complexity the project administration can be performed by a project administrator (e. g. a contract expert) or by a project management assistant. In small projects and projects of medium complexity the role of project administrator is usually performed by the project manager.

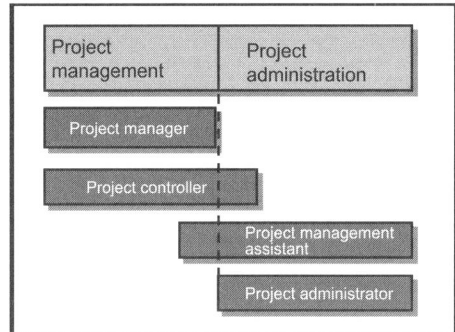

Fig. E1.36: Assignment of project administration tasks to
different role performers

Budgeting project administration

The costs of project administration are to be considered in the project budget. Personnel costs as well as material costs incur.

E 2 Design of the business process "Project management"

The business process "Project management" is to be designed in accordance with the specific requirements of a project. The use of project management methods and of standard project plans, of project communication forms, of the project infrastructure, of project consultants and project management coaches, and of project management checklists, is to be managed.

E 2.1 Use of project management methods

The use of project management methods in projects is to be laid down in the organizational guidelines of project-oriented companies. In accordance with the structure of the project management process the project management methods to be used in the project start, in project coordination, in project controlling, in resolving a project discontinuity and in the project close-down, are to be differentiated. As regards the use these methods, a distinction is to be made between "must" and "can" use. Decisions regarding the use of "can" methods and regarding the degree of detail in the use of methods are to be made according to the project. Further the use of project management methods can be differentiated for projects and small projects (see Figure E 2.1).

Methods for the project start	small projects	project
project planning		
project scope planning		
• project objectives plan	must	must
• objects of considerations plan	can	must
• work breakdown structure	must	must
• work package specifications	can	must
project scheduling		
• project milestone plan	must	must
• project bar chart	can	must
• CPM schedule	can	can
project resources, project costs, project income		
• project resource plan	can	can
• project cost plan	must	must
• project income plan	can	can
designing the project context relationships		
• project environment analysis	must	must
• business case analysis	can	must
• project – other projects analysis	can	must
• pre- and post-project phase analysis	can	must
• project presentations, project vernissage	can	can
designing the project organization		
• project assignment	must	must
• sub-project assignment	can	can
• project organization chart	must	must
• project role descriptions	must	must
• project responsibility matrix	can	can
• project communication plan	must	must
• project rules	can	must

developing the project culture		
• project name	must	must
• project logo	can	can
• project-specific "social" events	can	can
project risk management and project discontinuity management		
• project risk analysis	must	must
• project scenario analysis and alternative planning	can	can

Methods for project coordination	small project	project
• to-do lists	must	must
• meeting minutes	must	must
• work package approval certificate	must	must

Methods for project controlling	small project	project
project controlling reports		
• project progress reports	must	must
• earned value analysis	can	can
• project trend analyses	can	can
• project score card (plus interpretations)	must	must
project controlling		
• to-do lists	must	must
adaptation of the project documentation		
• adaptation of the project management documentation	must	must

Methods for resolving a project discontinuity	small project	project
• definition of the project discontinuity	must	must
• planning immediate measures	must	must
• cause analysis	must	must
• planning alternative resolution strategies	must	must
• planning additional measures	must	must
• ending the project discontinuity	must	must

CHAPTER

E

Methods for the project close-down	small project	project
planning of measures		
• to-do list: Remaining work	must	must
• designing the environment relationships	must	must
• to-do list: Post-project phase	must	must
• adaptation business case analysis	can	must
know-how transfer		
• project close-down report	must	must
• special reports	can	can
• actual project management documentation	must	must
• project presentation	can	can
• articles in newsletters, on homepage, in journals	can	can
• exchange of experience workshop	can	can
assessment of performance		
• evaluation of project success	must	must
• assessment of the members of the project organization	can	must
symbolic actions in the project close-down		
• "social" end event	can	must
• closing project cost center	must	must
• project acceptance certificate	must	must

Designing the project management process	small project	project
project communication		
• kick-off meeting	must	can
• project workshop	can	must
• project team meetings	must	must
IT support		
• project management software	must	must

Fig. E2.1: Checklist: Use of project management methods (to be adapted for each company)

Each new project plan resulting from the use of a project management method is a model of the project and serves to construct the project reality. The use of several different project management methods enables the development of a management complexity that complies with the complexity of the project.

The quality of the project plans is to be ensured by applying multiple methods. The completeness of the project plans can only be ensured by relating the project management methods with each other and by cyclical revisions of the project plans. For example, insights from the project environment analysis can be incorporated in the work breakdown structure and/or in the project cost plan.

The degree of detail of project plans is to be determined in relation to the complexity of the project.

Project plans should be developed jointly by the project team in a project start workshop. Thereby the creativity of the team can be utilized and the identification of the project team members with the results is promoted. The initial development can be prepared by a small group of selected project team members. The use of moderation techniques ensures target-oriented and efficient teamwork. Visualization techniques promote communication in the project management process and support the documentation of the results.

Project plans are often understood to be instruments used exclusively for documentation. In fact, however, project plans are also instruments for decision-making (decision on alternative strategies), instruments of leadership (basis for agreements on objectives, establishing commitment) and instruments of communication. Adequate IT support (project management and graphics software) serves to design and communicate recipient-specific information.

In Chapter F project management methods to be used in the project management subprocesses are discussed.

E 2.2   Use of standard project plans

Standard project plans can be used for managing repetitive projects.

If a project-oriented company repeatedly performs certain types of projects (e.g. contracting projects of an IT company or product development projects of a pharmaceutical company), standard project plans can be developed for these types of projects. This kind of standardization represents an instrument of organizational learning and of knowledge management in the project-oriented company. Project plans which can be standardized are, for example, work breakdown structures, work package specifications, objects of consideration plans, milestone lists, project organization charts and project responsibility matrices.

The efficiency of the project management processes can be considerably increased by the adequate use of standard project plans. Standard project plans are to be adapted according to the respective project conditions, i.e. they are to be complemented, labeled project-specifically, etc.

E 2.3   Use of adequate project communication forms

In the project management process the communication forms of meetings between the project manager and individual project team members, team meetings and workshops can be combined. Project workshops are to be performed to assure the appropriate project management quality.

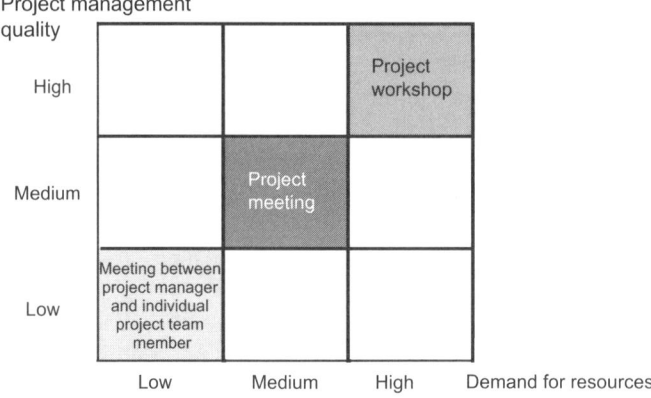

Fig. E2.2: Communication forms in the project

The objective of a meeting between the project manager and individual project team members in the project start process is to exchange information regarding the project and mutual expectations regarding the cooperation in the project. This general orientation forms the basis for the participation in project team meetings and in the project start workshop.

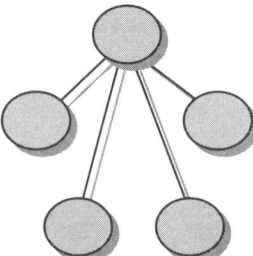

Fig. E2.3: Meetings between the project manager and
individual project team members: Communication structure

The objective of a "kick-off meeting" is that the project owner team and the project manager inform the project team about the project. It usually takes place in the form of "one-way communication", is 2 to 3 hours in length, with little opportunity for interaction.

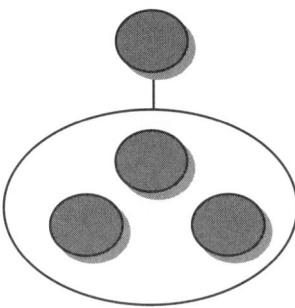

Fig. E2.4: Kick-off meeting: Communication structure

The objective of a project workshop is to jointly develop the "big project picture" in the project team. By the interaction of the team members in the workshop an important contribution to the development of the project culture is made. A project start workshop lasts 1 to 3 days, is moderated and generally takes place outside the usual workplace, possibly in a seminar hotel.

CHAPTER

E

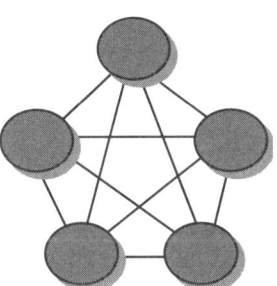

Fig. E2.5: Project workshop: Communication structure

The number of participants in a project workshop should not exceed 15. The project team members should attend the whole workshop, representatives of relevant project environments can attend the workshop selectively as guests. The most important results of the workshop should be presented to the project owner team at the end of the workshop. Involving representatives of relevant project environments and all members of the project organization in good time considerably contributes to project marketing.

Anti-theses for the project workshop
• We always moderate our problems ourselves.
• Workshops are frustrating, because afterwards things are still done the way the management wants.
• In workshops proper problems are often addressed by improper people.
• In workshops I feel like in kindergarten with wrapping paper, needle and marker.
• Holding workshops is a sign of weak leadership.

Fig. E2.6: Anti-theses for the project workshop

A checklist for designing a project start workshop is depicted in Figure E2.7.

Project start workshop: Design for one day
morning
• introduction: Ideas, objectives, agenda • information about the project and results of the pre-project phase • expectations of the project owner team, the project team, the partners, the suppliers, the consultants, etc. regarding the project • clarification of the project objectives and of the objects of consideration
afternoon
• completion of the project plans: Work breakdown structure, project cost plan, project risk analysis, etc. • review of the design of the project organization • presentation of the achieved results to the project owner team • planning of further course of action

Fig. E2.7: Checklist: Design of a one-day project start workshop

Fig. E2.8: Teamwork photograph

In projects of higher complexity a combination of several kick-off meetings and project start workshops with different target groups at different locations may be necessary.

The objective of regular controlling meetings of the project team is mutual information of the project team members on the project status and the agreement on further ways of proceeding. The objective of regular meetings of the project owner team is the information of the project owner team by the project manager on the project status and the taking of strategic decisions concerning the project.

E 2.4   Design of the project-related infrastructure

Professional project management requires the use of an appropriate information and communications technology (ICT) infrastructure as well as of an appropriate spatial infrastructure.

Especially in virtual project organizations with project team members working at different locations the planning of the software and telecommunications to be used in the project poses a challenge. The use of a uniform project management and office software is to be ensured, the appropriate hardware is to be provided. Decisions have to be made regarding the use of new communication tools, such as project management portals, collaboration software, telephone conferences and video conferences.

ICT tools		yes/no
Office-Software	• MS Word	
	• MS Excel	
	• MS PowerPoint	
	• etc.	
project management	• MS Project	
software	• Graneda Dynamics	
	• Super Project	
	• Time Line	
	• Primavera	
	• Project 2/Series X	
	• Project Scheduler	
	• etc.	
project management	• WelcomHome	
portals	• MS Project Central	
	• Proj-Net	
	• Project Office	
	• eProject – Enterprise 3.6	
	• etc.	
Telecommunications		
	• telephone conference	
	• video conference	
	• etc.	

Fig. E2.9: Checklist: Use of ICT in projects

A spatial infrastructure is to be planned and provided for holding meetings, for performing project presentations and, possibly, a project vernissage, as well as for creating a workspace for a Project Office.

E 2.5 Use of project management consultants and project management coaches

Projects, project managers and project teams represent new "objects" for consulting. Project management consulting can be defined as management consulting of a project, project management coaching is the management consulting of the project manager and/

or of the project team. Project management consulting and/or project management coaching serve to assure and increase the quality of the management of a project.

Due to the social complexity of projects the use of a project management consultant or a project management coach is recommended especially in the project start process. In the resolution of a project discontinuity project-external support may also be useful.

The decision regarding the use of a project management consultant or a project management coach should be taken jointly in the project team. As a project-external role, the role of the project management consultant and project management coach can be assumed either by an adequately qualified employee of the project-oriented company or by an external consultant.

E 2.6   Use of project management checklists

The use of checklists increases the efficiency in the project management process. The following checklists can be used:

- checklist: Use of project management methods,
- checklist: Design of a project start workshop,
- checklist: Design of a project controlling workshop,
- checklist: Agenda of a project owner team meeting,
- checklist: Use of ICT in projects,
- checklist: Table of contents of the project management documentation.

Some of these checklists were depicted above. As examples for further checklists for designing the project management process the "Checklist: Agenda of a project owner team meeting" and the "Checklist: Table of contents of the project management documentation" are depicted below.

1. Objectives
• develop a common view of the current project status • deviation analysis and discussion of the consequences • decisions • approval of the adapted project plans
2. Agenda
• introduction • minutes of the previous project owner team meeting • project progress report • decision on further course of action in the project • agreement on necessary measures
3. Participants
• project owner team • project manager • project management consultant (optional) • possibly selected project team members

4. Duration
•   2 - 3 hours

Fig. E2.10: Checklist: Agenda of a project owner team meeting

Project management documentation is to be differentiated from the project results documentation. Together they make up the project documentation, which should be structured in accordance with the work breakdown structure.

References

Watzlawick, P., (Wirklichkeit) Wie wirklich ist die Wirklichkeit? – München: Piper & Co. 1976

CHAPTER

E

F

Methods of project and programme management

Professional project management requires the application of project management methods. Following the systemic-constructivist approach of ROLAND GAREIS Project and Programme Management®, the use of project management methods makes it possible to cope with project complexity and project dynamics. The methods support communication in projects and give the members of the project organization orientation for their project work.

Project management methods can be differentiated either by the project management sub-processes or by the objects of consideration of project management (objectives, scope, schedule, resources, costs, income, risks, organization, culture, context).

The following methods are relevant for projects as well as for programmes. Programmes also require scheduling, cost planning, environment analysis, etc. To simplify the language, from now on "project management methods" will be used to mean project management methods as well as programme management methods.

Examples of project management methods are given in the case study of the project "Realization e-application". This case study serves to illustrate the methods of all project management sub-processes and ties the sections of Chapter F: Methods of project and programme management together. Specifics about the use of programme management methods are given in Chapter G: Programme management in more detail.

F Methods of project and programme management

Contents

F 1 Methods for the start process

F 1.1 Overview of the methods for the start process

The objective of the project start process is to establish a project as a social system. The communications necessary to do this are promoted and structured by the use of project management methods.

The project management methods to be used in the project start process can be differentiated by the objects of consideration of project management, i.e. objectives, scope, schedules, resources, costs, income, risks, organization, culture and context. The results of the application of methods are project management documents, i.e. project plans (e.g. work breakdown structure), graphics (e.g. project organization charts) or descriptions (e.g. expectations regarding the post-project phase).

Project management documents should be developed collectively by the project team during the project start workshop. To avoid spending too much time on basic questions in the project start workshop it is recommended that the project manager (and selected project team members) prepare preliminary versions of the project management documents. The project manager should also complete the project management documentation.

The project management methods to be used in the project start process can be selected with the help of the checklist in Figure F1.1.

Methods for the project start	small projects	projects
methods of project planning		
project scope		
• project objectives plan	must	must
• project objects of consideration	can	must
• work breakdown structure	must	must
• work package specifications	can	must
project schedules		
• project milestone plan	must	must
• project bar chart	can	must
• CPM schedule	can	can
project resources, project costs and project income		
• project resource plan	can	can
• project cost plan	must	must
• project income plan	can	can

Methods for the project start	small projects	projects
methods for designing the project context relationships		
• project environment analysis	must	must
• business case analysis	can	must
• project – other projects analysis	can	must
• pre- and post-project phase analysis	can	must
• project presentations, project vernissage	can	can
methods for designing the project organization		
• project assignment	must	must
• sub-project assignment	can	can
• project organization chart	must	must
• project role descriptions	must	must
• project responsibility matrix	can	can
• project communication plan	must	must
• project rules	can	must
methods for developing the project culture		
• project name	must	must
• project logo, picture mark	can	can
• project-specific "social" events	can	can
methods of project risk management and of project discontinuity management		
• project risk analysis	must	must
• project scenario analysis and developing alternative project plans	can	can

Fig. F1.1: Checklist: Project management methods to be used in the project start process

F 1.2 Methods for planning the objects of consideration and the objectives

The definition of the objects of consideration and the objectives of a project represent the basis for project planning in the project start process. Only knowledge of these enables any further planning, such as project scheduling or project cost planning.

F1.2.1 Objects of consideration plan

Definitions and examples

In a project various objects are "considered". Objects of consideration are, on the one hand, the (intermediary) results of a project and, on the other hand, additional objects which must be considered in order to obtain the results.

The results of the project "Realization e-application" are, for example, the installed software and hardware, but also the trained personnel and the adapted organization. To obtain these results it is possible that some form of self- or external financing are also objects to be considered[1)]

The objects of consideration of a project and its relationships can be depicted either graphically in a tree-structure or in the form of a list. The objects of consideration plan is a breakdown of the objects to be considered in a project.

✦ Realisation e Application	PROJECT OBJECTS OF CONSIDERATION
type of object	**object of consideration**
hardware	server
	PCs
	IT network
software	e-applikation user
	e-applikation administrator
	database
	office software
	IT interfaces
content seminars	content seminar 1
	content seminar 2
	content seminar 3
seminar participants	regular seminar customers
	new seminar customers
organization: roles	customer/user
	tutor
	administrator
	internet-provider
organization: business processes	seminar application process
	seminar administration process
	technical maintenance process

---

1) An example of the application of project management methods is given by the project management documentation of the project "Realization e-application" by ROLAND GAREIS CONSULTING. The objective this typical e-business project was to introduce an e-supported learning application for RGC seminars.

documents	business process documentations
	role descriptions
	manuals (user, administrator)
	training manuals
	source code documentation
	maintenance plan
personnel	tutors
	administration
financing	own financing
	external financing
marketing	brochures
	events
pilot operation	pilot operation

Version: 1.0	Date 01. 02. 2003	Author: GS

Fig. F1.2: Objects of consideration plan for the project "Realization e-application"

The objects of consideration plan is to be distinguished from the work breakdown structure. In the work breakdown structure the phases and work packages are depicted, not objects.

Objects of consideration specification

The objects of consideration specifications serve to describe the quantity and quality of the individual objects of consideration. For repetitive projects, often standard object specifications can be developed. For contracting projects the object specifications are developed during the offer process. These become part of the contract when it is signed.

As an example, the specifications of two objects of consideration are illustrated in Figure F1.3.

Objectives of objects of consideration planning

The objective of objects of consideration planning is the development of a common view of the considered objects for the members of the project organization and for representatives of relevant project environments, such as customers, suppliers and partners. A common language for the designation of the objects of consideration is to be developed and their quantities and qualities are to be described in the object specifications.

The objects of consideration form the basis of the planning of the project objectives and the project scope, as well as for designing the project organization.

In the project start process the objects of consideration can either be analytically defined, or the objects of consideration can be derived from the project objectives plan and the work breakdown structure.

✪ Realisation ✪ e Application	**PROJECT** **OBJECTS OF CONSIDERATION** **SPECIFICATIONS**
Object	**Specification**
e-application user	software solution with following features: • seminar selection • seminar description • guest book • texts for seminar • case studies • examples • registration for seminars
e-application administrator	software solution with following features: • discussion forum • overview of the seminar documents • reservation list • seminar list/seminar data • new seminar documents • participant status • seminar registration
Version: 1.0	Date: 01. 02. 2003 Author: GS

Fig. F1.3: Extract of the objects of consideration specifications of the project "Realization e-application"

Designing the objects of consideration plan

1. Structuring of the objects of consideration

The relevant criteria for structuring the objects of consideration of a project are primarily the results to be achieved. In an IT project these are, for example, the software, the hardware, the personnel and the organization. In an engineering project these are, for example, the construction, the individual technical systems and their components, the personnel and the organization of the production company. In construction or engineering projects it is also possible to differentiate between locations. Differentiation according to location can be made by individual buildings and/or floors within a building.

The connection between the individual objects of consideration can be depicted in a hierarchical breakdown.

For repetitive projects standard objects of consideration plans can be developed and used. These standard structures are to be adapted to each specific project.

2. Depicting the objects of consideration

The objects of consideration can be depicted in the form of a tree structure or in the form of a list. To ensure a clear overview, the objects of consideration should be hierarchically structured and coded.

Since the objects of consideration are objects and not activities, an object-oriented designation (with nouns) should be used.

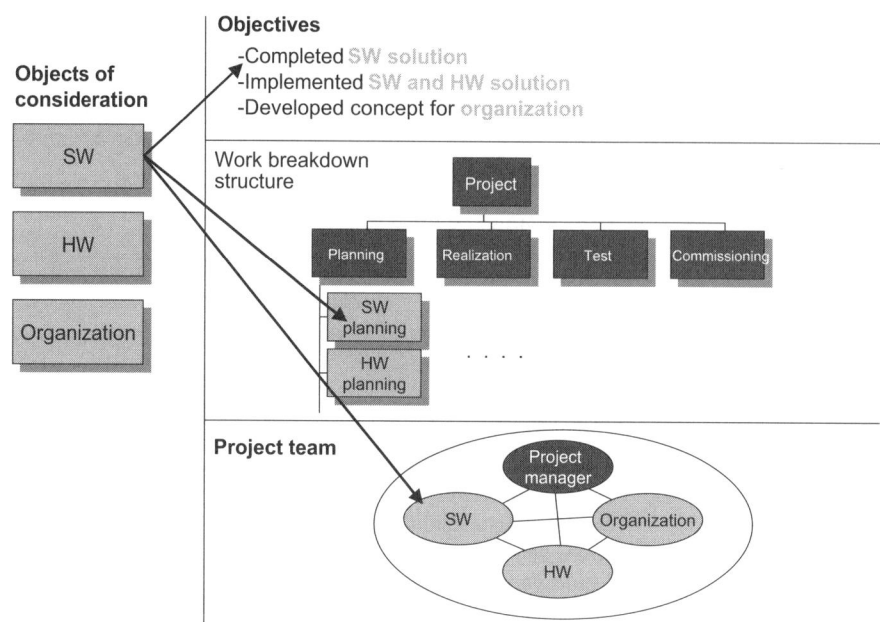

Fig. F1.4: Connection between the objects of consideration and other project plans

3. Developing specifications for the objects of consideration

Specifications of objects of consideration are only to be developed for selected objects of consideration. For repetitive projects standard object specifications can be used.

F1.2.2   Project objectives plan

Definitions and examples

Projects can be viewed as goal-determined organizations. In projects, above all, the objectives being related to content, schedule and budget are to be realized. In the project objectives plan the project content-related objectives are specified. To plan the remaining project objectives, additional planning methods are used, such as project scheduling or project cost planning. These planning results are summarized in the project assignment.

The project content-related objectives can be defined as the desired situation after the project close-down.

✪ Realisation e Application	PROJECT OBJECTIVES PLAN	
main objectives		
• e-application tested in pilot operation introduced for 3 RGC seminars as per specification (use, administration and maintenance processes) • adequate organization, 3 e-application tutors and e-application administrators trained • marketing implemented (presentation at events, hints in brochures) • financing for the application of the e-application secured • therefore reductions in the durations of the seminars and increase in the teaching quality (more information beforehand, tests) • prerequisite for increase in customer loyalty developed		
additional objectives		
• gaining experiences for further e-projects • further developing cooperative relationship with supplier		
Non-objectives		
• establishing new seminars • continuous operation and maintenance activities		
Version: 1.0	Date: 01. 02. 2003	Author: GS

Fig. F1.5: Project objectives plan of the project "Realization e-application"

CHAPTER

F

Objectives of project objectives planning

The main objective of project objectives planning is the development of a "holistic" project view. A project owner is not interested in sub-optimization through the realization of partial objectives, but in a holistic problem solution. Only the considering of all "closely-coupled" objectives ensures a holistic project view.

The installation of the software and hardware in the framework of the project "Realization e-application", for example, remains only a sub-optimization, if not all the personnel and organizational prerequisites for the use of the e-solution are provided. The development of these prerequisites is an objective to be realized in the project which is closely connected to the objectives of software and hardware installation.

The project objectives are to be operationalized as much as possible by a corresponding quantification. Only then it is possible to control whether or not the project objectives have been achieved and to evaluate the project success.

Developing the project objectives plan

1. Differentiating between main, additional and non-objectives

Main, additional and non-objectives can be differentiated in the project objectives plan. The main objectives of a project are, for example for the project "Realization e-application", the installed software and hardware as well as the achievement of the personnel and organizational prerequisites. Additional co-objectives could be process-related objectives, such as the development of a supplier into a regular supplier or the further development of know-how in the management of e-projects. A non-objective could be the establishment of new seminars.

The definition of non-objectives can contribute to the clarifying of the project objectives. Discussing the non-objectives enables project owner and project team to bring up and clarify anything which may be poorly defined.

In the project objectives plan it is possible to differentiate between the basis plan and objectives which have been adapted during periodic project controlling.

2. Quantifying the project objectives

Project objectives are to be quantified as much as possible in order to make their fulfill-ment measurable. For example, the objective "install hardware" can be quantified by specifying the essential components and their numbers.

The quantification can be performed in relation to the objects of consideration. These can serve as a checklist when examining the completeness of the project objectives.

The project boundaries must be respected during the quantification. The project objec-tives plan represents the results which a project should achieve, and not the results ex-pected in the post-project phase. For example, in the project "Realization e-application" the prerequisites for the increased customer loyalty are to be developed. The improve-ment itself can only be observed in the post-project phase.

F 1.3 Methods for planning the scope

The project scope can be planned in a work breakdown structure and in work package descriptions. The basis for developing a work breakdown structure are the objects of con-sideration plan and the objectives of the project.

F1.3.1 Work breakdown structure

Definitions and examples

The work breakdown structure (WBS) is a model of a project. It depicts the scope to be fulfilled in the framework of a project and segments the scope into sub-units which can be planned and controlled, so-called work packages. The project scope can be repre-sented graphically in a tree structure and can be listed in a table.

The work breakdown structure is neither a schedule, nor a cost or resource plan. The work breakdown structure is also not an organization chart of the project. It is the com-mon structural basis for the scheduling, cost and resource planning. In the work break-down structure the work packages to be assigned to the project team members are de-fined. Furthermore, it is the basis for the project filing system.

The work breakdown structure is a central communication instrument in project man-agement. The work breakdown structure is to be developed by the project team and to be approved by the project owner. Teamwork ensures, on the one hand, the necessary diversity and creativity and, on the other hand, the acceptance of the common solution. The work breakdown structure develops a common project understanding among the members of the project organization and contributes to the unification of the project's language and develops commitment among the team members.

The work breakdown structure is a relatively stable project management method since changes in schedule, costs or resources are not reflected in it. On the other hand, knowledge gained from other project management methods (such as project environment analysis or project cost planning) has to be integrated in the work breakdown structure. Improved project information also requires an updating of the work breakdown structure. It is supposed to always represent the latest status of the project scope.

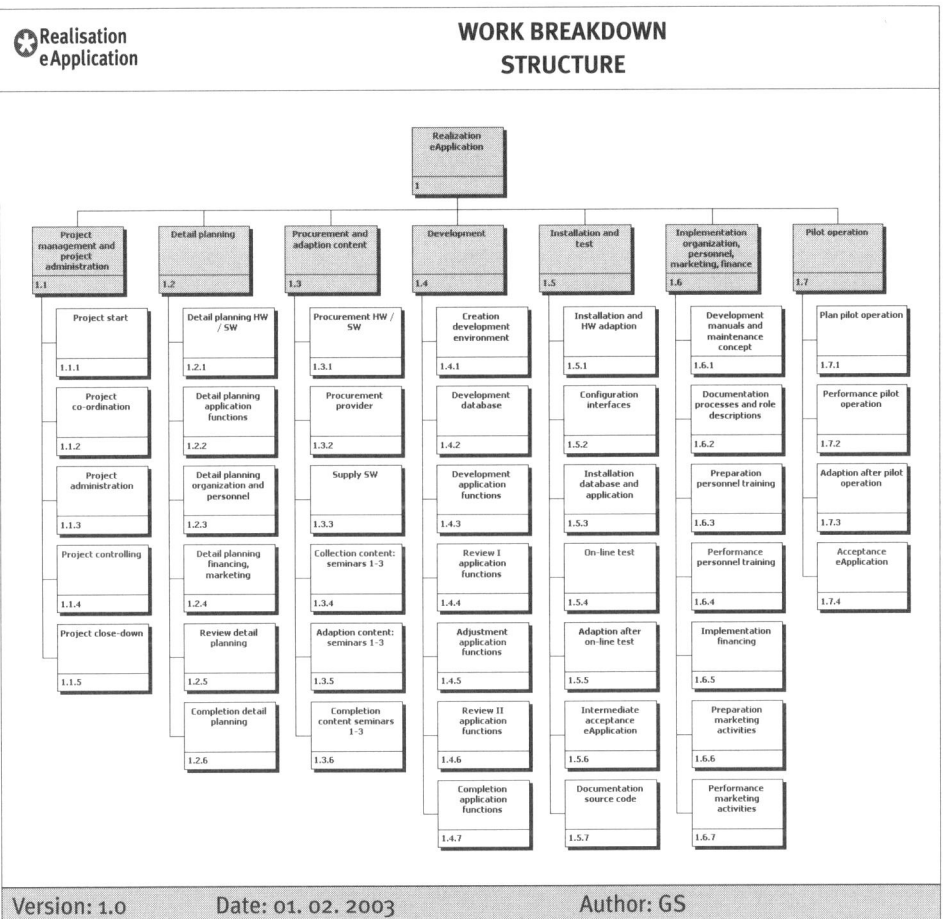

Fig. F1.6: Rough work breakdown structure of the project "Realization e-application"

Objectives of work breakdown structuring

The objective of work breakdown structuring is the segmentation of the project scope into work packages which can be planned and controlled. The entire scope should be depicted in the most complete manner possible already during the project start process. This way a common view of the project is developed for the members of the project organization and for the representatives of relevant project environments. The work break-down structure provides a contribution to the common language, to agreements and to developing commitment within the project.

Furthermore, the work breakdown structure represents the basis for scheduling, resource and cost planning. In addition, it should enable a clear assignment of work packages to project team members. A process-oriented structure is to serve as a basis for controlling the progress of the project.

Developing a work breakdown structure

1. Listing and grouping of work packages

In new, unique projects it is recommended that the work packages to be fulfilled be col-lected and listed. This can take place in the form of a brainstorming session. The listed work packages can be aggregated to groups.

In repetitive projects work breakdown structures can be standardized. Standard work breakdown structures for different types of projects can be used as the basis for devel-oping project-specific work breakdown structures.

2. Defining possible structuring criteria

On one hand, the objects of consideration of a project can be used for the structuring of the work breakdown structure. The result of the planning of the objects of consideration is a structure with many levels.

For these objects of consideration different business processes are to be performed. These functions are also structuring criteria.

Proven functions for the structuring of these processes are, for example, problem-solving cycles, work sequences, basic company functions and management tasks.

standard functions
problem-solving cycles
• gathering information, defining alternatives, evaluating alternatives, making decisions
work sequence
• conceptioning, planning, preparing, performing, follow-up
company functions
• procuring, warehousing, transporting, producing, selling, administering, man
management tasks
• planning, organizing, controlling

Fig. F1.7: Functions as structuring criteria for work breakdown structuring

3. Structuring the project scope

A work breakdown structure is developed through a step-by-step horizontal and vertical structuring of the project scope. The segmentation is made on the basis of functions (business processes) and objects of consideration. In order to develop a detailed work package segmentation it is necessary to combine several structuring criteria.

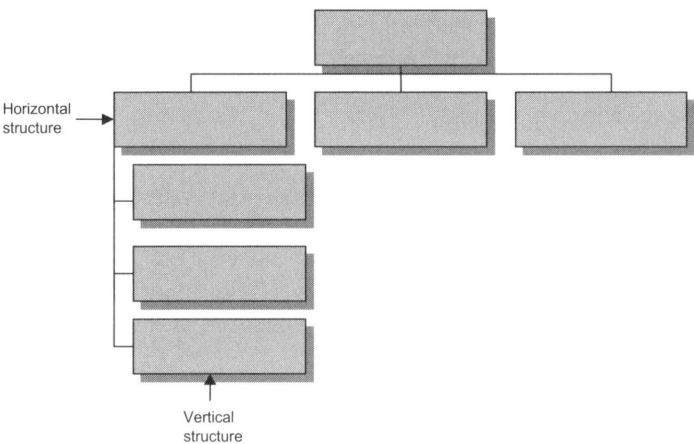

Fig. F1.8: Horizontal and vertical segmentation in work breakdown structuring

Although the work breakdown structure is not a sequencing structure, it is recommended to structure the work breakdown structure in a process-oriented way. A process orientation leads to project phases.

Each structuring criterion applied, as well as the phase scheme for the second level of the work breakdown structure, can be broken through or added to. Thus, for example, the phases planning, preparing, performing and follow-up of an event project can be expanded to include the marketing and project management, which are to be fulfilled parallel to these project phases. The positioning of marketing as a "phase" at the second work breakdown structure level is due to its assigned importance.

Project management is always to be represented as a function at the second level of the work breakdown structure and to be segmented into the sub-processes project start, project coordination, project controlling and project close-down.

The following work breakdown structure levels are structured by considering one phase (or group of work packages) each, either according to functions or objects of consideration. When structuring by objects the appropriate level of objects of consideration is to be selected. If possible, the same structuring criteria are to be used repeatedly in order to facilitate understanding and communication of the work breakdown structure.

4. Defining the adequate level of detail of the work breakdown structure

The work breakdown structure, as a central instrument in project management, is to be as little detailed as possible, but as much as necessary.

The detailing of the work breakdown structure is to be made until work packages which can be planned and controlled have been arrived at. These work packages represent the basis for the agreements between the project manager and the project team members and/ or suppliers, partners and customers. The project team members may further segment the individual work packages in order to fulfill their work

The level of detail of the work breakdown structure is also dependent upon the state of information in the project. In the start process, therefore, early project phases can usually be more exactly structured than later phases.

5. Work package names and codes

The names of the work packages have the character of "labels", which must be understandable and communicable to all members of the project organization.

The work package names initiate associations for the project team members during the project performance about what was discussed during the development of the work breakdown structure and for which a common understanding was reached. Even without the detailed work package specifications, they act as "mind catchers".

Each work package represents a function to be fulfilled within the project and is related to an object of consideration. It is, therefore, to be named in relation to function and the corresponding object of consideration (for example "Procurement of hardware and software").

Even if the development of a work breakdown structure is a creative process, the names of the work packages must be clear and consistent. The same terms are always to be used for identical functions or objects of consideration.

The coding of work packages serves, on the one hand, to clearly identify them and, on the other hand, to enable their sorting and selection. The code of a work package may consist of a combination of the function and the object of consideration.

There are work packages which refer to an object of consideration at a higher level (such as "Procurement of hardware and software") and work packages which refer to single objects of consideration (such as "Performance of pilot operation"). The work packages which refer to single objects of consideration can be coded using an object-specific code.

The number of digits used in the function-related portion of the code refers to the work breakdown structure level at which the work package lies. The coding enables the sorting and selecting of work packages according to function (e.g. all "procurement" work packages) and according to object (e.g. all work packages related to "hardware").

The function-related portion of the code is automatically assigned by many standard project management software packages as the work package is entered.

6. Depicting the work breakdown structure

Work breakdown structures can be represented graphically as a tree structure and in the form of a list.

The tree structure has the advantage that the relationships between the work packages are visualized. Furthermore, it supports the graphic depiction of the communication function of the work breakdown structure. The form used depends on the size of the project and the purpose. For larger projects different parts of the work breakdown structure may be selected for different addressees. The top management, for example, would only be interested in a rough plan of the entire project.

Tips for work breakdown structure planning
• process-oriented structuring • structuring according to the phases to be performed, not company divisions or departments • for reasons of communication and coding use no more than seven or eight work packages to segment a phase • use at least two work packages to segment a phase • use the corresponding objects of consideration as segmentation criteria • wherever possible, use the same segmentation criteria and uniform terminology • organize diversity in the project team in order to ensure the creativity required for the development of the work breakdown structure

CHAPTER

F

F1.3.2   Work package specifications

Definitions and examples

Work package (WP) specifications are quantitative and qualitative descriptions of the scope to be fulfilled in the framework of a work package. Work package specifications can also specify the method of measuring progress.

Work package specifications should provide orientation for the project team. Work package specifications are to be developed by the responsible project team members. An alignment of the interfaces between work packages can be made in small groups of project team members afterwards. The decision about the content of the respective work package specification should be taken by the project manager and the responsible project team member.

Work package specifications serve as a basis for agreements. They are especially important for internal projects. They increase transparency in project work.

The work package specifications represent a part of the project management documentation in the project handbook.

Objectives of the development of work package specifications

Objectives of work package specifications are to establish clear boundaries between individual work packages and to recognize interfaces between work packages. In a work package specification the work package results and the means to measure the progress of the work packages are defined.

By means of work package specifications the contents and the results of the individual work packages are separated from each other. Work package specifications save further detailing of the work breakdown structure. The development of work package specifications gives orientation for the work of the project team members and ensures commitment.

✪ Realisation ❂ e Application				WORK PACKAGE SPECIFICATIONS	
WBS code:	1.2.2	WP label:	detailed planning application functions		progress
WP content					
• definition of user groups					20%
• description of user and administrator functions					50%
• description of the user interface					80%
• development of a "script" for the essential functions					100%
WP results					
• "Script" of essential application functions, considering the defined user groups					
• 3 designs for the user interface					
WBS code:	1.5.4	WP label:	online test		progress
WP content					
• preparation of test data					10%
• preparation of forms for generating a protocol of the test					20%
• performance of online test					80%
• development of test protocol					100%
WP results					
• test protocol					
WBS code:	1.7.2	WP label:	Perform pilot operation		progress
WP content					
• performance of 3 pilot seminars with application					-
• recording functional errors, optimization potential					30%
• getting feedback from customers and administrators					50%
• correcting functional errors immediately					90%
• documenting optimization potential					100%
WP results					
• pilot operation performed for three seminars					
• documentation of optimization potential					
Version: 1.0		Date: 01. 02. 2003		Author: GS	

Fig. F1.9: Examples of work package specifications of the project "Realization e-application"

Developing of work package specifications

1. Selecting the work packages for which work package specifications will be developed

Basically, only those work packages are to be specified whose content and results are not clear. For new, unique projects it is recommended that a contribution to the clarification of the project content be made by specifying many work packages. For repetitive projects standard specifications can be used.

2. Specifying the work packages

Forms can be used for the development of work package specifications which should contain the description of the scope to be fulfilled, the work package results and possibly the method of measuring progress. The work package schedule or the work package costs are not part of the work package specification, but may possibly be used for internal assignments.

The description of the results of the respective work packages should also state the form of the results.

A description of the method of measuring progress should give information about intermediary results, for example at 30%, 50% or 70% progress.

F 1.4   Methods of scheduling

F1.4.1   Planning the project scheduling

Before a schedule can be developed, the part of the project to be scheduled, the planning depth and the planning method to be used must be determined.

Scheduling can be done for either an entire project or for individual project phases. Different scheduling methods can be used for different planning objects. Thus, for example, a bar chart can be used for the entire project and a CPM schedule for a project phase.

The basis for scheduling is the work breakdown structure. Work packages of different levels of the work breakdown structure can be planned in regard to their timing. A differentiation can be made between rough and detailed planning. In case the work breakdown structure does not have a deep enough structure for a detailed scheduling, then the individual work packages must first be broken down. Thereby "activities" as elements of the detailed scheduling come about.

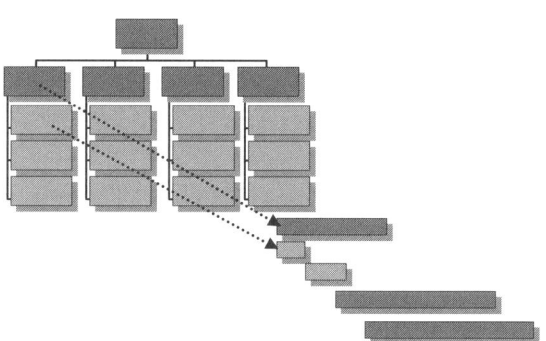

Fig. F1.10: Connection between the work breakdown structure and detailed scheduling

Tips for selecting the planning depth
• Use as little detail as possible, but as much as necessary.
• Use only as much detail as is necessary for controlling.
• Build hierarchies of schedules: Rough schedules for the whole project and detailed schedules for individual project phases.

The project logic and the project dates can be planned by means of a list of dates, a bar chart and/or network planning methods, such as CPM (critical path method). These methods compete with each other, but they also complement each other. The CPM is the most complicated and the list of dates is the simplest of the methods. Milestone planning is a specific form of listing dates. IT-supported use of one of the more highly developed methods also produces the corresponding results of less developed methods.

The efficient use of one or more scheduling methods is dependent upon the complexity and dynamics of the project.

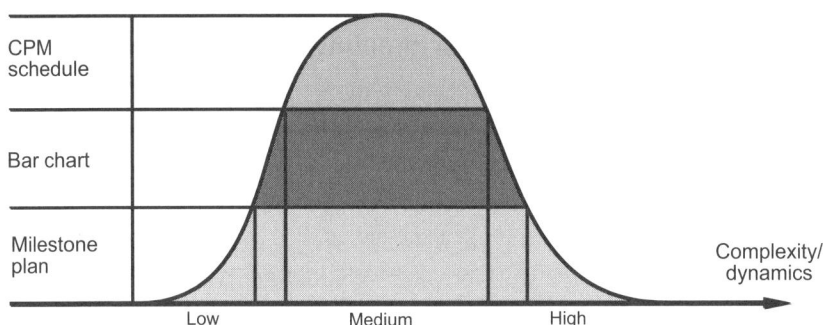

Fig. F1.11: Use of scheduling methods in relation to project complexity

The information requirements for the application of the different scheduling methods vary, as do the results that can be achieved with them.

method	information requirement
milestone plan	• list of work packages • dates of milestones
list of dates	• list of work packages • start date and/or end date of the work package
bar chart	• list of work packages • duration of the work packages • placement in time of the individual work packages
CPM schedule	• list of work packages • duration of the work packages • technological and resource dependency between work packages

Fig. F1.12: Information requirements for different scheduling methods

F1.4.2   Milestone plan

Definitions and examples

A milestone plan lists the dates of central events in the progress of a project, so-called project milestones. A project milestone plan contains no more than 8 or 9 milestones.

Project milestones are often connected to a symbolic event in a project, such as ground breaking and house-warming in a construction project.

The milestones are related to the start and end events of work packages.

✪ Realisation e Application	PROJECT MILESTONE PLAN	
WP-Code	milestone	planned date
1.1.1	project assigned	01. 02. 2003
1.2.6	detailed planning completed	27. 02. 2003
1.3.3	SW delivered	21. 03. 2003
1.4.3	draft application functions developed	04. 04. 2003
1.4.7	application completed	29. 04. 2003
1.5.6	intermediary approval of application made	20. 05. 2003
1.6.4	training performed	26. 06. 2003
1.1.5	project approved	30. 08. 2003
Version: 1.0	Date: 01. 02. 2003	Author: GS

CHAPTER

F

Fig. F1.13: Milestone plan for the project "Realization e-application"

Objectives of milestone planning

The objective of milestone planning is to determine the dates of important project events. Those provide orientation for the members of the project organization.

Developing a milestone plan

1. Selecting the object of scheduling and the planning depth

The object of scheduling of the milestone plan is always the entire project. The milestone plan is the roughest of all possible scheduling methods.

2. Listing the milestones and assigning dates to them

Per project 8-9 milestones should be defined. "Project assigned" and "Project approved" are obligatory milestones. For each milestone planned dates are to be determined. These can be adapted during periodic project controlling.

Tips for milestone planning
• define 8-9 milestones per project
• event-oriented definition of the milestones
• reference to start or end events of work packages

F1.4.3   List of dates

Definitions and examples

The list of dates is a listing of dates for work packages. The dates refer to the start and/or end event of the work packages.

✪ Realisation ✪ e Application	PROJECT LIST OF DATES		
work package		start date	end date
code	name		
1.1.1	project start process	01. 02. 2003	10. 02. 2003
1.2.1	detailed planning HW + SW	11. 02. 2003	15. 02. 2003
1.2.2	detailed planning application functions	11. 02. 2003	20. 02. 2003
1.2.3	detailed planning organization and personnel	11. 02. 2003	15. 02. 2003
1.2.4	detailed planning financing, marketing	11. 02. 2003	15. 02. 2003
1.2.5	review detailed plans	21. 02. 2003	22. 02. 2003
1.2.6	completion detailed planning	23. 02. 2003	27. 02. 2003
1.3.1	procure HW + SW	26. 02. 2003	20. 03. 2003
1.3.2	procure provider	28. 02. 2003	28. 02. 2003
1.3.4	collect content: Seminars 1-3	21. 02. 2003	02. 03. 2003
Version: 1.0	Date: 01. 02. 2003	Author: GS	

Fig. F1.14: Form for a milestone plan

Objectives of developing a list of dates

The objective of developing a list of dates is the planning of the start and end dates for the work packages, and to provide orientation for the members of the project organization.

The duration of the work packages and the logical relationships between the work packages are not explicitly planned and documented in the list of dates.

Developing the list of dates

1. Selecting the object of scheduling and the planning depth

The object of scheduling of a list of dates can be either the whole project or one or more project phases. A list of dates can be developed, for example, for the phase "Testing of software" in an IT project. A list of dates is a detailed scheduling method.

2. Listing the work packages to be planned and assigning dates to them
them

The work packages to be planned must be listed. With the use of IT this list can be developed by sorting and selecting the work packages by the work breakdown structure code. Planned dates are to be determined for each work package. Either the start and/or the end dates of a work package can be planned. The dates have to be adapted during periodic project controlling.

F1.4.4   Bar chart

Definitions and examples

The bar chart is a graphic representation of the project and/or a project phase which depicts the work packages as time-proportional bars.

Knowledge about the placement in time of the work packages is the prerequisite for the development of a bar chart. An explicit planning of the technological (and resource) interdependencies between the work packages does not take place.

CHAPTER

F

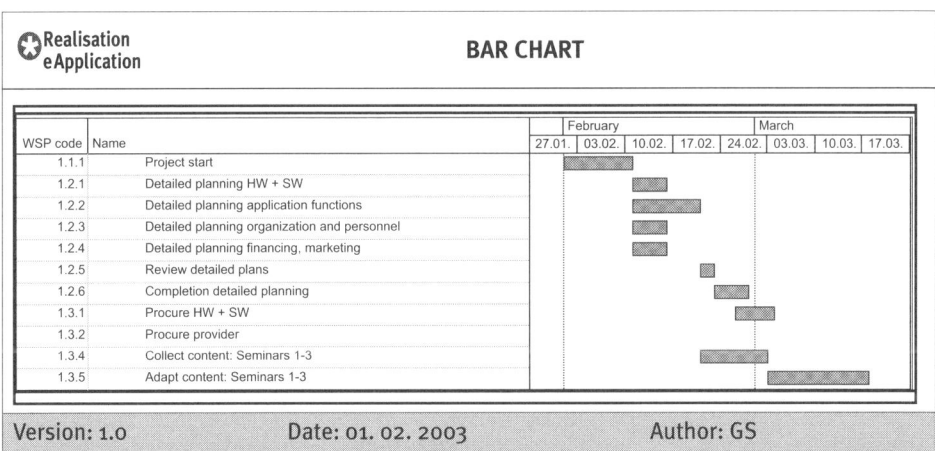

Fig. F1.15: Rough bar chart for the project "Realization e-application"

Objectives of bar chart planning

The objective of bar chart planning is the scheduling of the position in time of the work packages to provide orientation for the members of the project organization.

The bar chart is an important communication instrument in project management.

Developing a bar chart

1. Selecting the object of scheduling and the planning depth

The object of scheduling of the bar chart can be the entire project and/or one or more project phases. A bar chart can be used as a rough or as a detailed scheduling method. For the entire project a rough bar chart, and for individual project phases detailed bar charts can be developed.

2. Estimating the work package duration

Knowledge of the duration of the work packages is a prerequisite for developing a bar chart. The duration can be estimated either globally or analytically. In either case, the bases of the estimation are dates gained through experience.

Tips for estimating work package duration
• Estimate all durations in the same time unit.
• The work day will usually be the time unit used; for rough planning a week can be used as a time unit.
• The work package duration is to be differentiated from the pure working time.
• Estimate the probable duration without adding a reserve.
• The basis for the estimation is the assumption of a normal, usual resource employment for the performance of a work package. This "normal duration" creates minimal work package costs.
• The estimation is made by those responsible for the performance of the work package.

3. Graphic representation of the bar chart

The work packages are listed in the vertical line of the bar chart. In the horizontal line, which corresponds to a time axis, a time-proportional bar is drawn in for each work package.

In a networked bar chart the logical dependencies between individual work packages are defined and drawn in.

The development and visualization of bar charts is supported by project management software.

✪ Realisation e Application	ESTIMATING WORK PACKAGE DURATION	
Work package scope		
• develop the content for 3 seminars (self-assessment, pre-reading text, case studies, multiple choice test) • for each seminar the same scope of work		
Productivity per resource combination		
• work team with 2 content experts: 1 seminar in 5 days • work team with 3 content experts: 1 seminar in 4 days		
Duration estimate		
• decision: Use of work team with 2 content experts • calculation: 3 seminars x 5 days per seminar = 15 days		
Version: 1.0	Date: 01. 02. 2003	Author: GS

Fig. F1.16: Estimating the duration for the work package "Develop content – seminar 1-3" analytically

F1.4.5   Critical path method (CPM) schedule

Definitions and examples

The CPM schedule is a graphic representation of a project or part of a project which visualizes the position in time and the duration of the work packages as well as their relationships to each other.

Since the CPM uses mostly "activities" and not "work packages" as planning units, in this section the term "activities" will also be used. The CPM is considered a network planning method. Therefore, often the term "network" is used for the CPM schedule.

In network planning it can be differentiated between the "arrow diagramming" and "precedence diagramming" methods. Since, in practice, the precedence diagramming method is the one mostly used this method will be discussed here. In the precedence diagramming method the activities are shown as nodes and the relationships between the activities as arrows.

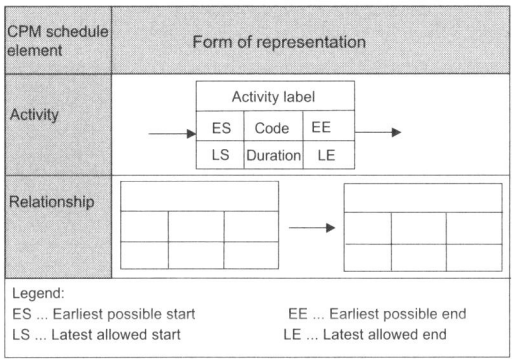

Fig. F1.17: Representation of CPM schedule elements

Basically, the logical relationship between two activities can be represented as end-to-start, start-to-start, end-to-end or start-to-end.

Kind of relationship	Relationship between
Normal sequence	End of one task and start of another task
Start sequence	Start of one task and start of another task
End sequence	End of one task and end of another task
Leap sequence	Start of one task and end of another task

Fig. F1.18: Possible logical relationships between activities

The relationships between the activities can be established by the following questions:

- Which events are prerequisites for the start (or the end) of the activity in question?
- For which events is the start or the end of the activity in question a prerequisite?

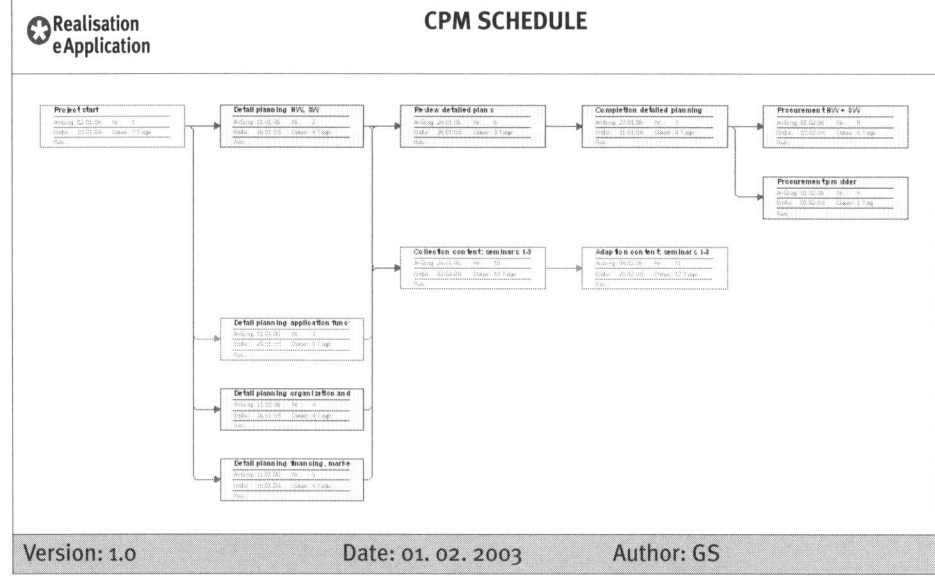

Fig. F1.19: Extract of the CPM schedule of the project "Realization e-application"

This CPM schedule can also be represented as a networked bar chart.

Code	Project name
1.1.1	Project start
1.2.1	Detail planning HW, SW
1.2.2	Detail planning application functions
1.2.3	Detail planning organization and personnel
1.2.4	Detail planning financing, marketing
1.2.5	Review detailed plans
1.2.6	Completion detailed planning
1.3.1	Procurement HW + SW
1.3.2	Procurement provider
1.3.4	Collection content: seminars 1-3
1.3.5	Adaption content: seminars 1-3

Version: 1.0 Date: 01. 02. 2003 Author: GS

Fig. F1.20: Networked bar chart for the project "Realization e-application"

As opposed to bar chart planning, the CPM technique can differentiate between planning the logic and scheduling. Because of the possibility of depicting the logical dependencies between activities, a network can be developed independently from any scheduling assumptions.

Objectives of CPM scheduling

The objective of CPM scheduling is the planning of logical relationships between activities and the position of the activities in time in order to provide orientation for the members of the project organization.

Through its relatively complicated visualization, the CPM schedule is to be seen as a communication instrument only for project management experts. It is, however, the basis for developing other communication instruments, such as the networked bar chart, bar chart or milestone list.

Developing a CPM schedule

1. Selecting the object of scheduling and the planning depth

The object of scheduling a CPM schedule can be the entire project and/or one or more project phases. A CPM schedule can be used as a rough or as a detailed scheduling method. For the entire project a rough CPM schedule, and for the individual project phases detailed CPM schedules can be developed.

2. Listing the activities to be planned

For the development of a detailed CPM schedule the work packages are to be broken down into activities which meet the following criteria:

- the activity is performed without interruption,
- the resource requirement is constant throughout each time unit and
- there is a proportional relationship between the output of the activity and the duration of the activity.

3. Developing the logic of the CPM schedule

The relationships between the activities of a project should be planned on the basis of technological dependencies, not resource dependencies. The consideration of scarce resources takes place in a later optimization step.

Important parts of the logic of the CPM schedule are developed in the project team. The use of visualization techniques (e. g. post-its on flip charts) promotes communication in the project team. Subsequently, results can be implemented by means of project management software.

Schedules of already performed projects can be used as orientation tools for the development of new plans. For repetitive projects it is possible to use standardized schedules.

4. Estimating the duration of activities

As with bar charts, the duration of an activity is estimated either analytically or globally.

5. Calculating the CPM schedule

The dates for the activities and the total project duration are calculated with the use of project management software on the basis of simple formulas. It is also possible to make the calculation manually

The calculation of dates identifies extreme dates for each activity, i.e. the earliest possible and/or the latest allowed dates. If there is a difference between the latest allowed and the earliest possible date for an activity, then this activity has a (total) float, i.e. a time reserve.

If the float of an activity is zero, then this activity is considered critical. The extension of the duration of a critical activity will extend the duration of the entire project. The chain of critical activities is called the critical path of the project.

6. Optimizing the CPM schedule

In case the duration of the CPM schedule turns out to be too long, it may be possible to shorten it by overlapping activities on the critical path.

When all the possibilities for overlapping have been exhausted and additional time savings are required, it should be considered whether an acceleration is possible by contracting out or by increasing the deployment of resources.

The desired date to start the performance can be set between the earliest possible and the latest allowed date per activity. These decisions are to be made considering the use of floats of other activities.

Tips for developing a CPM schedule

- CPM schedules for entire projects should have a maximum of 500 activities
- For presenting purposes the entire plan can be divided into partial plans with 100 to 200 activities
- The clarity of the CPM schedule can be aided by arranging the activities according to objects of consideration
- The legibility of the CPM schedule can be improved by colored marking of (technologically) related activities

---

F 1.5   Methods of budgeting

F1.5.1   Project budgets for internal and external projects

Depending on the type of project, the project budget will include either only project-related costs or project-related costs and income.

Internal projects cannot be assigned any income, therefore budgeting is limited to the planning of the project costs. In external projects services are performed for a third party in return for payment. The income received is to be planned in an income plan and compared to the project costs in the project profit calculation. Services which are performed in (external) projects, are, for example, complex customer contracts for construction, engineering, ICT companies or event organizers.

According to cost accounting terminology external projects are profit centers and internal projects are cost centers. One associated challenge for the project-oriented organization is the establishment of temporary profit centers and cost centers.

In internal as well as in external projects one should aspire to the most holistic view of the project costs possible. This means that not only the costs that lead to disbursements, but also the opportunity costs should be planned. Furthermore, project-related costs of potential partners and of the customer (in external projects) should also be taken into account.

In planning the project budget it must be minded that only those costs and income are taken into account which accrue during the project. Costs and income relating to the post-project phase are to be considered in a business case analysis of the investment which is initialized by the project.

F1.5.2   Project cost plans

Definitions and examples

A cost plan is a model of a project from the cost perspective. Cost plans can be developed for individual work packages, for individual objects of consideration, for internal assignments and for the entire project.

The integration of project scope, project schedule and project costs requires a uniform structural basis for their planning. This structural basis is the work breakdown structure. Planning units for the cost planning, therefore, are the work packages of the work breakdown structure.

The individual work packages of the work breakdown structure are those units for which planned costs are planned and actual costs are recorded. Cost structuring according to the work breakdown structure provides the possibility for the work package costs to be grouped according to various criteria and to be influenced at different levels of aggregation.

Internal assignments are developed by grouping work packages according to areas of responsibility. Calculating the costs of internal assignments enables cost responsibility accounting. For the realization of a cost controlling and a cost responsibility accounting the structure of the cost plans must coincide with the cost structures used for cost recording.

Summing the costs according to objects of consideration enables visualization of the cost consequences of changes to the scope.

These structural relationships of the project cost planning are shown in the following figure.

The planning and controlling of project costs can be performed either by means of project management software or with database or spreadsheet software.

A detailed view of the cost plan for the project "Realization e-application" is depicted in Figure F1.22.

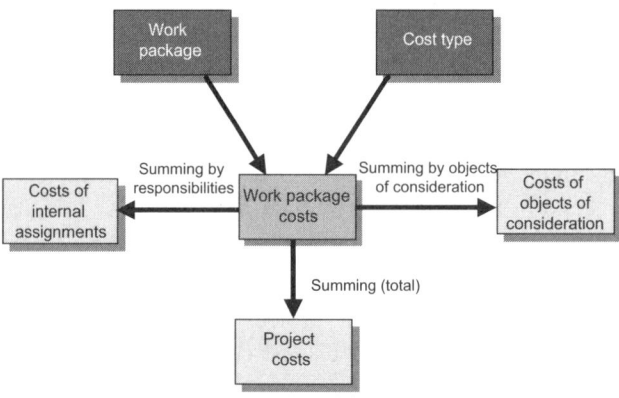

Fig. F1.21: Structural relationships of project cost planning

✪ Realisation eApplication	**PROJECT** **COST PLAN**				

Work package		Cost type	Planned quantity	Price	Planned costs
WBS	**Label**				
1.1.1	Project start	Personnel	5 PD	€ 400,00	€ 2.000,00
1.1.2	Project coordination	Personnel	18 PD	€ 400,00	€ 7.200,00
1.1.3	Project administration	Personnel	6 PD	€ 240,00	€ 1.440,00
		Material	1	€ 1.000,00	€ 1.000,00
1.1.4	Project controlling	Personnel	6 PD	€ 400,00	€ 2.400,00
1.1.5	Project close-down	Personnel	4 PD	€ 400,00	€ 1.600,00
Phase 1.1	**Project management and project administration**				**€ 13.640,00**
1.2.1	Detailed planning HW + SW	Personnel	1 PD	€ 400,00	€ 400,00
		Services	1	€ 1.000,00	€ 1.000,00
1.2.2	Detailed planning application functions	Personnel	2 PD	€ 400,00	€ 800,00
		Services	1	€ 2.000,00	€ 2.000,00
1.2.3	Detailed planning organization and personnel	Personnel	3 PD	€ 400,00	€ 1.200,00
1.2.4	Detailed planning financing, marketing	Personal	3 PD	€ 400,00	€ 1.200,00
1.2.5	Review detailed plans	Personnel	2 PD	€ 400,00	€ 800,00
		Services	1	€ 500,00	€ 500,00
1.2.6	Completion detailed planning	Personnel	1 PD	€ 400,00	€ 400,00
		Services	1	€ 1.000,00	€ 1.000,00
Phase 1.2	**Detailed planning**				**€ 9.300,00**
1.3.1	Procure HW + SW	Personnel	2 PD	€ 400,00	€ 800,00
1.3.2	Procure provider	Personnel	1 PD	€ 400,00	€ 400,00
1.3.3	SW delivery	Material	1	€ 3.000,00	€ 3.000,00
...	... usw				

Version: 1.0	Date: 01. 02. 2003	Author: GS

Fig. F1.22: Detailed view of the cost plan of the project "Realization e-application"

Objectives of project cost planning

Project cost plans serve the planning and documentation of the project costs. They provide a basis for decisions (e. g. deciding whether to perform a project or not, or deciding on the offer price for an external project), and they enable controlling.

Furthermore, they form the basis for the assessment of project success and for the agreement of objectives between the project owner and the project team.

Developing project cost plans

1. Selecting the relevant planning depth

The project cost plan is structured according to the work breakdown structure. It must be decided which level of the work breakdown structure the cost plan should relate to. In most cases a relatively high work breakdown structure level will be sufficient for the requirements of cost planning and cost controlling. The work packages of the lowest work breakdown structure level or the scheduling activities usually provide too much detail for project cost planning.

2. Planning the work package costs

The cost types which are to be considered for each work package must be determined. It is usually enough to limit the cost types to personnel costs, material costs, costs for services from suppliers and "other" costs.

For each cost type the (plan) quantity is to be determined. By multiplying the plan quantity by the plan price the costs are determined.

Determining the plan quantities for the work packages presupposes a qualitative and quantitative planning of the resources to be used. This planning step corresponds to an integrative consideration of the project resources and the project costs.

The work package costs are to be determined by the project team members fulfilling the individual work packages. Central calculation departments cannot be assigned the job of a detailed project cost planning, since they do not have the process-specific know-how to be able to determine either the relevant plan quantities or the plan prices. The early involvement of those departments charged with the fulfillment of the individual work packages in the calculation also creates a basis for motivation which facilitates the agreement on the costs.

3. Calculating the project costs

By summing the work package costs the "production costs" of a project can be determined. By taking into account additional overhead costs, the project costs are calculated according to the full costing approach.

4. Determining the cost of the objects of consideration

By summing the costs of all the work packages related to a particular object of consideration the cost of the object can be determined. The prerequisite for the necessary (IT-supported) sorting of work packages is to be developed through the work breakdown structure coding.

5. Agreeing on objectives

Determining the costs of internal assignments and their use as target figures for the project team members enables a cost responsibility accounting. The planned costs for the entire project can be the basis for the agreement of objectives between the project owner and the project team.

F1.5.3 Project cost histograms and project cost curves

Definitions and examples

Project cost histograms and project cost curves are graphic representations of the project costs over time. In project cost histograms the project costs are shown by period. In project cost curves the project costs are accumulated over time. This planning step corresponds to an integrative consideration of the project costs and the project schedule.

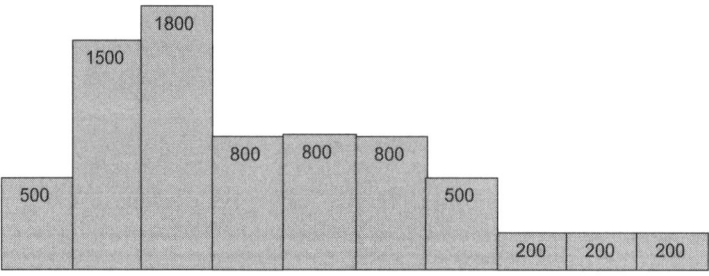

Fig. F1.23: Project cost histogram

The costs during early and late project phases are normally lower than in phases of high productivity. This results in a typical s-curve of the project costs.

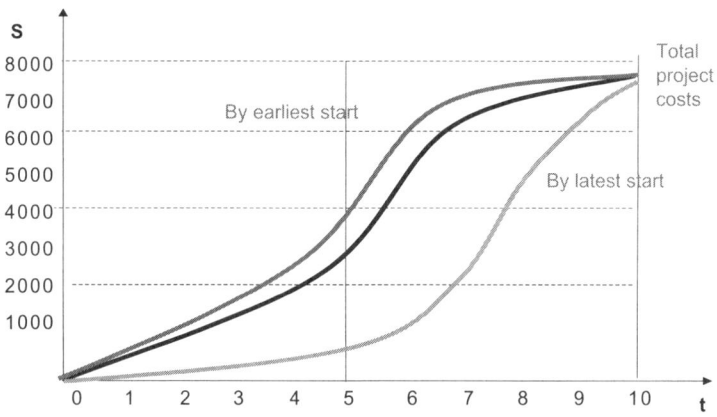

Fig. F1.24: Accumulated project costs

Objectives of the project cost histogram and project cost curve

The objective of developing a project cost histogram and a project cost curve is the visualization of project costs occurring over time.

This creates the possibility of optimization according to cost planning and/or bookkeeping objectives. For example, it is possible to move costs from one bookkeeping period to another by rescheduling work packages.

Furthermore, the basis for the planning of project-related cash outflows and for the development of project progress curves is established.

Developing project cost histograms and project cost curves

1. Calculating project costs per period

The bases for the calculation of the project costs over time are the work package dates and the work package costs. The dates of the work packages are visualized in the project schedule. The work package costs result from the project cost planning.

With regard to the occurrence of costs in the periods of the work package duration, certain assumptions are to be made. In case there is no explicit information about how the costs vary over time in a work package, it should be assumed that the costs are proportional to the work package duration. Most project management software packages enable moving costs to the start, to the end and/or to distribute them evenly across the duration of the work package.

2. Developing graphics and optimization

In project cost histograms the costs per period are totaled and depicted. In project cost curves the costs per period are depicted as accumulated. The necessary calculations and visualizations are performed by project management software.

F1.5.4   Project income plan

Definitions and examples

The development of a project income plan is only relevant for external projects, since internal projects cannot have income.

Project-related income can, on the one hand, be payments from customers for services, but also sponsoring or subvention payments. In most cases there are few payments which are easy to plan. The timing of the payments depends on the payment agreements which have been made. Furthermore, in external projects interest income may accrue.

In the following example the project income for the project "pm days '03" is shown. The objective of this project of RGC was the organization of a project management event in Austria. The income arose out of the fees paid by participants and exhibitors at the event, as well as from sponsoring.

Objectives of project income planning

The objective of project income planning is to plan project-related income and to assign it to the project. That way the responsibility of the project team for securing the project-related income is apparent and a basis for the project success calculation is developed.

Developing a project income plan

1. Listing the types of income

The different types of income which can be attributed to the project are to be listed.

2. Planning the income

For the individual income types the amount and the timing of the income is to be planned and depicted in an income plan.

WBS code	Income	
1.4.2	Sponsors	€ 18.150,00
1.4.2	Exhibitors	€ 18.500,00
1.4.2	Other income	€ 3.200,00
1.7.1	Participants	
	- Research conference	€ 9450,00
	- Practice conference	€ 87.850,00
	- Expert seminar	€ 11.000,00
	Total	€ 148.150,00

Fig. F1.25: Income plan of the project "pm days '03"

F1.5.5   Financial project success

Definitions and examples

The financial project success can be defined as the difference between the project income and project costs, whereby the term "project costs" is defined differently in the full costing and in the contribution method.

By applying full costing the project success is calculated as the difference between the project income and "full" project costs, which also contains a share of the fixed (overhead) costs of the organization performing the project. A positive project success is referred to as a project profit, the negative project success as a project loss. By applying the contribution method project success is calculated as the difference between project income and the variable project costs. The resulting project success is referred to as "contribution". This project-related contribution serves to cover the fixed costs of the organization performing the project and to realize a profit for the organization.

Objectives of project success calculation

project success. By comparing project costs and project income the influence factors of the project success become apparent. Not only a minimization of costs but also a maximization of income contributes to the optimization of the project success.

Developing a project success calculation

1. Calculating the project success

It must be decided whether the project success is to be calculated on the basis of full costing and/or contribution costing. Then the project success is to be calculated as the difference of project income and project costs.

2. Agreeing on objectives

The calculated project success is to be interpreted in regard to its suggestibility and its risk, and possibly used as a basis for project-specific agreements.

WBS code	Phase	Costs
Costs		
1.1.	Project management	€ 40.000,00
1.2.	Planning	€ 700,00
1.3.	Preparing for first announcement	€ 9.000,00
1.4.	Preparing for final announcement	€ 12.000,00
1.5.	Review and final marketing	€ 50.000,00
1.6.	Preparing for pm days '03	€ 5.000,00
1.7.	Performing pm days '03 and follow-up	€ 9.000,00
		€ 119.748,00
Income		
1.5.	Sponsors	€ 18.150,00
1.5.	Exhibitors	€ 18.500,00
1.7.	Other income	€ 3.200,00
1.7.	Participants	
	- Research conference	€ 9.450,00
	- Practice conference	€ 87.850,00
	- Expert seminar	€ 11.000,00
		€ 148.150,00
Contribution		
		+28.402,00

Fig. F1.26: Project success calculation of the project "pm days '03"

F 1.6   Resource planning

Definitions and examples

A project resource plan is a tabular and/or graphic representation of the requirements of a resource for a project over a period of time.

In project resource planning not all the resources of a project are planned, just the so-called "bottleneck resources". Bottleneck resources are those resources which are scarce for the project and therefore have an influence on the achievement of the project objectives. Bottleneck resources can be, for example, personnel, machinery, finances, material, storage space, etc. Personnel bottleneck resources are to be differentiated by qualifications which are not interchangeable. In an engineering department, for example, one can differentiate between a design engineer and a draftsman. It is possible to consider individuals as bottleneck resources, rather than groups with certain qualifications. In projects which initialize infrastructure investments (construction, engineering, IT investments), the financing often represents a considerable bottleneck resource.

A basis for project resource planning is already developed in the determination of the plan quantities in project cost planning.

F1.6.1 Project resource plan (general)

Objectives of planning the project resources

The offer and demand on a bottleneck resource can be represented in project resource histograms and in project resource curves. On the basis of comparison of the resource requirements with the currently available resources, the project-related over- and/or under-coverage of project resources can be determined. This may lead to an increase or a reduction in the offer of resources, or to compensation by shifting work packages.

CPM-supported project resource plans can be developed for different time positions of the work packages.

Since planning project resources is limited to the consideration of bottleneck resources, this is no instrument for general resource planning in the project-oriented organization. It is, therefore, not a substitute for a department-related personnel planning.

Project resource plans are to be developed for each scarce project resource. The following procedure can, therefore, be run through many times for different project resources.

Developing project resource plans

1. Assigning the bottleneck resources to the work packages

Each bottleneck resource for which a project resource plan is to be developed is to be defined. The work packages of the work breakdown structure are the carriers of the resource requirements. It must be determined for each work package whether the bottleneck resource is required.

2. Estimating the resource requirement during each period per work package

For each bottleneck resource the requirement is to be estimated per work package per planning period. These plan quantities are either directly available as results of the project cost planning or can be derived from them.

As a rule, it can be assumed that the resource requirement of a work package per period remains the same.

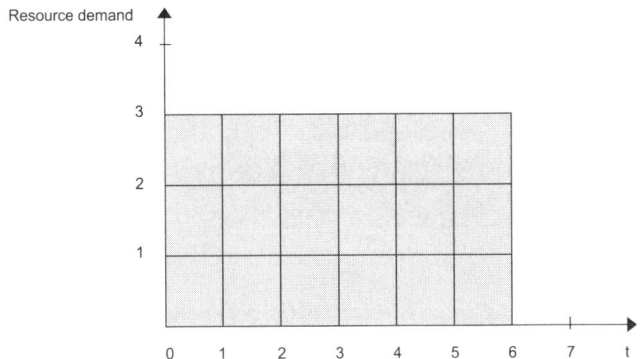

Fig. F1.27: Unchanging resource requirement of a work package per period

3. Developing a resource histogram and a resource curve

By summing the requirements for a bottleneck resource of work packages in a project per period a resource histogram can be developed. The accumulation of the periodic resource requirement leads to a resource curve. In CPM-supported project resource planning resource plans can be developed for the earliest and latest positions of the work packages.

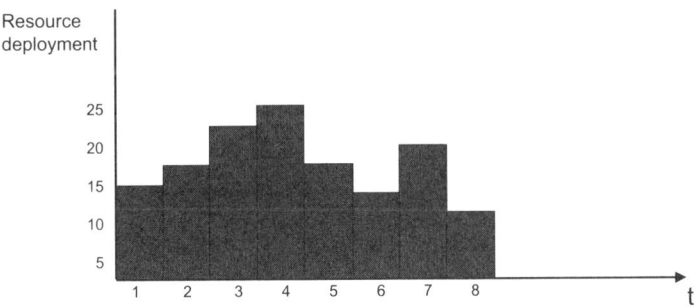

Fig. F1.28: Resource histogram

4. Determining the available amount of a bottleneck resource in an organization

The available amount of a bottleneck resource in an organization is to be determined and compared to the resource requirement. The available resource quantity can remain the same over the duration of the project or vary in different periods. The resource requirement of a project per period can be smaller or larger than the available resource quantity. Over- and under-coverage can be determined.

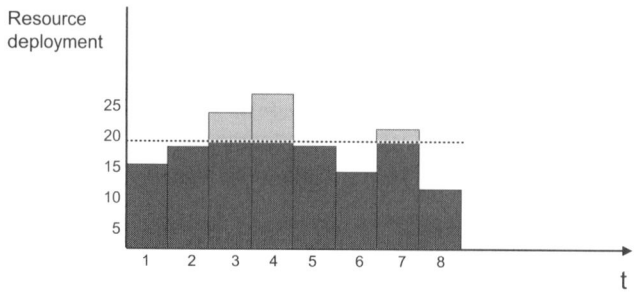

Fig. F1.29: Over- and under-coverage of the available bottleneck resource

5. Making optimizations

With resource bottlenecks a reworking of the project plans is necessary. By shifting work packages and/or by changing the logic or by dividing work packages it may be possible to ensure the most even use of a bottleneck resource and to even out the bottlenecks.

Furthermore, possibilities to increase or reduce the offer of resources must be taken into account. Possible measures for increasing the offer of personnel resources are, for example, the establishment of incentives for increased performance, overtime or weekend work (in limited amounts) or the use of temporary workers.

6. T-supported resource planning

Most standard software packages for project management have a module for project resource planning. This enables the differentiation between resource requirements at the start, at the end or constantly over the duration of a work package.

F1.6.2   Project finance plan

Objectives of planning the project financing

Financing can be a bottleneck resource in projects. To plan project financing it is necessary to differentiate between project-related disbursements and project costs.

The data from the project cost planning are the basis for determining the cash disbursements. The differentiation between costs and disbursements is shown in Figure F1.30.

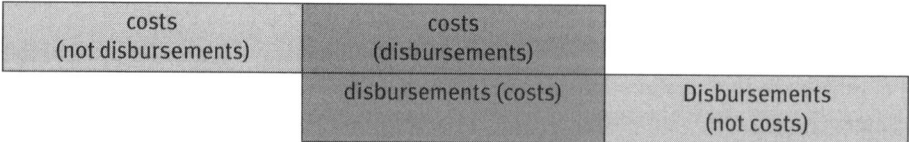

Fig. F1.30: Differentiation between costs and disbursements

The largest part of all project costs leads to disbursements. Examples of cash and non-cash costs are project-related personnel and material costs, travel costs, etc. Disbursements which are not effective in the respective period are, for example, investment payments for machinery. Non-disbursement costs however are, for example, interest, depreciation or risk surcharges.

The timing of a disbursement is usually dependent upon the timing of the costs. It should be taken into account that where costs accrue continually over time and proportionally to production, disbursements are made at discrete points in time and are not always proportional to production (for example down-payments for deliveries or services).

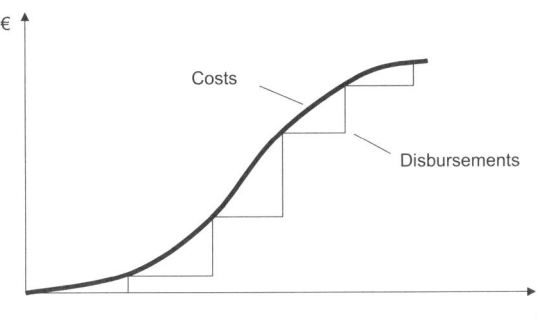

Fig. F1.31: Accumulated costs and disbursements over time

By definition, project-related receipts only occur in external projects. In internal projects the planning of the cash inflow is not applicable. In external projects the receipts occur as set forth in the contract.

reason for cash inflow	date	cash inflow in % of the total amount	Cash inflow in € 1,000
contract concluded	1. 2. 200x	30% down payment	EUR 15,000
delivery of machines	17. 10. 200x	30% payments	EUR 15,000
installation completed	15. 11. 200x	30% payment	EUR 15,000
hand-over	3. 12. 200x	10% final payment	EUR 5,000

Fig. F1.32: Planning of a cash inflow in a contracting project of an engineering company

Project financing plans are tabular and/or graphic representations of the project-related disbursements and receipts over time. The cash inflow and outflow of a project can be represented in a histogram per period and accumulated in a curve. In a CPM-supported project finance planning histograms and curves can be developed for different time positions of the work packages.

Project finance plans are used for project-related liquidity planning. By calculating the cash surplus per period the requirement for and/or the availability of financing can be planned. A project-related financing requirement exists when there is a disbursement surplus in an individual period. This is calculated out of the difference between the project-related disbursements minus the project-related receipts.

The financing requirement may be financed within the project. In this case the interest costs developed by the financing requirement should be considered as calculatory costs of the project. In case the project-related receipts exceed the disbursements then there is a receipt surplus. Receipt surpluses can be invested. The interest income should be considered as calculatory income of the project.

Developing project financing plans

1. Defining the planning depth

In order to control project liquidity and to optimize the project-related interest costs and income, financing plans for the entire project should be developed. The project financing plans can be developed at a relatively high work breakdown structure level. Per project phase, however, different work breakdown structure levels can be considered for the determination of the project-related cash flows.

2. Determining the project-related cash flow

Starting from the time position of the work packages, specific assumptions about the accrual of cash flows can be made. For example, with supplier services the schedule of down payments, partial payments and final payments must be considered. These are determined by the agreement of payment terms and cannot be directly deduced from the cost accrual.

3. Planning the cash inflow

In external projects the project-related cash inflow is to be planned. The amount of the receipts and their due dates are usually set forth in the customer contract.

4. Developing a cash flow histogram and a cash flow curve

Out of the graphic representation of the cash in- and outflows in histograms and curves the project-related financing requirement and/or financing surplus becomes apparent.

5. Calculating the project-related interest costs and income

Assuming corresponding calculatory interest rates for interest on the positive and/or negative cash flow surpluses, project-related interest costs and/or income can be calculated.

6. IT-supported project finance plann

Most standard software packages for project management enable the specification of cash flows as specific resources. In case the consideration of additional payment dates is necessary, they may be defined as milestones in the schedule.

F 1.7   Methods for designing the context relationships

F1.7.1   Project and project context

The project context is the setting in which the project is performed. The dimensions of the project context are the company strategies, other projects of the project-oriented organization, the pre- and the post-project phase of the project, relevant project environments and the business case of the investment initialized by the project.

In different analyses the project context can be determined and strategies and measures for designing the project context relationships can be planned.

Understanding of the project context provides members of the project organization with an orientation for their actions. A corresponding outward orientation of the project organization is ensured.

Tips for project context analyses

- Perception of a project in its context
- Communicate the "big project picture" through the project context analyses
- Project management as relationship management: The communication with relevant project environments can be specifically designed on the basis of the project environment analysis
- Performance of project marketing tasks inside the project, inside and outside the project-oriented company
- Assurance of outward orientation in the project already during the project start process

---

F1.7.2 Project and company strategies analysis

Definitions and examples

In the analysis of the relationships between the company strategies and the project under consideration it should be clarified whether the strategies of the company caused the performance of the project and in which form and to what extent the project contributes to the realization of the company strategies. Furthermore, it should be established whether or not the project influences the company strategies.

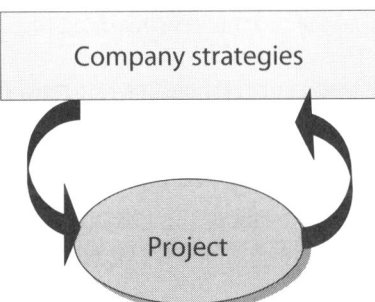

Fig. F1.33: The relationships between a project and the company strategies

External projects should always contribute to ensuring market share and relatively short-term financial success. Internal projects serve, above all, the long-term survival of the company through product, organizational and personnel developments and through investments in the infrastructure of the company.

The strategies of a company can either be explicitly planned and documented or they may exist only implicitly. In this case the relevant assumptions about the company strategies are to be made and documented.

✪ Realisation e Application	PROJECT AND COMPANY STRATEGIES ANALYSIS
company strategies	contribution of the project to realization
continual quality improvement	• further development of the quality of the seminars
growth	• increase in turnover for the profit center "Seminars" through a more attractive offer
innovation leadership	• use of new technologies (e-business) for the performance of seminars
cooperative relationship with suppliers	• further work with regular supplier
Version: 1.0	Date: 01. 02. 2003 Author: GS

Fig. F1.34: Project and company strategies analysis of the project "Realization e-application"

Objectives of the project and company strategies analysis

The objective of the project and company strategies analysis is to find the connections between the objectives of a project and the company strategies and to document them. This analysis enables the understanding of a project's sense.

The members of the project organization as well as the representatives of relevant environments have a right to be informed about the connections. The communication of the contribution of a project to the implementation of the company strategies promotes the acceptance of a project and contributes to ensuring the relevant management attention, for example in regard to the provision of required resources.

Developing the project and company strategies analysis

1. Listing relevant company strategies

It is to be determined whether explicit and documented company strategies are available. If this is not the case, assumptions about the implicit strategies are to be made. Each company strategy which is relevant for the project under consideration is to be listed.

2. Analyzing the relationships between the project and the company strategies

The relationships between the project objectives and the explicitly or implicitly existing company strategies are to be analyzed and documented. In case such connections cannot be made then either the company strategies are to be adapted or the performance of the project put into question.

3. Communicating the relationships between the project and the company strategies

The relationship between the objectives of the project under consideration and the company strategies is to be communicated to the members of the project organization and to the relevant environments in the corresponding form.

Questions for the analysis of the relationships between a project and company strategies
• Which company strategies will be realized by the project? • How does the project influence the (further) development of the company strategies? • How great is the importance of the project for the company? • What priority does the project have in the company? • How large is the share of the project turnover on the total turnover of the company?

F1.7.3   Project and other projects analysis

Definitions and examples

Relationships between a project and other projects can be synergetic or competing. Dependencies can arise between projects in regard to objectives, the methods used, the resources used and necessary intermediary results.

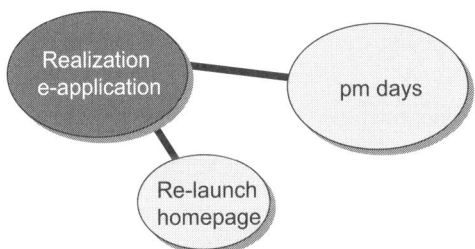

Fig. F1.35: Relationships of the project "Realization e-application" to other projects at RGC

✪ Realisation e Application	PROJECT AND OTHER PROJECTS ANALYSIS	
other projects	relationships	measures
re-launch homepage	• common user interface and technology	• use the same programmer
pm days	• marketing the new e-supported seminar application during the pm days	• plan and prepare a presentation of the e-application
Version: 1.0	Date: 01. 02. 2003	Author: GS

Fig. F1.36: Analysis of the relationships of the project "Realization e-application" to other current projects

Dependencies in terms of results can refer to the aspired end results and/or to intermediary results. Intermediary results of other projects can be prerequisites for further project work of the project under consideration. For example, the implementation of a new software package within an IT project can be a prerequisite to further work in an organizational development project.

Resource synergies can result, for example, in the performance of several projects with the same supplier. Resource competition can arise, above all, around the availability of "scarce" experts as project team members.

Objectives of the project and other projects analysis

The objective of the analysis of the relationships between a project and other projects is to determine whether there are synergies or conflicts in regard to the objectives, the results to be achieved and/or the methods and resources to be used. The result of the analysis forms the basis for using the synergies and for avoiding conflicts.

Developing the project and other projects analysis

1. Listing the projects to consider

Projects having relationships to the project under consideration are to be listed. The basis for this selection can be information from the project portfolio database and from the members of the project organization.

2. Analyzing the relationships between the project under consideration and other projects

Potential relationships between the project under consideration and the selected projects in regard to the objectives, the results to be obtained and/or the methods and resources to be used are to be analyzed.

3. Coordinating with other projects

The measures necessary to use the synergies or to avoid conflicts between projects are to be defined and communicated.

F1.7.4   Pre-project and post-project analysis

Definitions and examples

Projects often have long histories. Information about the concrete reasons which led to a project, about actions and decisions before the formal project start are important for understanding a project.

Actions and decisions during the pre-project phase can no longer be influenced by the project team. But through the project history factors which can promote or inhibit project success become apparent.

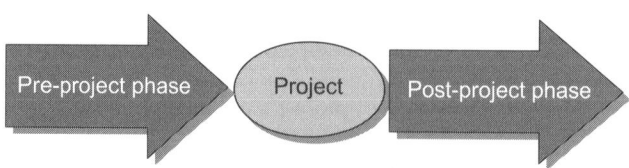

Fig. F1.37: Time dimension of a project

Expectations regarding the post-project phase could be, for example, the application of the project results for reference purposes, the establishment of a long-term customer relationship, career advancements for the project team members, etc. Expectations regarding the post-project phase can be influenced.

Information about the pre- and the post-project phase should be exchanged by the members of the project organization during the project start process. In this way a uniform level of information can be ensured.

Not only the objectives to be realized in a project and the project scope but also the project history and the expectations regarding the post-project phase influence the design of a project. For example, the project history influences the selection of the project team members, the definition of project values, the planning of communication forms, the establishment of the methods to be used, etc. Potentials for the realization of expectations regarding the post-project phase are to be created in the project.

Objective of the pre- and post-project phase analysis

The objective of the analysis of the pre-project phase and the post-project phase is to make activities and decisions from the pre-project phase as well as expectations regarding the post-project phase transparent. This creates the basis for an appropriate structuring of the project.

Developing the pre- and post-project phase analysis

1. Analyzing the existing documentations and decisions from the pre-project phase

Existing documents (protocols, plans, descriptions, etc.) which were developed during the pre-project phase are to be analyzed for relevant information. Possibly interviews with promoters of the project idea and with decision-makers are to be made. Consequences for the structuring of the project can be derived and documented.

2. Analyzing the expectations regarding the post-project phase

From the existing documents and from interviews the expectations regarding the post-project phase are to be analyzed. Consequences for the structuring of the project can be derived and documented.

✪ Realisation e Application	PRE-PROJECT AND POST-PROJECT PHASE ANALYSIS
Description of the results of the pre-project phase	
Environment relationships relevant for the project • IT supplier • internet provider	
Decisions/events relevant for the project • software selection: Oracle Web DB 3.0 (portal) • selection of IT supplier • acceptance of important application functions (on the basis of specification)	
Documentations relevant for the project • Documentation of the project "Conception e-application" • Specification of the e-application	
Expectations regarding the post-project phase	
Further development of the environment relationships • further cooperation with IT supplier • customer loyalty through new seminar quality	
Measures for the post-project phase • use the e-application for further seminars • evaluate the user feedback; further develop the contents • develop the English version of the e-application	
Use of the experiences gained • use the project management documentation for training and consulting purposes • use the know-how for other e-projects	
Version: 1.0	Date: 01. 02. 2003 Author: GS

Fig. F1.38: Analysis of the pre-project and the post-project phase of the project "Realization e-application"

CHAPTER

F

3. Communicating to the members of the project organization

The results of the analysis are communicated to the members of the project organization in the project start workshop.

Questions for the pre- and post-project phase analysis
• What concrete reasons led to the project? • Who promoted the performance of the project? Who inhibited it? • What data and documentation already exist, which can be used? • What decisions have already been made which must be taken into account? • What similar projects have already been performed in the organization? • What actions and decisions are necessary after the project? • Which follow-up projects can/should/must ensue?

F1.7.5   Project environment analysis

Definitions and examples

The (relevant) social environments of a project can be considered its social context.

The project environment analysis considers the relationship of a project to its relevant environments. It is assumed that the relevant environments of the project cannot be (directly) changed. Therefore, the project environment relationships are considered, they can be designed. The management of these relationships is a project management task.

"Relevant" for a project are those environments which can have an effect on the project success. Relevant project environments can be differentiated into project-internal and project-external environments. Project-external environments are, for example, customers, suppliers, banks but also other business units and departments of the project-oriented company. The project team and the project manager can be considered as (project-internal) environments because their relationships to the project significantly influence its success.

CHAPTER

F

The project environment analysis can be represented either as "cloud graphic" or as "segment graphic". These graphics are to be complemented by the description of the project environment relationships. The segment graphic is to be used when a large number of relevant environments is to be considered.

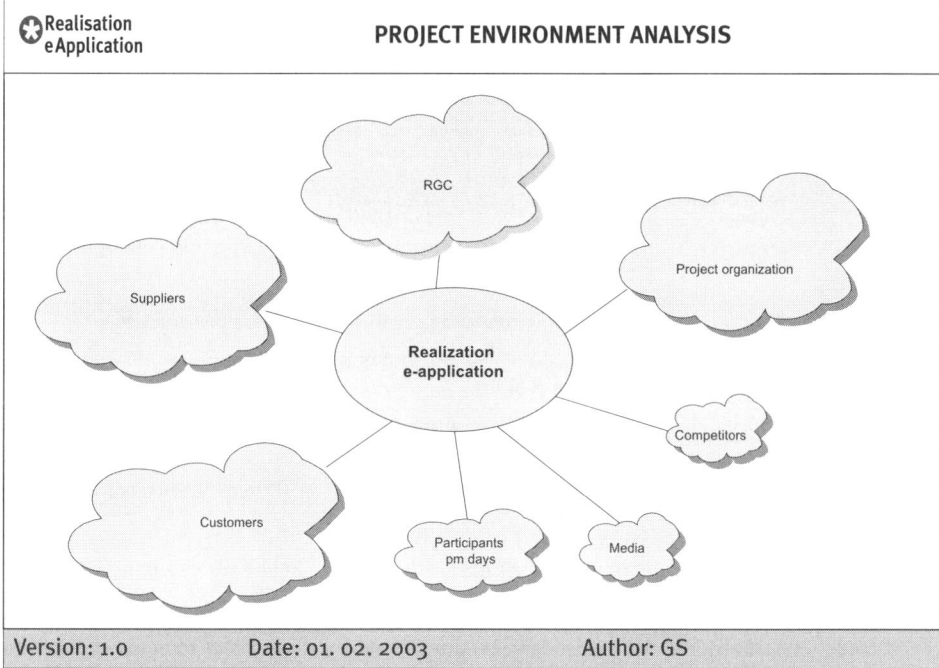

Fig. F1.39: Project environment analysis for the project "Realization e-application" (cloud graphic)

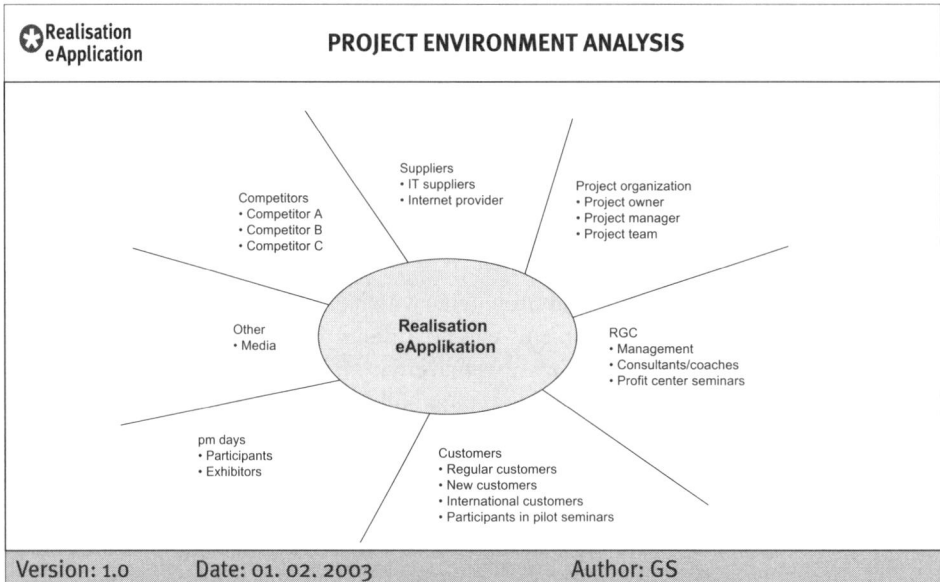

Fig. F1.40: Project environment analysis for the project "Realization e-application" (segment graphic)

Realisation e-Application PROJECT ENVIRONMENT RELATIONSHIPS		
relevant environment	**relationship**	**measures**
RGC	• vested interest in success	• make qualified resources available
project organization	• IT supplier with responsibility for whole project	• common project ownership by RGC and IT supplier
competitors	• pressure from competition	• ensure competitive edge
pm days participants	• interest in new e-application	• good presentation
participants pilot seminars	• additional services, not yet mature product	• preliminary info about pilot situation
IT supplier	• large interest in product (reference)	• ensure qualified resources
internet provider	• no interest in product	• info about project
Version: 1.0	Date: 01. 02. 2003	Author: GS

Fig. F1.41: Project environment relationships for the project "Realization e-application"

The project environment analysis should be developed by the project team as part of the project start workshop. The common construction of the relevant project environments is important. Selected representatives of relevant environments can be involved. The communication of the results to the outside, especially the communication of the evaluation of the relationships, is to be performed selectively.

Objectives of the project environment analysis

One objective of the project environment analysis is to ensure an outward orientation in the project. The project environment analysis is an instrument of project marketing.

Understanding the environment in which a project exists provides members of the project organization with an orientation for their behavior and actions.

The amount of documentation in the project environment analysis can be limited to the graphic representation of the relationships of the project to only a few environments or, if required, also contain detailed results of analysis (mutual expectations, potentials and conflicts, strategies and actions).

Developing the project environment analysis

1. Constructing the relevant project environments

A list of relevant project environments must be developed. The definition of the relevant project environments represents a social construct. The project team defines those environments as relevant for the project which it perceives as such and upon which it can agree.

A project environment is to be differentiated into several environments if different strategies and actions for the design of the relationships between the project and the individual environments are necessary (e.g. further differentiating the environment "Banks" into the environments "Commercial bank" and "Control bank".

2. Evaluating the project environment relationships

The significance of individual relevant environments for the project and the amount of interaction between the project and each environment can be analyzed and evaluated. The relative significance of individual environments and the intensity of their interactions can be analyzed in the form of a list.

environment	significance 1 ... very high 5 ... very low	interaction 1 ... very often 5 ... very seldom
project owner	1	3
project team	1	1
company management	3	5
customer	1	3
supplier	4	3
media	5	5

Fig. F1.42: Analysis of the significance and the interaction with relevant project environments

In a project environment graphic the significance and the intensity of the interaction be-tween the project and each environment can be expressed in symbols. The size with which a relevant environment is represented symbolizes the significance of the environ-ment. The significance of an environment for the project is dependent upon the possi-bility of it exerting influence on the project's success. The distance with which an envi-ronment of the project is represented should reflect the intensity of the interaction be-tween the project and the environment.

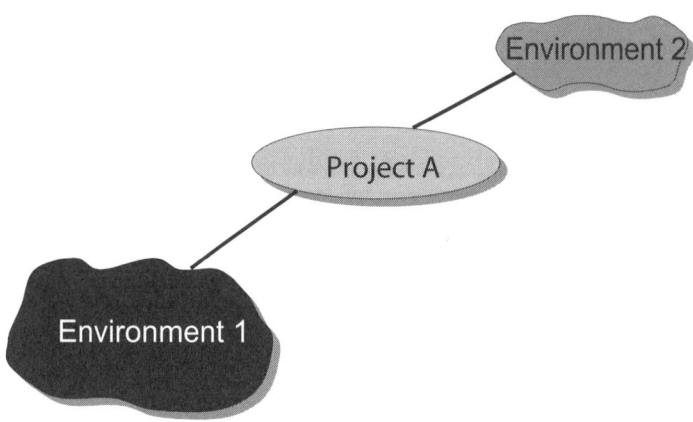

Fig. F1.43: Graphic representation of the project environment analysis

The evaluation of the significance of an environment and the intensity of the interaction relate to a particular planning or control day. The project environment relationships change in the course of the project and therefore require "social" controlling.

3. Analyzing individual project environment relationships

Important decisions in a project are guided by the expectations project environments have toward a project. The qualification of the relationship between a project and each environment can take place on the basis of the description of the mutual expectations. It is possible to differentiate between process- and results-related expectations and/or manifest and latent expectations. Fears are also expectations, namely negative expecta-tions.

When in a project the expectations of a relevant project environment toward the project are analyzed then expectations regarding expectations are formulated. Even when these expectations are not the real expectations of this environment, they still guide the be-havior in the project.

The advantage of the method of formulating the expectations is that project team mem-bers put themselves in the position of a social environment and thereby learn to under-stand their point of view. Potential conflicts between the project and an environment can be recognized as structural (and not necessarily as personnel-related).

Project-internal environments have strong process-related expectations and tend to have a common history and future. The relationships of a project to project-internal environments can, therefore, be influenced in many ways. Project-external environments, on the other hand, have strong result-related expectations regarding the project and have, above all, a common present. The relationships of a project to its external environments can, therefore, only be influenced in a limited way.

The project environment relationships can be globally evaluated with the help of graphic symbols (+ for positive, − for negative, +/- for ambivalent) or analyzed by describing conflicts and/or potentials..

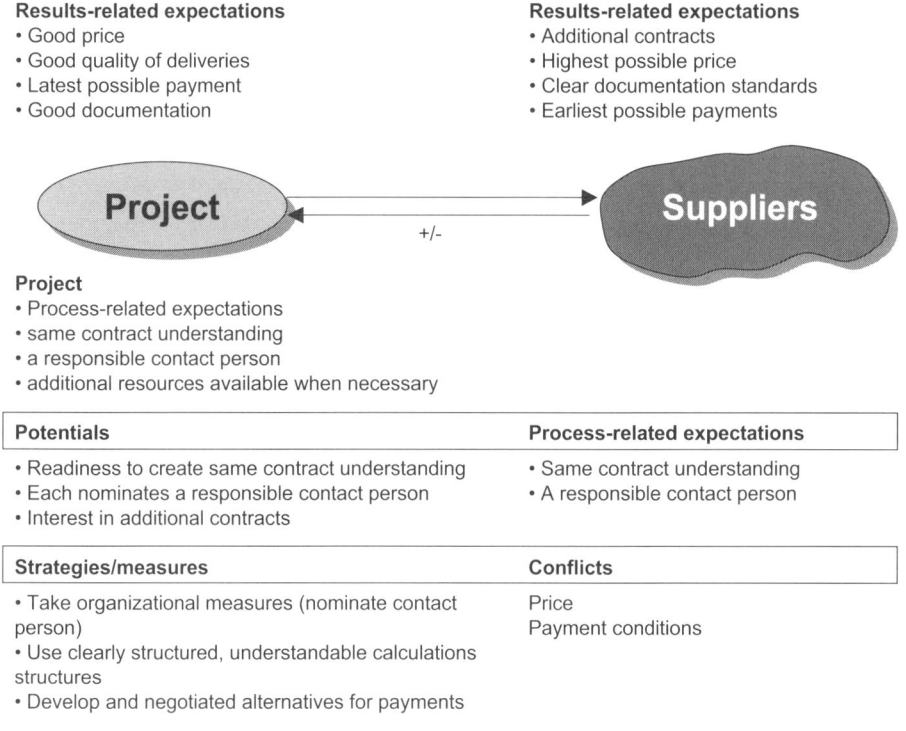

Results-related expectations
• Good price
• Good quality of deliveries
• Latest possible payment
• Good documentation

Results-related expectations
• Additional contracts
• Highest possible price
• Clear documentation standards
• Earliest possible payments

Project
• Process-related expectations
• same contract understanding
• a responsible contact person
• additional resources available when necessary

Potentials	Process-related expectations
• Readiness to create same contract understanding • Each nominates a responsible contact person • Interest in additional contracts	• Same contract understanding • A responsible contact person

Strategies/measures	Conflicts
• Take organizational measures (nominate contact person) • Use clearly structured, understandable calculations structures • Develop and negotiated alternatives for payments	Price Payment conditions

Fig. F1.44: Analysis of a project environment relationship

4. Analyzing the relationships between relevant project environments

Since the relationship of an environment to the project is also dependent on its relationship to other project environments, the relationships between project environments can also (selectively) be considered. However, this measure substantially increases the complexity of the analysis.

5. Planning strategies and measures for the design of the project environment relationships

On the basis of the analysis of potentials and conflicts concrete strategies and measures for the design of the project environment relationships can be planned and communicated.

F1.7.6 Adaptation of the business case analysis

In the project start process the assumptions made in the business case analysis are to be verified and, if necessary, adapted. Thereby the first business case controlling is completed.

During periodic project controlling further business case controlling is to be made. Thereby the project organization also assumes responsibility for optimizing the business case.

The method of the business case analysis is described in Chapter I2: Assigning a project or a programme, as it is to be developed during the project assignment process. It represents the basis for the investment decision.

F 1.8 Organizational design

Objectives of designing a project organization

The traditional forms of project organization, the new elements for the design of project organizations, the possibilities for the design of project organization charts and the different project roles were described in Chapter C. Methods for teamwork and for leading in projects were shown in Chapter D.

These are the concepts upon which the following methods for designing the project organization are based. The most important of these methods are:

- project assignment and sub-project assignment,
- project organization chart,
- project role definitions and project role descriptions,
- project responsibility matrix,
- project communication plan and
- project rules.

The process of designing a project organization

Designing a project organization is a creative process. By using an adequate organization for the performance of a project a project-oriented company creates a competitive advantage for itself. The adequate project organization is to be designed in the project start process. It may be possible to use project plans from the project assignment process for this (e.g. a draft of the project organization chart).

The project organization is also part of "social" project controlling. The project organization lives; possibly additional project team members will be required, the frequency of the project team meetings may be changed, the project rules added to, etc. Such adaptations in the design of the project organization take place in order to increase the efficiency of the project. They are the results of reflections in the project team.

A fundamental change in the project organization becomes necessary if a project discontinuity is to be resolved.

F1.8.1   Project assignment

Definitions and example

A project assignment is a summary of the objectives agreed between the project owner team and the project team. The project assignment should contain the following information:

- project start date, project end date,
- project objectives, project non-objectives,
- project phases,
- project costs, project income,
- project owner team, project manager, project team members,
- relationships to other projects, and
- relevant project environments.

Additional and more detailed methods for agreements between the project owner team and the project team are the project management documents of the project start process.

Objectives of the project assignment

The project assignment serves to document the objectives agreed on in the project start process between the project owner team and the project team. The documentation promotes commitment to the agreed objectives.

Developing the project assignment

The result of the project assignment process, i.e. setting the project boundaries and defining the project context as well as preparing the first project management documents (work breakdown structure, milestone plan, project organization chart, project objectives plan and project environment analysis), forms the basis for the project assignment document.

The detailing and completing of the project planning and the involvement of additional experts as project team members may lead to an adaptation of the project assignment in the project start process. The cyclical formulation of the project assignment promotes the quality of the agreed objectives.

The project assignment should be generated in written form. The project owner team should assign the project team, not only the project manager.

✳ Realisation e Application	**PROJECT ASSIGNMENT**	
project start date: 01. 12. 2003	project end date: 30. 08. 2003	

project objectives:	project non-objectives:
• Through pilot operation tested e-applicati-on introduced for 3 RGC seminars as per specification (use, administration and maintenance processes). • Adequate organization implemented, 3 e-application tutors and e-application admi-nistrators trained. • Marketing implemented (presentation at events, hints in brochures). • Financing for the application of the e-appli-cation secured. • Experiences gained for further e-projects. • Cooperative relationship with supplier further developed. • Therefore reduction in the durations of the seminars and increase in the teaching qua-lity (more information beforehand, tests). • Prerequisite for increase in customer loyal-ty developed.	• establishing new seminars • continuous operation and maintenance activities

project phases:	project costs: external EUR 5,800	project income:
• project management and project administ-ration • detailed planning • procurement and adaptation content • development • installation and test • implementation organization, personnel, marketing, financing • pilot operation		

project owner team: Huber, Fischer	project manager: Schubert

project team members: Schubert, Hofer, Meier

relationships to other projects: Re-launch homepage, pm days

relevant project environments: RGC, project organization, competitors, pm days participants, participants in pilot seminars, IT supplier, internet provider

——————————— project manager	——————————— project owner team

Version: 1.0	Date: 01. 02. 2003	Author: GS

Fig. F1.45: Project assignment form

F1.8.2   Assignment of sub-projects
Definitions and example
Several work packages can be assigned by a sub-project assignment to a member of the project organization. The assignment contains the objectives of the assignment, the work packages to be fulfilled, their costs and completion dates, as well as the project team members and project contributors responsible for the fulfillment of the work packages.

A sub-project assignment is a summary of the agreed objectives between the project manager and a project team member who is responsible for the group of work packages. Further project management documents provide an additional basis for the agreement of objectives between the project manager and the project team member.

Sub-project assignments are given to cooperative partners who are responsible for parts of the project. In the project "Realization e-application", for example, a sub-project assignment was given to the IT supplier for the development of the software solution.

Objectives of the sub-project assignment
The sub-project assignment serves to document agreements between the project manager and a project team member in regard to the fulfillment of a group of work packages. Documenting the agreement in the sub-project assignment ensures commitment and a clear regulation of responsibilities.

Developing a sub-project assignment
During the project start process those parts of the project are to be identified which should be assigned in the form of a sub-project assignment. For this, the objects of consideration plan can be used as a basis.

Those work packages which are the subject of a sub-project assignment are to be grouped, the costs and schedule objectives are to be documented and coordinated with the responsible project team member and/or sub-team.

A sub-project assignment should be made in writing. It can be assigned either to a project team member or to a sub-team. In either case the "contractor" is to be seen as a part of the project organization, since continuous coordination and controlling of the work packages in the sub-project assignment are necessary.

The use of this work package-related development of sub-project assignments ensures the compatibility of the project management structures used in the project and in the sub-projects for which cooperative partners are responsible.

F1.8.3   Project organization chart
Definition and example
A project organization chart is a graphic representation of the organizational structure of a project. In a project organization chart the roles of the project organization and their relationships are shown.

In the standard project organization chart of *ROLAND GAREIS Project and Programme Management*® the individual roles (project manager, project team member and project

✪ Realisation e Application	SUB-PROJECT ASSIGNMENT IT SUPPLIER	
sub-project contractor: Fischer	sub-project assignment number: XYZ	
start date: 11. 02. 2003	end date: 28. 08. 2003	
Sub-project assignment objectives: • ensure the operation of the e-applica-tion for RGC • provide the technological infrastruc-ture (server and network)	project costs: € xxx	

WBS code	work package
1.2.1	• detailed planning HW and SW
1.2.2	• detailed planning application functions
1.3.1	• procure HW and SW
1.3.2	• procure provider
1.4.1	• establish development environment
1.4.2	• develop database
1.4.3	• develop draft application functions
1.4.5	• revise application
1.4.7	• completion application
1.5.1	• install and adjust HW
1.5.2	• configure interfaces
1.5.3	• install database and application
1.5.5	• adapt after online test
1.5.7	• document source code
1.6.1	• develop manuals and maintenance concept
1.6.4	• perform personnel training
1.7.3	• adapt after pilot operation
employees: Hofer, Holzer, Sommer	
Version: 1.0 Date: 01. 12. 2003 Author: GS	

Fig. F1.46: A Sub-project assignment for the project "Realization e-application"

contributor) as well as team roles (project owner team, project team and sub-teams) are depicted. This depiction corresponds to an "empowered project organization".

The project organization chart for the project "Realization e-application" is shown in Figure F1.48.

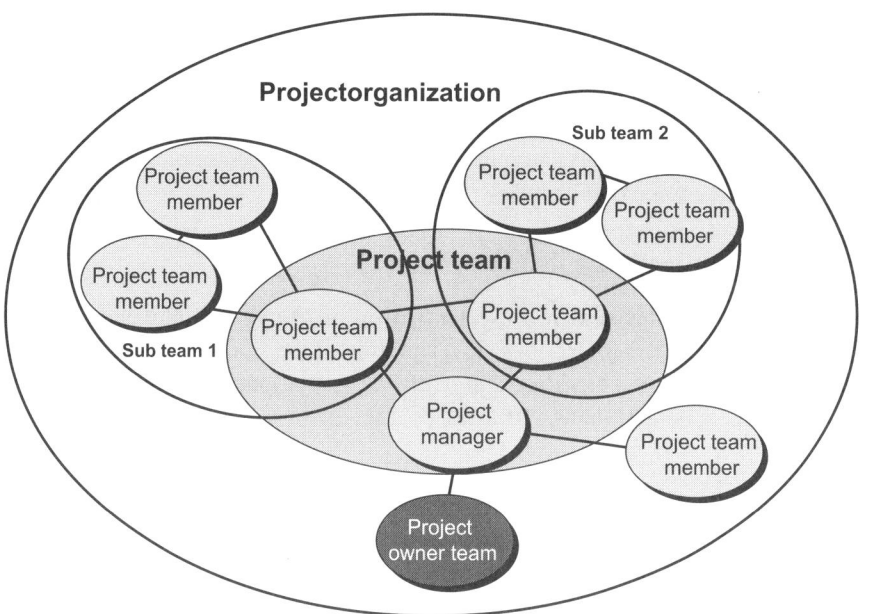

Fig. F1.47: Standard project organization chart

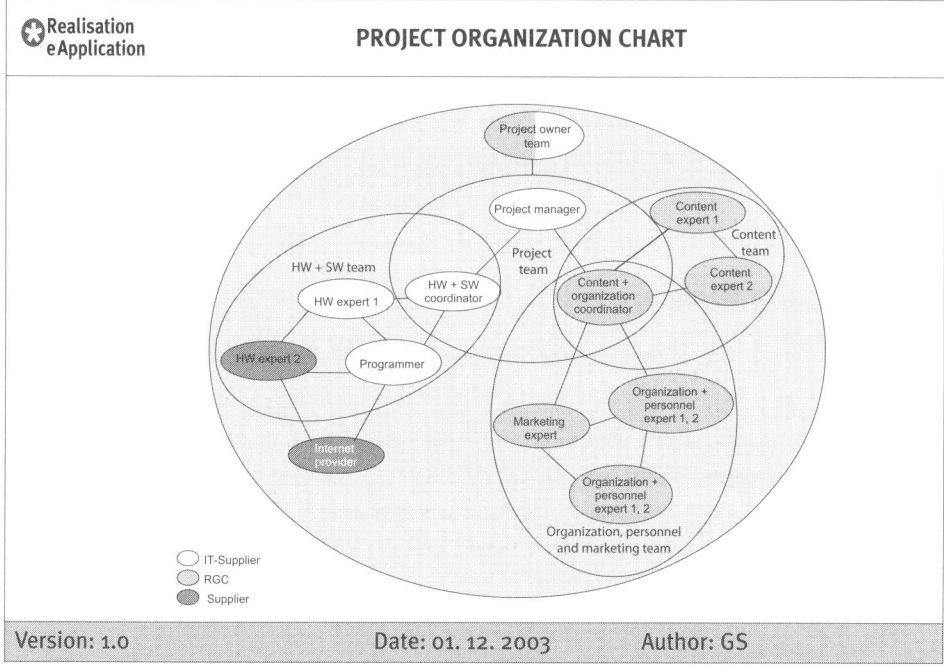

Fig. F1.48: Project organization chart for the project "Realization e-application"

Objectives of the project organization chart

The project organization chart should depict the organizational structure of a project. It should visualize the project roles and important relationships between the roles. All members of the project organization should gain orientation for the project work from it.

The project organization chart can be supplemented with further methods of organizational design, such as project role descriptions.

Developing a project organization chart

For the development of a specific project organization chart a standard chart (see Figure F1.47) can be used. The project-specific adaptation of the standard is made by defining the structure (number of individual and team roles, relationships between these roles) and by a project-specific labeling of roles.

The labeling of project roles is a creative act, since the job titles of the members of the project organization in the permanent organization should not be adopted, but new titles should be found to express their specific functions in the project. The role of project contributor Schmidt in the permanent organization of RGC is, for example, "Event manager". Her role in the project "Realization e-application" was "Content expert".

In the project organization chart relationships between project roles are only depicted as examples. The number of relationships and their intensities should be expressed. In order to document the quality and quantity of the different relationships the project organization chart is to be supplemented with the project communication plan. This plan visualizes, for example, who participates in which project meetings.

In addition to the project roles and their relationships to each other the names of the persons fulfilling the roles can also be visualized in the project organization chart.

F1.8.4   Project role definitions and project role descriptions

Definitions and example

A project role is an organizational element of a project. Expectations regarding a project role exist before someone takes over the role.

Project roles can be described through the representation of objectives, organizational position, the tasks to be fulfilled, decisional authority and the relationships of the role to relevant project environments. By describing project roles those who perform them attain clarity in regard to their tasks and their cooperation with other project role performers.

Persons have to be assigned to perform the different project roles. The defined roles of the project "Realization e-application" and the persons assigned to these roles are shown in Figure F1.49.

✪ Realisation ✪ e Application	**PROJECT ROLES**
project role	person performing the role
project owner team	Huber, Fischer
project manager	Schubert
HW + SW coordinator	Hofer
content and organization coordinator	Meier
content experts	Schmidt, Bauer
organization and personnel experts	Müller, Berger
marketing expert	Gruber
HW experts	Holzer, Winter
programmer	Sommer
internet provider	N. N.
sub-teams	
content team	Meier, Schmidt, Bauer
organization, personnel & marketing team	Müller, Berger, Gruber
HW + SW team	Hofer, Holzer, Winter, Sommer
Version: 1.0 Date: 01. 12. 2003	Author: GS

Fig. F1.49: Project roles and persons performing the roles of the project "Realization e-application"

Objectives of the project role definition and the project role description

An objective of the project role definition is the identification of all project roles necessary for the successful performance of a project. Care should be taken that the definition of the project roles is complete. All work packages of a project are to be fulfilled by the defined project roles.

The defined project roles are to be described so that the person performing the role is clear as to his/her position in the organization, tasks and formal authority. In this way all members of the project organization are given orientation for their project work.

The definition of the project roles is the basis for the development of a project organization chart and for the personnel allocation within the project.

Defining project roles and project role descriptions

The definition of project roles can be differentiated by project team members, project contributors and sub-teams. An example for a project team member of the project "Realization e-application" is the "SW + HW coordinator", an example for a project contributor is the "HW expert" and an example for a sub-team is the "SW + HW team".

Descriptions of project roles can be standardized. In Chapter C3 standard role descriptions for all project roles were given. Although there are basically unchanging expectations toward project roles the standard role descriptions are to be adapted for each specific project.

The project role descriptions should be adapted by the project team. The adaptations should consider jointly achieved results as well as the communication process in clarifying the expectations toward each role.

F1.8.5   Project responsibility matrix

Definitions and example

The work packages to be fulfilled by project roles and representatives of project environments can be distributed with the help of a project responsibility matrix. A responsibility matrix is a method for planning and documenting the tasks, the project roles and project environments involved in the fulfillment of individual work packages.

Responsibility matrices are matrix representations. In the rows of the matrix the work packages are listed and the columns show the project roles and the project environments. In the fields of the matrix the functions to be performed are depicted..

Objectives of the project responsibility matrix

With the help of the project responsibility matrix the distribution of the work packages to project roles and project environments can be planned and agreed on. In the responsibility matrix the form of the collaboration of the members of the project organization in fulfilling individual work packages is planned.

The responsibility matrix is also an instrument for conflict management. Possible conflicts regarding the fulfillment of individual work packages can be identified early and dealt with before a conflict occurs.

The responsibility matrix is an integration instrument. The results of the work breakdown structure, the role definitions and the project environment analysis are brought together in the planning of the work package distribution. The development of the responsibility matrix serves to verify the completeness and the level of detail of the work breakdown structure, the role definitions and the project environment analysis.

Developing a project responsibility matrix

The work breakdown structure, the defined project roles and the relevant project environments form the bases for the development of the responsibility matrix.

The work packages to be considered in the development of the responsibility matrix are to be selected. In order to keep the planning complexity at a minimum, work packages for which the organizational responsibilities are basically clear, and for which no cooperation between the roles is necessary, should not be considered. It is recommended to concentrate on important work packages and those where the organizational responsibility is unclear.

The assignment of functions in regard to the fulfillment of work packages to project roles and project environments is made by

- listing the selected work packages as rows,

| ⊛ Realisation / ⊛ eApplication | PROJECT RESPONSIBILITY MATRIX | | | | | | | | | | | | | |

Roles and environments

WBS code	WP label	project owner team	project manager	HW + SW coordinator	content + organization coordinator	content expert	organization/personnel expert	marketing expert	HW experts	programmer	internet provider	sub-team content team	sub-team organization, personnel, marketing	sub-team SW and HW team
1.2.1	detailed planning HW + SW		C							I	I			P
1.2.2	detailed planning application functions	D	C										C1	P
1.2.3	detailed planning organization and personnel	I	C									C1	P	
1.2.4	detailed planning financing, marketing	I	C									C1	P	
1.3.1	procure HW + SW													

Version: 1.0   Date: 01. 02. 2003   Author: GS

Legend:
C ...   coordination, initialization
P ...   performance
C1 ... contribution, consulting
D ...   Decision
I ...   Information

Fig. F1.50: Extract of the responsibility matrix of the project "Realization e-application"

- listing the project roles and project environments as columns and
- entering the functions in the intersection points of the matrix by means of codes.

To identify the functions different codes can be used. It is recommended that meaningful letters rather than number codes be used. In order to make the responsibility matrix easy to understand, no more than six or seven codes should be used.

It is apparent from the analysis of the rows of the responsibility matrix which cooperations are necessary for the fulfillment of individual work packages. On the basis of the analysis of the columns the project-related work load on the individual project roles can be determined.

The communication in the project team during the development of the responsibility matrix about the boundaries of the individual work packages and about the possible organizational responsibilities is an important objective of the development process. Potential conflicts that become visible during the process can be dealt with in the project team. Persons performing important roles must, therefore, be involved in the development process.

For repetitive projects standard responsibility matrices can be developed and used. These standards are to be adapted to the specifics of each project.

F1.8.6   Project communication plan

Definitions and example

Objectives of verbal project communication are to inform the members of the project organization, to support decision-making and to solve conflicts in the project. The forms of verbal project communication are

- meetings between the project manager and project team members and/or between project team members,
- project meetings (e.g. of the project owner team, of the project team, of sub-teams),
- project workshops (e.g. the project start and the project close-down workshop) and
- project presentations.

These different forms of project communication were already described in Chapter E. Project meetings and project workshops are central leadership instruments which have an important integrative function in projects.

In a project communication plan meetings and workshops used in a project are planned with regard to their type, objectives, participants and frequency. In Figure F1.51 the communication plan for the project "Realization e-application" is depicted.

Objectives of the project communication plan

With the help of the project communication plan the objectives, participants and frequencies of project meetings and project workshops are planned and agreed. Through the use of the different forms of meetings and workshops the energy in the project can be directed.

Developing the project communication plan

To meet different objectives different project meetings and project workshops with different participants and different frequency are required.

The different communication forms are to be labeled specifically (e.g. project owner team meeting, project team meeting, etc.). Participants in the project communication can be members of the project organization or representatives of relevant project environments.

To give orientation to all participants the dates of the project meetings and workshops should be agreed medium-term. Additional meetings can be called if required.

Realisation e Application	PROJECT COMMUNICATION STRUCTURES		
meeting	content	participants	frequency
project start workshop	information on the project objectives, the project structures, strategies, organization, developing the "big project picture"	project team	1x in the project start process
project owner team meeting	project status and project preview, context information, current problems, strategic decisions	project owner, project manager, selected project team members	every 2-3 weeks
project team meeting	project status and project preview, review, definition of project strategies, operational decisions	all project team members, possibly selected representatives of environments	every 2-3 weeks
sub-team meeting	problem-solving	sub-team members, possibly project manager and selected representatives of environments	1x per week
project close-down workshop	reflection, feedback, dissolving the social system, clarifying remaining tasks	project team and project owner	1x in the project close-down process
Version: 1.0	Date: 01. 02. 2003		Author: GS

Fig. F1.51: Communication plan for the project "Realization e-application"

F1.8.7 Project rules

Definitions and example

In addition to the general rules of the project-oriented company, project-specific organizational rules are to be developed. These project rules can apply to

- the expected behavior in the project,
- use of ICT,
- project documentation and
- project marketing.

✪ Realisation e Application	**PROJECT RULES**

expected behavior in the project

- punctuality at meetings
- no mobile phones, no phone calls during project meetings
- no sending of deputies to meetings
- joint responsibility for results
- those present have authority to make decisions
- etc.

use of ICT

- The following version of MS Office is to be used in the project: 2000.
- The following version of MS Project is to be used in the project: 2000.

project documentation

- The central file for all project documents is on the server: Data
- All project-related correspondence is to be copied to the project manager
- The project-related correspondence (letters, fax, e-mail) will be given a consecutive number
- The project letterhead is to be used without exception
- etc.

project marketing

- The project name "Realization e-application" is to be used on every project document
- Font and color of the project name: Tahoma, 12 point, italic, dark red
- etc.

Version: 1.0	Date: 01. 02. 2003	Author: GS

Fig. F1.52: Extract of the project rules of the project "Realization e-application"

Objectives of the project rules

Project rules serve to give the members of the project organization orientation for their project work. The project rules enable their work in the project to be efficient and co-operative.

Developing project rules

A draft of the project rules is to be made by the project team during the project start process and agreed with the project owner team. On the basis of regular reflection of the project team about the process of working together in a "social" controlling the project rules are to be adapted as required.

The project rules are to be documented and included in the project handbook.

F 1.9 Project culture development

Objectives of the development of a project culture

The fundamental objectives of the deliberate development of a project culture were already described in Chapter D3: Project culture. Primarily, the development of a project-specific identity should serve to secure competitive advantages for the performance of the project. Competitive advantages are created by defining clear project boundaries, by securing the recognition of the project and by promoting the identification of the members of the project organization with the project. The methods of project culture development also form the basis for professional project marketing.

Important methods for realizing these objectives, i.e.

- project name and project logo,
- project values and project mission statement,
- project slogans and project-related anecdotes,
- project-related artifacts, project language and a project room, and
- project-related events

are described below.

The process of developing a project culture

The development of a project culture is a process that requires time and energy. Since projects are usually defined for dealing with short- to medium-term business processes, the deficit in time must be compensated by an increased use of resources and by appropriate communication forms. This is possible, for example, with the performance of the project start workshops. It can be developed by using the results of the project assignment process, such as project name and drafts of project management documents.

Controlling of the project culture is performed during project controlling. Elements which give identity to the project, such as the project name, the project values, etc., should not be changed during controlling. Potential adaptations to the project culture are the result of reflection processes in the project.

A fundamental change in the project culture is required for resolving a project discontinuity.

F1.9.1   Project name, project logo and project color

A project name serves to make a project and all project-related information recognizable and assignable.

The project name should enable the identification of the project type and promote associations with the project objectives. For example, conception projects are to be differentiated from realization projects. The project names "Conception e-application" and "Realization e-application" identify a chain of product development projects.

Project names should be short, but no abbreviations should be used which outsiders would not understand. A contract number "A 2003-12", for example, is not a good name for a contracting project.

The project name "Route 66" for a vacation trip by airplane from Vienna to New York, continuing by hire car to Miami, is a bad choice. For the association with Route 66 is that of an east-west trip, not a north-south trip through the USA.

The use of a project logo seems to presuppose time-consuming design work. It is, however, rarely necessary to create a picture as a project logo. Often it will suffice to use the project name as word picture, e.g. written in italics and in color.

The choice of a project color supports the recognizability of a project. In the chain of projects "Conception ABS", "ABS" and "ABS optimization", for example, the continuity of the chain of projects is also ensured through the retention of ABS (antibiotic strategies) as an element of each project name and through an uniform use of blue and turquoise as project colors. But in order to communicate the independence of the projects the two colors were used in inverse in the project "ABS optimization". Examples of different colored documentation from both projects are shown in Figure F1.53.

 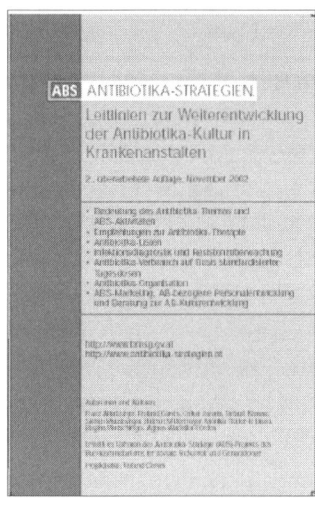

Fig. F1.53: Cover of the book "Guidelines for further development of the antibiotic culture in Austrian hospitals" (1st issue as a result of the project "ABS", 2nd issue as a result of the project "ABS optimization")

F1.9.2   Project values and the project mission statement

Project values provide benchmarks for what is considered good, valuable and desirable in a project. They determine the behavior of the members of the project organization consciously and unconsciously and provide orientation for their actions. Project values are therefore an important leadership instrument in projects.

In defining project values the following questions can be used as an aid:

- What is specific about the project results?
- What is important and what is not important to the project team?
- What differentiates the project from other projects?

In the definition of the project values results-related and process-related statements can be differentiated. These central values of the project "Realization e-application" are depicted in Figure F1.54 as an example.

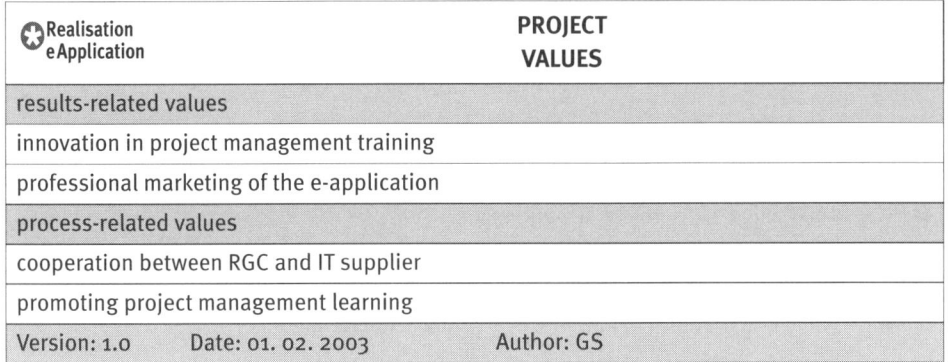

Realisation e Application	PROJECT VALUES
results-related values	
innovation in project management training	
professional marketing of the e-application	
process-related values	
cooperation between RGC and IT supplier	
promoting project management learning	
Version: 1.0 Date: 01. 02. 2003	Author: GS

Fig. F1.54: Values of the project "Realization e-application"

Central values can be communicated by documenting them in a "Project mission statement". As an example the mission statement of the project "IPMA '90" for the organization of the world congress of the IPMA – International Project Management Association is depicted in Figure F1.55. In this project mission statement the project values were not formulated into full sentences, but were defined as pairs of terms. The values were interpreted in meetings and in presentations, their implementation took place in the daily project work. The value pair "mind and body", for example, was implemented by a challenging congress programme on the one hand, and by light lunches and relaxing Tai Chi exercises, which attended to the physical well-being of the congress participants on the other hand.

mission statement of the project "IPMA '90"
significant in contents and socially stimulating
tradition-minded and future-oriented
mind and body
science and practice
international and innovative
(learning) experience and chance

Fig. F1.55: Mission statement of the project "IPMA '90"

F1.9.3 Project slogans and project-related anecdotes

Project slogans should communicate what is important in the project and/or what is especially important in a project phase. Project slogans can be formulated for an entire project or for individual project phases. The slogans which were used in the individual phases of the project "IPMA '90" are depicted in Figure F1.56. These slogans were used by the project manager to give orientation to the project teams. For marketing the world congress additional product-related slogans were formulated.

date	phase	slogans for the project "IPMA '90"
12/88	marketing	Total marketing!
10/89	design	No more creativity! Love for the details and for high production quality!
05/90	final preperation	Have fun, everything is decided!
06/90	performance	IPMA´90 – A High Touch Project!

Fig. F1.56: Slogans for the project "IPMA '90"

The image of a project can also be developed through project-related anecdotes. An anecdote about a cooperative customer meeting can support the creation of a project-related customer relationship.

F1.9.4 Project-related artefacts, project language and project rooms

Project management documents, such as a project brochure, the project organization chart and the project plans, can be considered as project-related artefacts.

The professional content as well as the graphic arrangement of these documents communicates the project culture. Great importance is attached to "symbolic project management" (see also Chapter D3). The project language, i.e. project-specific terms and labels, is also communicated by the project management documents.

A project room represents an "organizational home" for a project. The project room does not have to be available to the project all the time, it is sufficient for providing an identity when the same room is always available for project meetings. The decoration of the project room with selected artefacts of the project, such as print-outs of the work breakdown structure and the project organization chart, supports the identification with the project.

F1.9.5 Project-related events

Project-related events, such as outdoor weekends or project vernissages, are further elements for the development of the project culture. The use of such events is dependent on the size and strategic importance of the project, since they incur additional costs. But even in small projects it is recommended for the project team to mark the end of the project start workshop with a social event, for example.

F 1.10 Risk management

F1.10.1 Definitions and example

Definition: Project risk

A project risk can be defined as the possibility of a negative or positive deviation from a project objective.[2] Deviations regarding the project scope, the project schedule, the project costs and the project income are considered.

The use of project management methods, such as the work breakdown structure, project cost plans and the project environment analysis in the project start process and in the project controlling process, (implicitly) contribute to avoiding negative deviations from the objectives and to promoting positive deviations. Because of their relative uniqueness, complexity and dynamics projects have many risks. It is recommended in addition to this implicit risk management also to perform an explicit project risk management.

Occasions for project risks

Projects are dynamic. They develop over time. As shown in Chapter B1.6, developments in projects can be classified as either continual or discontinual. Occasions for developments in projects are self-organizing processes and interventions of relevant project environments. These can lead to changes in the project structures and thereby pose project risks.

Project risk management takes into account continual project developments, while the management of discontinuities in projects accounts for discontinual developments.

Types of project risks

The types of project risks can be differentiated as follows (see Figure F1.57).

Time-wise, risks can be differentiated into risks of the project and risks of the post-project phase. In project risk management only project-related risks are to be considered. The risks of the post-project phase are objects of consideration of the business case analysis.

Definition: Project risk management

Project risk management is a project management task. It includes the identification and evaluation of risks, the planning and performing of risk management measures (avoidance and/or promotion of risks and provision for risks) and risk controlling (see Figure F1.58). Project risk management considers deviations from the planned

---

2) In the practice of project management it can be observed that positive deviations from objectives are not understood as risks and are therefore not considered in risk analysis. Thereby important potentials in projects are lost

criterion	type of risk	criterion	type of risk
project level	• project risk • project phase risk • work package risk	cause	• risk from the market • risk from organization • etc.
object level	• object risk • component 1 risk • etc.	reach	• isolated risk • coupled risks
function	• engineering risk • procurement risk • etc.	measurability	• measurable risk • non-measurable risk
area	• technical risk • commercial risk • legal risk	insurability	• insurable risk • non-insurable risk

Fig. F1.57: Differentiating risks in engineering projects

project scope, project schedule, project costs and project income at the level of the work packages, the project phases and the overall project.

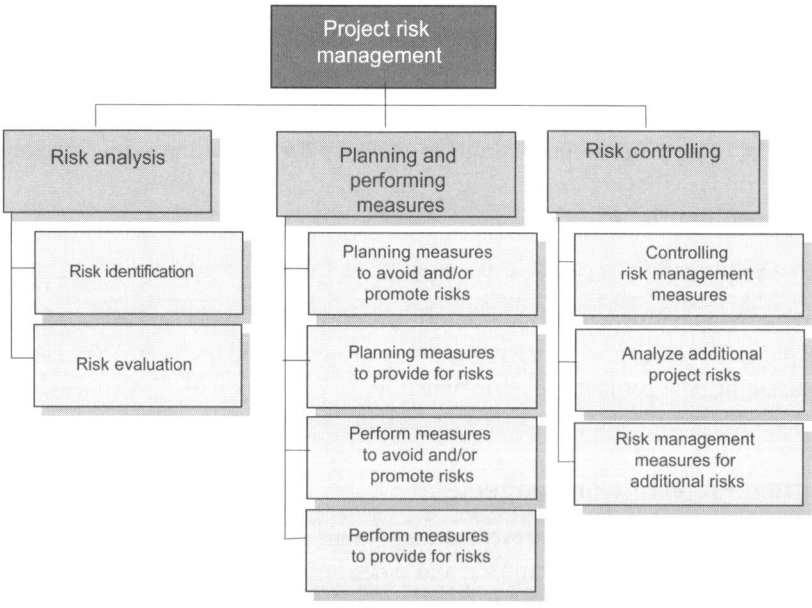

Fig. F1.58: The tasks of project risk management

To differentiate between measures for risk avoidance and provision for risk, the following serves as an example. Possible measures to avoid the risk of a car accident are not to drive a car, to select a known route and to drive carefully. Possible measures for providing for the situation that the car accident occurs are to get accident insurance, to fasten the seat belts and to drive a car with airbags.

The risk analysis and the planning of risk management measures performed for the project "Realization e-application" are depicted in Figure F1.59.

✪ Realisation e Application		**PROJECT** **RISK ANALYSIS**		
WBS code	WP/phase	risk	characteristic/ description	measure
1.1	project management	quality of the project organization	unclear project roles; no common understanding of the form of cooperation	discussing the role descriptions, the cooperation
1.2	detailed planning	quality of specifications	not comprehensive enough	check the specification at project start with project owner team, adapt if required
1.3	procurement and adaptation content	content quality	improved seminar quality	ensure that potential for improvement is clear
1.4	development	deployed tool	new tool, not known to the developer	comprehensive training
1.5	installation and test	installation, test	interfaces not completely defined	clarify all interfaces
1.6.	implementation	administrator	too little attention in comparison to the users	definition of the administrator functions
1.7	pilot operation	administration	too much administrative effort	short-term adaptation, estimate effort
		acceptance	participants do not accept e-application solution	short-term adaptation, marketing measures

Interpretation of the risk management in the project "Realization e-application":

For risk management in the project "Realization e-application" there is an identification of important risks at the project phase level. No risk checklist is available. The project team identified possible risks in a brainstorming session on the basis of the work breakdown structure. The identified risks were described. No monetary evaluations and no estimations of occurrence probabilities were made. To manage the identified risks risk management measures were planned and performed.

Version: 1.0	Date: 01. 02. 2003	Author: GS

Fig. F1.59: Risk analysis and plan of risk management measures for the project "Realization e-application"

F1.10.2   Objectives of project risk management

The earliest, most complete identification of project risks, the minimizing of possible negative deviations and the optimization of possible positive deviations are objectives of project risk management.

The explicit application of project risk management is dependent on the project situation. Above all the scope, the complexity of the project, the relationships to the project environments and the importance of the project for the project-oriented company determine the requirement for risk management. It is also dependent on the possible extent of the costs, income and schedule deviations and their probability. Only in cases of large deviations and high probabilities an analytical project risk management is recommended (see Figure F1.60). For other situations a quick-and-dirty project risk management, as was used in the project "Realization e-application", is sufficient.

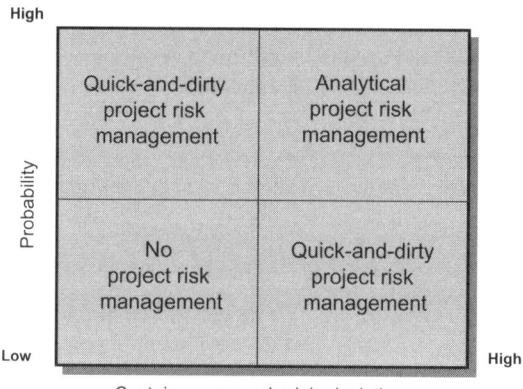

Fig. F1.6o: Forms of project risk management

In practice, risk management is used most often for repetitive contracting projects. This is where the most experience in risk analysis exists and contingencies are calculated as provision.

F1.10.3   Identification of project risks

Project risks can either be identified by work packages, objects of consideration or project environments. Since the planning and the controlling of schedules, costs and income are not made for objects of consideration or for project environments but for work packages, risk management is to be made on the basis of work packages. The objects of consideration plan and the project environment analysis can be used as additional project-specific checklists.

Risk checklists often display structures related to work packages, objects of consideration and project environments (see e. g. Figure F1.61). Such checklists are not compatible with other project management documents. The use of such checklists is therefore limited. In contrast to the use of such general checklists, it is recommended (for unique projects) that project plans be recognized and used as project-specific checklists.

Checklist for risk analysis
Project and general company policy
• Is it necessary to be especially considerate of the interests of consortium partners, customers, administrative bodies?
• Does the project have a considerable publicity effect?
• Is the contract value equal to the magnitude and interests of the company?
• Do any business relationships to the country or to the potential client already exist?
Project and sales policy
• Are the determined sales objectives, methods and route in agreement?
• Is the point of delivery located in a politically tense area?
• Are storage, transport and loading requirements cleared (also at destination)?
• Are entry and employment of assembly and commissioning personnel at the place of assembly assured?
• Are descriptions, instructions, operating manuals and maintenance manuals available in local language?
Project location
• Stable political circumstances?
• Favorable outlook for economic development?
• Good market potential?
Desired foreign participation
• Does the foreign partner wish to participate in the management?
• Does the foreign partner wish for another type of cooperation, such as financial contribution or cooperation of a company from the customer's country?
Competitive situation
• Are references and experiences sufficient?
• Are competitive companies to be expected (if yes, which ones?)
• Is a price war to be expected as a result of the competitive situation?
• Is the company technically and economically competitive?

Fig. F1.61: Extract of a risk checklist of an engineering company

CHAPTER

F

For a risk analysis only those phases and work packages are to be selected where risks are expected. It is not necessary to consider all phases and work packages.

When risks are dependent on each other there is a compound risk situation. In the case of a compound risk the risks must be considered together in order to avoid multiple evaluation (see Figure F1.62).

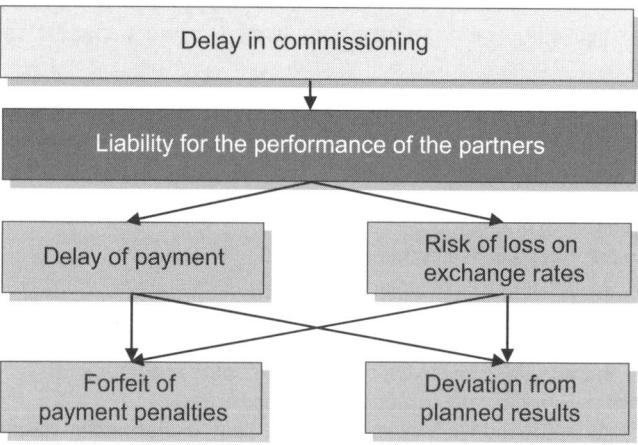

Fig. F1.62: Example of a compound risk situation

Evaluation of project risks

In risk evaluation the possible deviations from the objectives are evaluated qualitatively and possibly also quantitatively. Often a qualitative description of the risks is sufficient. A quantification of the project risks in the form of cost, income and schedule deviations is not always necessary.

The quantitative evaluations can be monetary for costs and income, and in time units for schedules. Evaluations can also be performed on other scales, such as (very) low, medium and (very) high (see Figure F1.63).

For the quantitative evaluation of risks occurrence probability estimations are necessary. An example of this type of risk evaluation is shown in Figure F1.64 "Engineering: Risk analysis and risk management measures". In this example the expected value of the risk is a product of the possible financial damage ("costs") and the probability that the risk will occur. The expected value is not the costs that would be incurred but a purely statistical value.

WBS code	WP label	Risk	Description positive deviation	Description negative deviation	Deviation	P
1.1	Project management	Relationship to RGC	A great deal of trust in services		-2	h
1.2	Prepare guidelines for project and programme management	Development process		Review processes; no common views	2	m
		Compatibility with concern		Compatibility problems with guidelines of the concern	1	n
1.3.4	Data collection of the projects	Data quality		Quality insufficient	2	m
		Collection process	Much information available, collection especially fast		-2	m
1.3.6	Workshop: Review with decision makers	Decision-making		Slow decision-making	1	m
1.3.8	Development of a concept for IT implementation	Compatibility of infrastructure		Standard tool not compatible with infrastructure	0	m
1.4.1	Establishment of roles, career paths, certifications	Acceptance		Resistance from the union representative, HR	1	m
1.4.3	Performance of assessment of PM personnel	Acceptance		Resistance from the union representative, HR	1	l
1.5	Development of concept for PM Office, PP Group	Development process		Differing interests of representatives of R&D, organization, engineering	1	h
1.6	Decision-making	Decision-making		Little interest in making decisions	1	2
...	etc.					

Fig. F1.63: Extract of a risk evaluation from an organizational development project (quantitative risk evaluation using a scale from -2 to +2; evaluating the probability using a scale with low, medium, high)

work package						
No.	label	risk	characteristic	costs in € 1,000	probabi-lity	Expected value in € 1,000
1120	design front axle	design effort	additional design effort for front axle because of large spring de-flection and small headroom cannot be done at the same time	90.–	80%	72.–
1160	design thread	design effort	additional design effort for thread, because required tractive force not possible	50.–	50%	60.–

2110	component delivery	delivery time	delay in delivery of components because no suitable supplier available	200.–	30%	60.–
3250	production rotary act	completion date	delay completion date of rotary act because too many rejects, tolerances cannot be kept	30.–	70%	21.–
	total expected value					213.–

Fig. F1.64: Engineering: Risk analysis and risk management measures

Calculating the project risk

The project risk can be calculated summing the expected values of the individual work package risks (see Figure F1.64). This results in the expected value for the project risk.

A differentiated calculation of the project risk can be made with the help of the probability theory, which enables depicting the project costs (or the project duration) as probability distributions.

For this, the work package costs and income and/or work package durations are defined as stochastic data. The costs (income) and/or durations are viewed as a random variable, which is described by means of probability distributions. These probability distributions are, as a rule, continuous distributions, because work package costs or durations can take on any value between certain limits.

Since statistical data for project planning is usually available only in a limited amount, probability data are, as a rule, identified through subjective estimates. These can be represented by means of standard probability distributions (such as normal distribution, distribution or uniform distribution). The advantage of the use of standard distributions is that only a few estimates are required in order to represent the whole distribution.

The standard distribution is easily determined on the basis of three estimated values: An optimistic, a most frequent and a pessimistic. For this, the probabilities that the optimistic value does not exceed the maximum and the pessimistic value does not fall short of the minimum have to be defined. These data suffice to calculate the expected value and the standard deviation of the random variable. The formulas are depicted below:

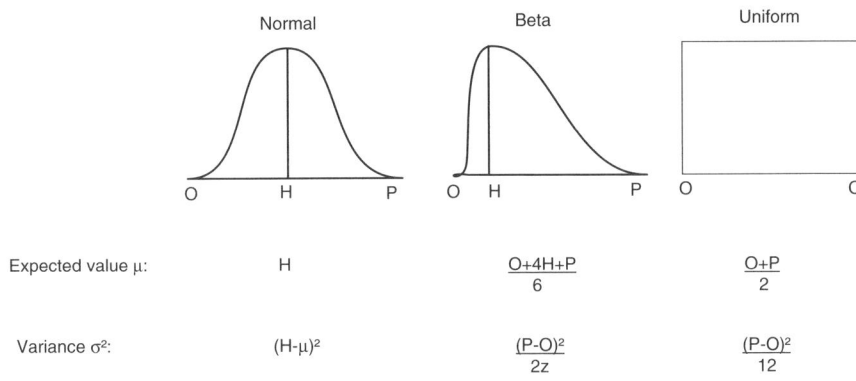

	Normal	Beta	Uniform
Expected value μ:	H	$\dfrac{O+4H+P}{6}$	$\dfrac{O+P}{2}$
Variance σ²:	$(H-\mu)^2$	$\dfrac{(P-O)^2}{2z}$	$\dfrac{(P-O)^2}{12}$

Legend:
O = Optimistic estimate
H = Most frequent estimate
P = Pessimistic estimate
z = Factor for converting the variance to values of the standard normal distribution

Fig. F1.65: The expected value and standard deviation of standard distributions

The expected value and variances of the project costs and/or project income and the project duration can be determined by simulation (Monte Carlo method). Both the variance and the standard deviation represent risk ratios.

The distribution of project costs, project income and/or project duration is approximately normally distributed.

From standard normal distributions the probability that a certain value of the project costs and/or the project income and/or the project duration will be exceeded or fall short can be assessed.

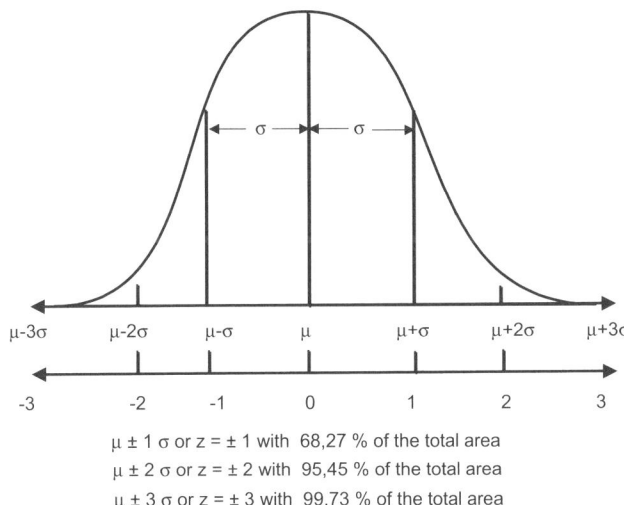

μ ± 1 σ or z = ± 1 with 68,27 % of the total area
μ ± 2 σ or z = ± 2 with 95,45 % of the total area
μ ± 3 σ or z = ± 3 with 99,73 % of the total area

Fig. F1.66: Standard normal distribution

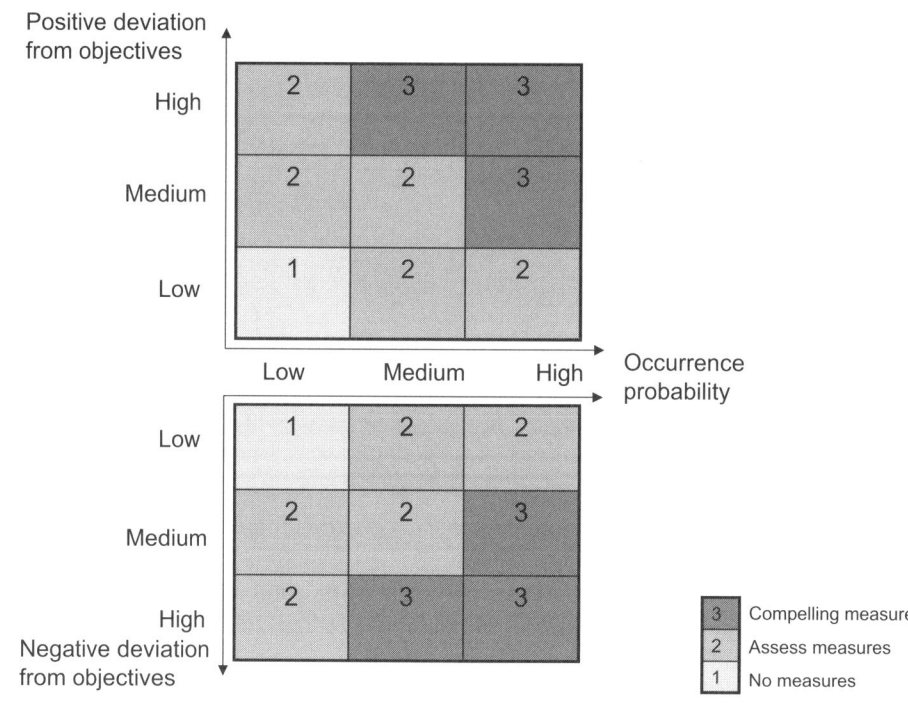

Fig. F1.67: Project risk matrix with standard strategies

F1.10.4   Risk management measures in projects

Risk management measures in projects should, on the one hand, avoid the occurrence of negative deviations from the objectives and/or their consequences through precautionary measures. On the other hand, the occurrence of positive deviations from the objectives and/or their consequences should be promoted. Risk management measures are to be planned and performed.

A prerequisite for planning risk management measures is the selection of the risks for which measures should be set and/or the definition of priorities between the risks under consideration. The selection of relevant risks can be made with the help of a project risk matrix. A project risk matrix with standard strategies is shown in Figure F1.67. The standard strategies relate to the different combinations of deviations from objectives and their probabilities.

Risk management measures can be differentiated into preventive (avoiding or promoting) and corrective (provision-making) measures. Preventive measures should avoid the occurrence of negative deviations. Corrective measures should diminish negative consequences in case a risk actually occurs (e. g. see Figure F1.68).

preventive measures	corrective measures
• market research and credit rating checks for supplier selection • use of experienced project personnel • use of approved methods and products • involvement of those affected by the project results • promotion of creativity • empowerment and coaching	• contracting measures (transfer of risk to customer, consortium or supplier) • risk contingency in the calculation • making reserves • insurance policies • create redundancy in the project structures (potential substitute personnel) • define an escalation model for problem situations

Fig. F1.68: Examples of preventive and corrective risk management measures

Avoiding and/or promoting and provision-making measures incur costs. Before the measures are realized, therefore, the costs and benefits are to be estimated. The benefits can come either from the elimination of a project risk or from a risk reduction.

Once decided upon, risk management measures are to be taken up in all the project plans, the additional work packages in the work breakdown structure, the additional costs in the project cost plan, etc.

⋯⋗ CASE STUDY F1: Project risk management in an engineering project

Project description:
- contract of an Austrian engineering company to construct a production hall with production equipment for a customer
- performance of the contract as general contractor
- objects of consideration: Construction, technical building equipment, production equipment
- duration: 2.5 years
- budget: Approx. € 100 Mio.
- The scope of work is visualized in the following work breakdown structure:

Project risk management in the project start process
- Based on a draft of the project risk management from the offer project, project risks were identified and evaluated in the project start process. The probability, the possible cost and scheduling effects were estimated and the expected value was calculated. Furthermore, preventive measures were planned and the remaining risks were estimated (see Figure F1.70).

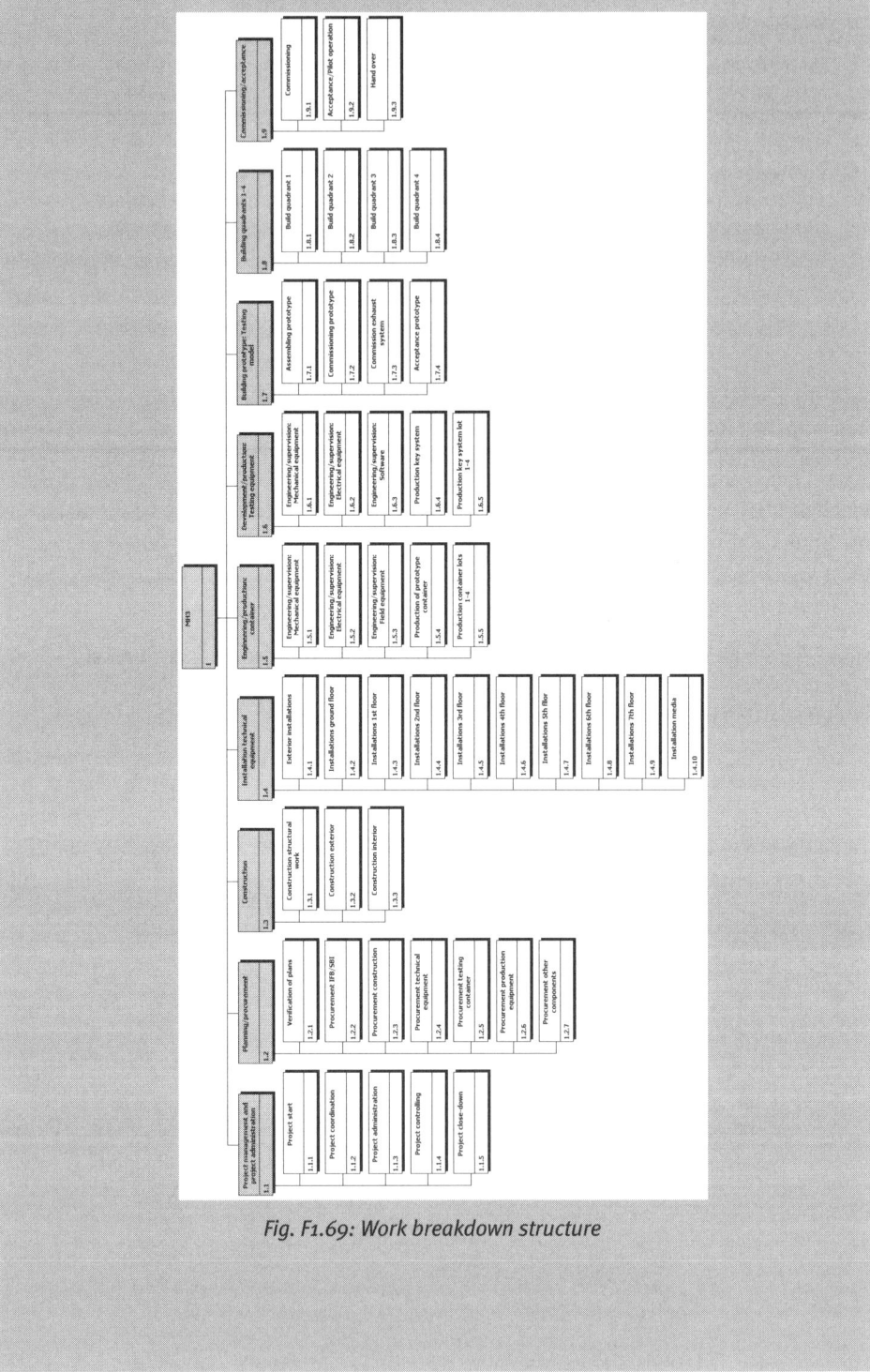

Fig. F1.69: Work breakdown structure

#	Type of Risk	Evaluation before measures [%]	[T€]	[t]	Preventive measures	Probability after measures	Cost deviation after measures	Schedule deviation after measures	Expected costs after measures	Expected duration after measures
1	Professionalism of project management	0%	0	0	Timely and complete provision of documentation	50%	1000		500	
2	General planner relationship	0%	0	0	No measures possible	60%				
3	Customer relationship	0%	0	0	Develop contract text for subcontractors according to the main contract	40%	500	50		20
4	Resources (own)	0%	0	0		40%	1000		400	
5	Information, communication	0%	0	0	Schedule risks and cost risks are carried by customer	10%	2000	10	200	1
6	Procurement success, quality of bids	50%	50	0		50%	4000	10	2000	5
7	Quality of the documentation	10%	500	0	Determine selection criteria for subcontractors; obtain information		500	60		
8	Weather, chemistry	30%	135	0	No additional measures possible	50%	100	50	50	25
9	Loss, damage	10%	300	0		25%	1000		250	
10	Warranty	50%	250	0	Warranty transferred to customer	80%	10		8	
11	Composition of the ground	25%	75	0	Design freeze, change management	10%	1000		100	
12	Construction supplier relationship	0%	0	0	Verification of plans	10%	1500	40	150	4

Fig. F1.70: Risk management for an engineering project in the project start process

CHAPTER

F

275

Project risk management in the project controlling process
During the project controlling of the engineering project it was recognized that the risk analysis made during the start process did not meet the professional requirements of the project. The strengths and weaknesses of the project risk management performed during the project start process shown in Figure F1.71 were analyzed. It was decided that the risk analysis should be revised and the risk management measures planned again.

Strengths	Weaknesses
• Discussion on risks has taken place • Detailed, analytical consideration • Differentiated cost and schedule deviations • Consideration of remaining risks	• Structure: Mixed scope, environments, objects • Completeness uncertain • No positive deviations • No provision-making measures • Differing depth of analysis • No supporting graphs or charts • No correlation between risks considered • "100% risks"

Fig. F1.71: Strengths and weaknesses of the project risk management of the project start process

In the following Figure F1.72 the project risks are assigned to the individual project phases. Possible negative and positive deviations from the objectives and avoiding and/or promoting and provision-making measures are described. An evaluation of the risks was not yet performed.

Risk # new	WBS code	WP label	Type of risk	Description of positive deviation	Description of negative deviation	Avoiding/promoting measures	Provision-making measure
1	1.1	Project management	Professionalism of project management		Competence of the team insufficient	Signature rules; consulting + training	
2			Customer relationship	Good relationship, quick decision-making	Missing planning documents; slow decision-making, claims	Social events, communications politic, invitation to team meetings	Contract definition; claim management
3			Resources (own)		Resource bottlenecks	Procure external resources; establish priorities by management	
4			Information, communication		Information flow incomplete; inadequate communication culture	Infrastructural measures (uniform communication platform: DB system, EDP support)	
5	1.2	Planning/procurement	Procurement success, quality of bids	Extensive procurement successes	Bids incomplete	Market analysis, professional tendering	Risk budget
6		6	Quality of the documentation		Plans incomplete	Provision of complete documentation in time	
7	1.3	Construction	Weather, chemistry	Especially good weather conditions, no impediments	Bad weather		Insurance
8			Loss, damage		High loss, damages		
9			Warranty		Will be claimed		Transferred to customer
10			Composition of the ground		Pole hits mineral spring		
11			Construction supplier relationship		Bankruptcy, additional charges	Establish selection criteria for subcontractors; obtain information	
12	1.4	Installation technical equipment	Loss, damage		High loss, damages		

Fig. F1.72: Identification of the project risks and description of risk management measures

The subsequent quantitative evaluation of the risks, differentiated by cost and schedule deviation, is shown in Figure F1.73.

Risk #	Evaluation before/after measures					Evaluation before/after measures				
	Occurrence probability	*Cost deviation (K€)*	*Schedule deviation (days)*	*Expected costs*	*Expected duration*	*Occurrence probability*	*Cost deviation (K€)*	*Schedule deviation (days)*	*Expected costs*	*Expected duration*
1	90%	1.000		900		50%	1.000		500	
2	60%	1.000	50	600	30	40%	500	50		
3	90%	1.000		900		40%	1.000		400	
4	90%	2.000	10	1.800	9	10%	2.000	10	200	1
5	50%	8.000	10	4.000	5	50%	4.000	10	2.000	5
6	80%	500	60	400	48		500	60		
7	50%	100	50	50	25	50%	100	50	50	25
8	90%	1.000		900		25%	1.000		250	
9	80%	167		133		80%	10		8	
10	10%	1.000		100		10%	1.000		100	
11	15%	1.500	40	225	6	10%	1.500	40	150	4
12	90%	1.000		900		25%	1.000		250	

Fig. F1.73: Quantitative evaluation of the risks

CHAPTER F

The changes in the project risks before and after the risk management measures are visualized in Figure F1.74.

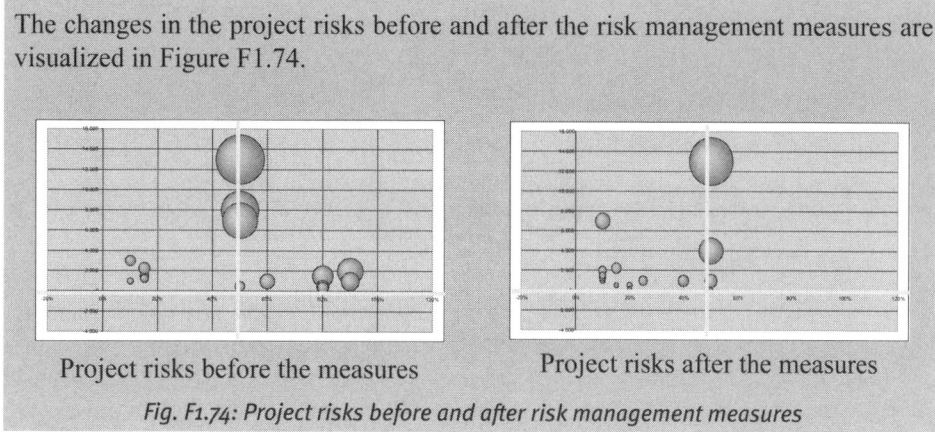

Project risks before the measures   Project risks after the measures

Fig. F1.74: Project risks before and after risk management measures

F1.10.5   Project risk controlling

The controlling of project risks is a task in the project controlling process. It includes controlling the risk management measures, the identification of new risks, the evaluation of newly identified risks, the adaptation and evaluation of old, still active risks and the planning and performing of new risk management measures. Methods for project risk controlling are described in Chapter F3.

F1.10.6   The organization of project risk management

The project manager and the project team are responsible for project risk management. The assistance of customer and supplier representatives broadens the view and creates possibilities when agreeing on risk management measures. The results of the risk analysis and the risk management measures are to be reviewed with the project owner team.

For the performance of risk management measures and controlling the success of the measures and/or the remaining risks the project manager can delegate responsibilities to "project risk owners". However, usually, it will be the responsibility of the project manager.

- Risk management requires corresponding communication forms in order to ensure the necessary experience and creativity for risk identification and risk evaluation and for planning risk management measures. Teamwork and the performance of risk management workshops are recommended.

Fig. F1.75: Participants in a project risk management workshop of an Austrian engineering company

Brainstorming and the interaction of the workshop participants substantially contributes to the complete listing of the risks of the project.

To ensure this completeness risk checklists can be used. For evaluating individual risks, for evaluating the project risk and for project risk documentation, standard software is available. Such software packages can, for example,

- define costs, income, resources and durations as stochastic variables,
- make Monte Carlo simulations,
- visualize probability distributions for the start events of work packages, for the project duration and/or for project costs and project income, and
- for example, develop a "criticality index" for non-critical activities of the CPM schedule.

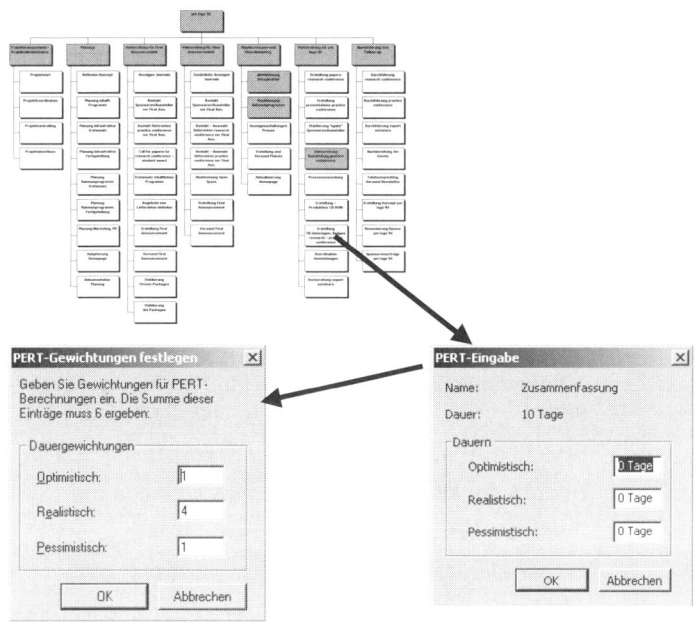

Fig. F1.76: Screenshot: Definition of the probability distribution for the duration of a work package

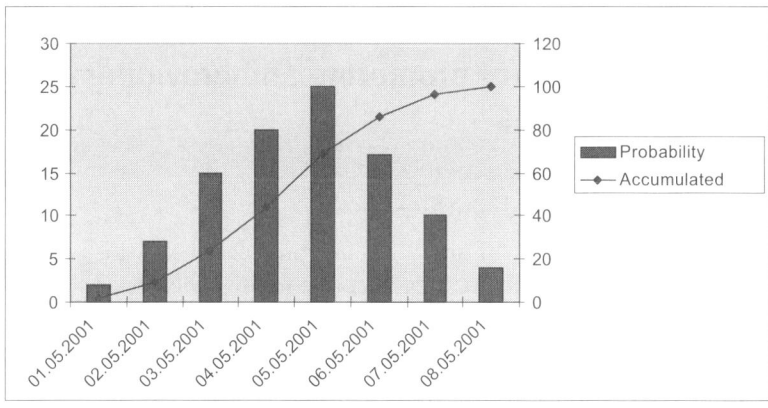

Fig. F1.77: Screenshot: Probability distribution of a project end date

Standard project management software, such as MS Project, has simple functions for project risk management. Additional functions are offered in special project risk management software packages (such as "Opera" as risk management module for the project management software "Open plan", both by Welcome Software Technology). Selected software packages for risk management are listed in the following table.

Software package	supplier	homepage
@RISK	Palisade Inc.	www.palisade.com
Cobra	Welcom	www.welcom.com
Contingency Calcula-tor TM	Westney Consultants Internati-onal, Inc.	www.westney.com
Crystal Ball®	Desicioneering	www.desisioneering.com
Decision Explorer	Banxia Software Ltd.	www.scotnet.co.uk/bankxia/de-main.html
Definitive Scenario	TerraMar Informasjonsystemer AS	www.definitivesoftware.com
DPL	ADA Decision Support Menlo Park	http://harker.coolware.com/kcoo-ley/ada/index.html
Expert Coice	Expert Coice Inc.	www.expertchoice.com
Futura	Futura International Group	www.futura-da.fi
Monte Carlo	Primavera Systems Inc.	www.primavera.com
Precision Tree	Palisade Inc.	www.palisade.com
Riskman	CCC Software Professionals Oy	www.riskmantools.com
Risktrak	Risk Services & Technology Inc.	www.risktrak.com
Riskexpert	Marotz Inc.	www.marotz.com

Fig. F1.78: Selected risk management software

F 1.11   Avoiding and/or promoting and providing for discontinuities

F1.11.1   Definitions and examples

A project discontinuity is a phase of instability in a project which leads to a change in the project identity (see Chapter B1.6). In projects the following types of discontinuities can be differentiated:

- a project crisis, based on an existential threat to the project,
- a project chance, based on new potentials of the project and
- a structurally caused change in the project identity, caused by a foreseeable change in the project.

Project crises and project chances come surprisingly, in spite of possible measures aimed at their avoidance or promotion. Structurally caused identity changes are predictable and therefore also planable.

The frequent occurrence of discontinuities in projects depends on the complexity and dynamics of the project. A project discontinuity can emerge because of changes in the relationships of a project to its relevant environments or because of self-organizing processes in the project. Causes for a project crisis can be, for example, a new company strategy, the loss of an important project partner or legal changes by regulatory authorities.

Examples of project crises, project chances and structurally caused identity changes in projects are shown in Figure F1.79.

Examples of project crises
• The reorganization of a service area of a city administration had to be interrupted because fundamental organizational decisions, which had to be taken outside the project by politicians and the city management, were still not made.
• The reactor accident at Chernobyl led to the endangerment of personnel from an Austrian engineering company at a construction site in nearby Shlobin. Crisis management measures at the construction site and in Austria for family members of the site personnel were to be made.

Examples of project chances
• The "e-application" developed by RGC for project management seminars was also interesting for use by the PROJEKTMANAGEMENT GROUP from the University of Economics and Business Administration, Vienna, in the course "Project management" and in the university training course "International project management". The project "e-application" was therefore substantially enlarged in scope during its performance. It was redefined as a cooperative project between RGC and PMG.
• During the performance of a logistics project for the Ministry of Defense of an eastern European state instead of the planned mainframe solution a new, recently introduced multiple server technology appeared to be more suitable. The IT supplier and its customer jointly decided use the new technology. For this purpose the structures of the project had to be substantially changed.

Examples of structurally caused changes in the identities of projects
• In the "ABS (antibiotic strategies) project", after the development of the antibiotic guidelines, the Federal Ministry for Social Security and Generations undertook to implement these guidelines with a road-show and counseling in Austrian hospitals. In order to gain the necessary outward orientation after a phase of creativity and inward orientation, a restructuring of the project was necessary. (The case is described in Chapter E1.6.)
• The necessary re-orientation in an engineering project at the beginning of the construction phase and after the phases of technical planning, procurement and production should be considered as a structurally caused identity change.

Fig. F1.79: Examples of discontinuities in projects

F1.11.2 Objects of consideration, tasks and objectives of the management of project discontinuities

Objects of consideration of the management of project discontinuities

For a professional management of project risks the methods of traditional project risk management are not sufficient. Project risk management presupposes the continuous development of projects. In addition to these continuous developments, discontinuities management also accounts for discontinuous developments in projects. Thereby the bandwidth of possible developments to be considered broadens. In project risk management the possible negative and positive deviations from the objectives are considered. In the management of project discontinuities extreme developments within a negative project scenario ("worst case") and a positive project scenario ("best case") are considered.

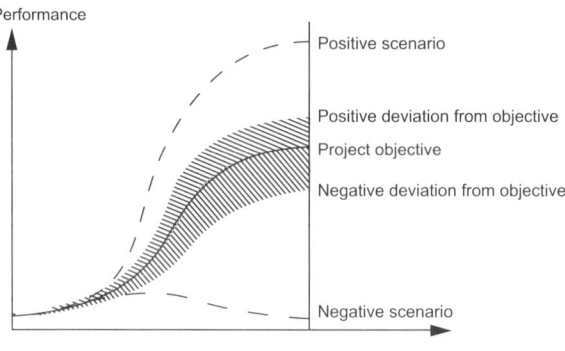

Fig. F1.80: Objects of consideration of project risk management and of the management of project discontinuities

Tasks and objectives of the management of project discontinuities

Through the use of professional project management the avoidance of project crises and/ or the promotion of project chances and provision for project discontinuities are implicitly performed. For example, a clear definition of project objectives, the design of an adequate project organization, the explicit development of a project culture, the adequate use of project planning methods and the conscious performance of reflections in the project all contribute to avoiding project crises.

In spite of this, many projects have discontinuities and many are stopped (see Figure F1.81).

The logic of failure
• There are more and more projects, but fewer and fewer successful projects (Atos Origin)
• More than 50% of all IT projects fail (GPM – Gesellschaft für Projektmanagement)
• More than 70% of all ERP-projects do not reach the defined project objectives (Gartner)
• 30% of all IT projects are stopped (Standish Group)
• 50% of all IT projects exceed the project budget by 90% (Standish Group)
• The results of IT projects fulfill only 40% of the defined functions (Standish Group)

Fig. F1.81: The logic of failure in IT projects

An increase in project success seems to be possible with the help of an explicit management of project discontinuities. This includes, on the one hand, the avoidance of project crises and/or the promotion of project chances as well as providing for project discontinuities and, on the other hand, the resolution of a project discontinuity (see Figure E1.13).

Strategies and measures for avoiding project crises and/or for promoting project chances are to be planned and implemented. The objective of providing for project discontinuities is to develop provision-making strategies and measures for discontinuities. Thereby an efficient resolution should be enabled in the event that a discontinuity arises.

For the management of project discontinuities time and room for reflection in the project are necessary. "Many project leaders are too reactive – they revel in fire-fighting and crises management, instead of balancing these skills with the more productive, strategic approach. One thing is clear – you should always create small but significant periods of reflection time. Without this you will never even realize, if your priorities are upside down."[3]

The avoidance of project crises and/or the promotion of project chances and the provision for project discontinuities are not sub-processes of project management, they are tasks which are to be performed in the project start process and in the project controlling process. The resolution of a project discontinuity represents its own sub-process of project management which is dealt with in Chapter E1.6.

CHAPTER

F

F1.11.3 Methods for avoiding and/or promoting project discontinuities

Early warning methods

Objectives of the use of early warning methods in project management are:

- early identification of project risks which may lead to project discontinuities,
- analysis of the causes and the relationships between these project risks, and
- evaluation of the extent of possible project discontinuities.

The early identification of potential project discontinuities enables the development of appropriate measures in time.

Methods for early warning are project ratios, project-related indicators and methods for identifying "weak signals" (poorly defined, unstructured information). Usually, project controlling ratios are used which relate to the project's past. This past-related information makes it possible to describe the project status but not to anticipate project discontinuities. For the identification of potentials for project discontinuities newer methods of early warning are better qualified, such as the scenario technique. Different early warning methods and their characteristics are shown in Figure F1.82, differentiated by "generation".

---

3) *Briner, et Al.* (Project Leadership) 1990,

generation	method	characteristics
first generation	• ratio analysis (such as project costs, project income, project duration, cash flows, etc.) • trend analysis	• use of quantitative information • oriented to the past • consideration of negative deviation from objectives
second generation	• indicators (such as fluctuation in the members of the project organization, commitment to the project, customer relations, etc.)	• use of quantitative and qualitative information • oriented to the past • consideration of negative and positive deviations from objectives • problems in identifying the indicators which show changes in the project over time • definition of the bandwidth inside which the indicators may move without being considered critical • assumption of the cause-effect principle
third generation	• scenario technique • environmental monitoring • feedback diagrams • discontinuity questionnaire • cross-impact analysis • sensitivity analysis	• use of quantitative and qualitative information • oriented to the future • consideration of negative and positive deviations from objectives • consideration of "weak signals" • high sensitivity for possible problems

Fig. F1.82: Early warning methods for the identification of potential project discontinuities

From a constructivist view, the assumption that potential project discontinuities are to be identified objectively by project ratios cannot be upheld. Instead, in a process of reality construction weak signals are to be identified and interpreted as potentials for project discontinuities. It is, therefore, important to give special attention to the design of early warning processes. Involving the members of the project organization and representatives of relevant environments, as well as moderating their communications, influences the results and their acceptance.

Central differences between traditional early warning and new approaches are summarized in Figure F1.82.

traditional early warning	early warning "new"
• discontinuities can be recognized by objective criteria • the future can be predicted on the basis of past trends • it is important to predict the correct future • the past is given sense retrospectively	• discontinuities are not "recognizable", they must be constructed • discontinuities cannot be predicted on the basis of information from the past • crisis potentials can be defined based on the construction of possible futures • the future is given sense anticipatively

Fig. F1.83: Traditional early warning versus new approaches of early warning

The scenario technique is an early warning method of the third generation which appears to be especially relevant for project management. It is briefly described here[4].

Scenario technique

A project scenario is a possible future state of a project. In applying the scenario technique in project management possible development alternatives of a project are conceptualized. In addition to the planned scenario at least a "worst-case" and "best-case" scenario are also defined and described. As a model a scenario funnel is used. As starting point of the funnel the date of planning is used. The possible project scenarios stretch across the funnel, which is limited by the extreme scenarios.

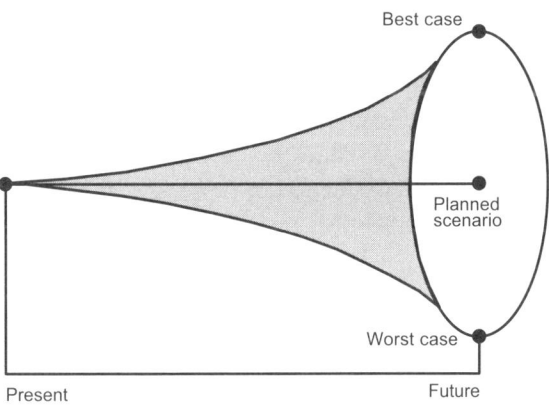

Fig. F1.84: Scenario funnel

The objective of the scenario technique is to develop alternative consistent pictures of the future. Not only the desired and/or probable scenario ("planned scenario") is defined but thinking in possible alternatives is promoted. Through the development of project scenarios the perception of the project becomes more complex. The occurrence of an

---

4) Descriptions of further early warning methods can be found, for example, in: Konrad, L., (Strategische Früherkennung) 1991, Simon, D., (Schwache Signale) 1986, Reibnitz, U. von, (Szenario-Technik) 1992

extreme scenario represents a discontinuity for the project, which can be prepared for with the help of the scenario technique.

The application of the scenario technique in project management includes the following steps:

- Identifying influencing factors
  Important factors are to be identified which can influence the project development. These influencing factors can be project environment relationships or project structures, such as the project organization, the project culture or the milestone schedule. The project environment analysis is thus an important instrument for identifying influence factors.
- Forecasting possible developments of the influencing factors
  Possible changes in the project environment relationships and in the project structures are to be forecast. Positive and negative deviations of the project objectives resulting from the possible developments of the influencing factors are to be described.
- Defining the project scenarios
  By bundling possible developments of influencing factors project scenarios can be defined. The combination of several possible positive changes results in the best-case scenario, several negative changes lead to the worst-case scenario. For communication purposes each of the defined project scenarios should be given a name.
- Describing the project scenarios
  The consequences of the individual project scenarios on the achievement of the business case objectives and the project objectives are to be described.
- Planning and implementing of strategies and measures for avoiding and/or promoting and for providing for the extreme scenarios
  The occurrence of an extreme scenario represents a project discontinuity. In order to avoid and/or promote and provide for this, strategies and measures are to be planned and implemented. The scenario technique is not limited to the definition and description of project scenarios but also includes the planning of strategies and measures.

The application of the scenario technique and the resulting planning of alternative project strategies for the project "Realization e-application" are shown in Figures F1.85 and F1.86.

⊛ Realisation e Application	PROJECT SCENARIOS		
criterion	scenarios		
	best case	worst case	middle case
quality of the solution developed	all functions available as planned	main functions not available in corresponding quality, improvements costly	important functions available immediately, improvements possible short- to medium-term
effort for administrating the solution	low administration effort	medium administration effort	medium administration effort
acceptance of the solution by the seminar customers	high acceptance by the seminar customers	little acceptance by the customers	medium acceptance by the customers
financing solution	adequate solution found	no adequate solution found	adequate solution found
Version: 1.0	Date: 01. 02. 2003	Author: GS	

Fig. F1.85: Scenarios of the project "Realization e-application"

⊛ Realisation e Application	ALTERNATIVE RESOLUTION STRATEGIES		
criterion	scenarios		
	best case	worst case	middle case
seminar operation	immediate application for all RGC seminars	interruption of the pilot application, "traditional" performance of seminars	as planned
marketing	immediate planning and performance of a broad marketing initiative	redefinition of marketing strategy	as planned
training, organization	as planned	interruption of the organizational adaptations	as planned
financing	as planned	search for cooperation partners for financing	as planned
Version: 1.0	Datum: 01. 02. 2003	Ersteller: GS	

Fig. F1.86: Alternative resolution strategies for the project "Realization e-application"

An extensive example of the use of the scenario technique for an event project is shown in Case Study F2.

···→ **CASE STUDY F2: Crisis avoidance and provision for a crisis in an event project**

Problem situation

In April 1994 Projekt Management Austria (PMA) held a 2-day international conference on the topic "Crisis management". After an extensive direct mailing campaign of the conference brochures the registrations for the conference remained well behind expectations until the end of January 1994 (especially those of foreign participants).

Since the success of the conference was mostly dependent upon the number of participants, on February 2nd 1994 methods for crisis avoidance and provision for a crisis were deployed. A situation analysis was performed, possible scenarios were defined and described, crisis avoiding measures were determined and provision-making measures for the occurrence of a crisis were planned.

Situation analysis

On the basis of a project environment analysis (see Figure F1.87) it was determined that

- the international cooperation with foreign crisis management publishers and institutes had not brought the expected marketing success,
- "crisis management" represented for PMA a new product in a somewhat new market,
- the term "crisis" had negative connotations and managers were shy of attending a crisis management conference, and
- in the project team many project experiences and social networks were available.

Definition and description of project scenarios

Four project scenarios were defined and described using the following influencing factors: German- and English-speaking participants, paying participants, sponsoring, number of exhibitors, budget, financial contribution, programme content and consequences for PMA (see Figure F1.88 "Scenarios for the crisis management conference").

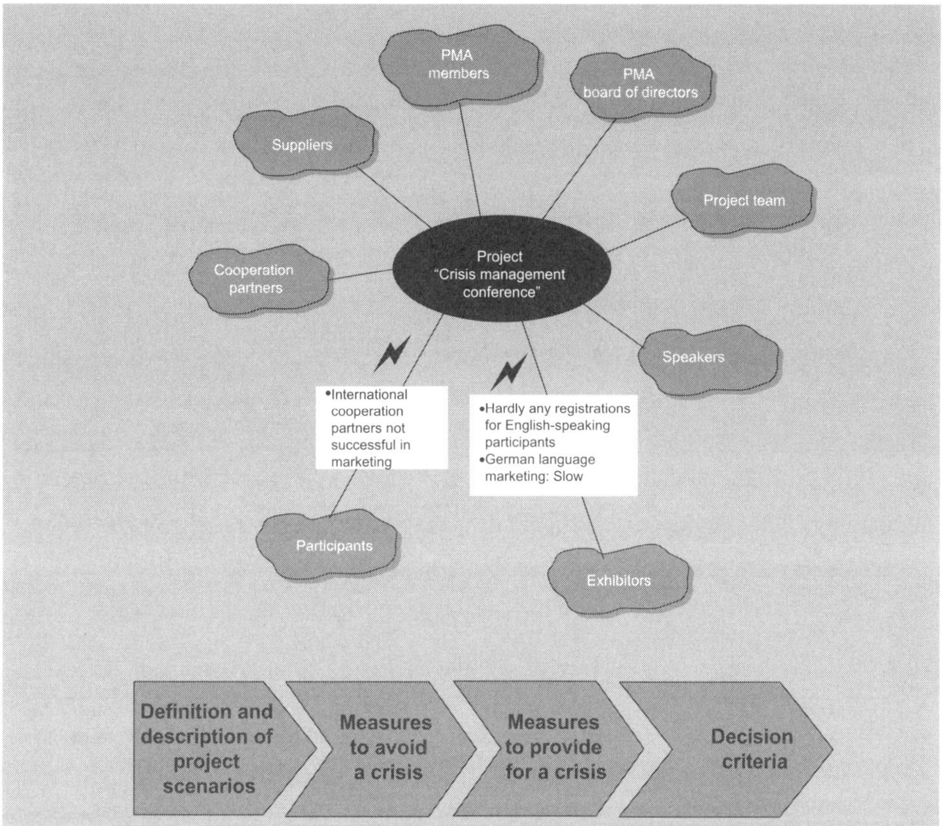

Fig. F1.87: Project environment analysis of the project "Crisis management conference"

PROJECT SCENARIOS				
Criteria	„Everybody happy"	„The flop"	„Flop in English"	„A happening"
German- & English-speaking participants	140 + 60 = 200	80 + 10 = 90	140 + 10 = 150	200 + 100 = 300
Paying participants	120	50	100	200
Sponsoring	ATS 300.000,-	ATS 280.000,-	ATS 280.000,-	ATS 350.000,-
Number of exhibitors	5	3	5	8
Budget	ATS 1,5 Mio.	ATS 0,75 Mio.	ATS 1,0 Mio.	ATS 1,8 Mio.
Contribution	ATS 350.000,-	ATS −140.000,-	ATS 100.000,-	ATS 800.000,-
Programme content	• Many participants • Adequate infrastructure • Good contacts • Satisfied partners	• Few participants • Empty rooms • Few contacts	• Visit OK • Little international flair	• Total networking
Consequences for PMA	• Good PR • Financial success	• PR ok • Financial loss	• PR ok • Financial success	• Great PR • Financial success „great"

Fig. F1.88: Project scenarios "Crisis management conference" (developed: February 2nd 1994)

Measures for crisis avoidance

The following measures for crisis avoidance were setFig. F1.88: Project scenarios "Crisis management conference" (developed: February 2nd 1994)

- reinforced marketing (additional mailing, reduced participation fee for university students, reinforced involvement of speakers and cooperation partners in marketing, telephone marketing),
- reinforced PR (press conference, press service, advertising in daily newspapers, radio and television),
- additional lectures,
- continuous review of the registrations.

Measures for crisis provision

For the different scenarios alternative plans were developed with measures to provide for a crisis, differentiated by programme, speaker, marketing and infrastructure.

Alternative plans	„Everybody happy"	„The flop"	„Flop in English"	„A happening"
Programme content	• As planned	• 2 instead of 3 parallel events • No evening event	• 2 instead of 3 parallel events	• Additional workshop
Speakers	• As planned	• Cancel speakers • Reinforced interactive work in groups	• Cancel some English-speaking speakers	• Invite some speakers to moderate additional workshops
Marketing	• After-sales marketing	• Pre-information to participants, exhibitors and speakers	• Pre-information to participants, exhibitors and speakers • After-sales-marketing	• Intensive after-sales marketing
Infrastructure	• As planned	• Reduce the lecture halls	• Reduce the lecture halls	• As planned, lots of turmoil

Fig. F1.89: Alternative plans for the crisis management conference

For all project scenarios the programme content was also adapted. It was planned that the scientific leader of the conference would present the case study given here during his opening speech, in order to inform the conference participants about the development of the project. At the same time the pressure to justify a not so successful event could perhaps also be reduced.

Criteria for decisions
Criteria were defined on the basis of which the decision for one of the alternative plans could be made (see Figure F1.90).

- Until March 23rd 1994 no more than 80 participants: Situation defined as "Flop"
- Until March 23rd 1994 no more than 20 English speaking participants: Situation defined as "Flop in English"

Fig. F1.90: Criteria for decisions

F1.11.4   Methods for providing for project discontinuities

The planning of alternative project strategies, the development of alternative project plans and the development of standards for resolving a project discontinuity are methods for providing for project discontinuities.

Developing alternative project plans

To provide for different project scenarios, alternative project strategies can be planned and alternative project plans (alternative work breakdown structures, schedules and cost plans) can be developed.

An example of a work breakdown structure of a research and development project in an Austrian engineering company in which two alternatives are planned (only one development cycle or two), is shown in Figure F1.91. In this project the productivity of a turbine should be increased on the basis of tests. If, during the tests in the first development cycle, the corresponding results cannot be achieved, a second cycle is already planned in the work breakdown structure. These two alternatives are also already included in the schedule and cost plans. However, no linear progression for the duration and costs is assumed, instead the necessary additional costs and duration are explicitly planned.

CHAPTER

F

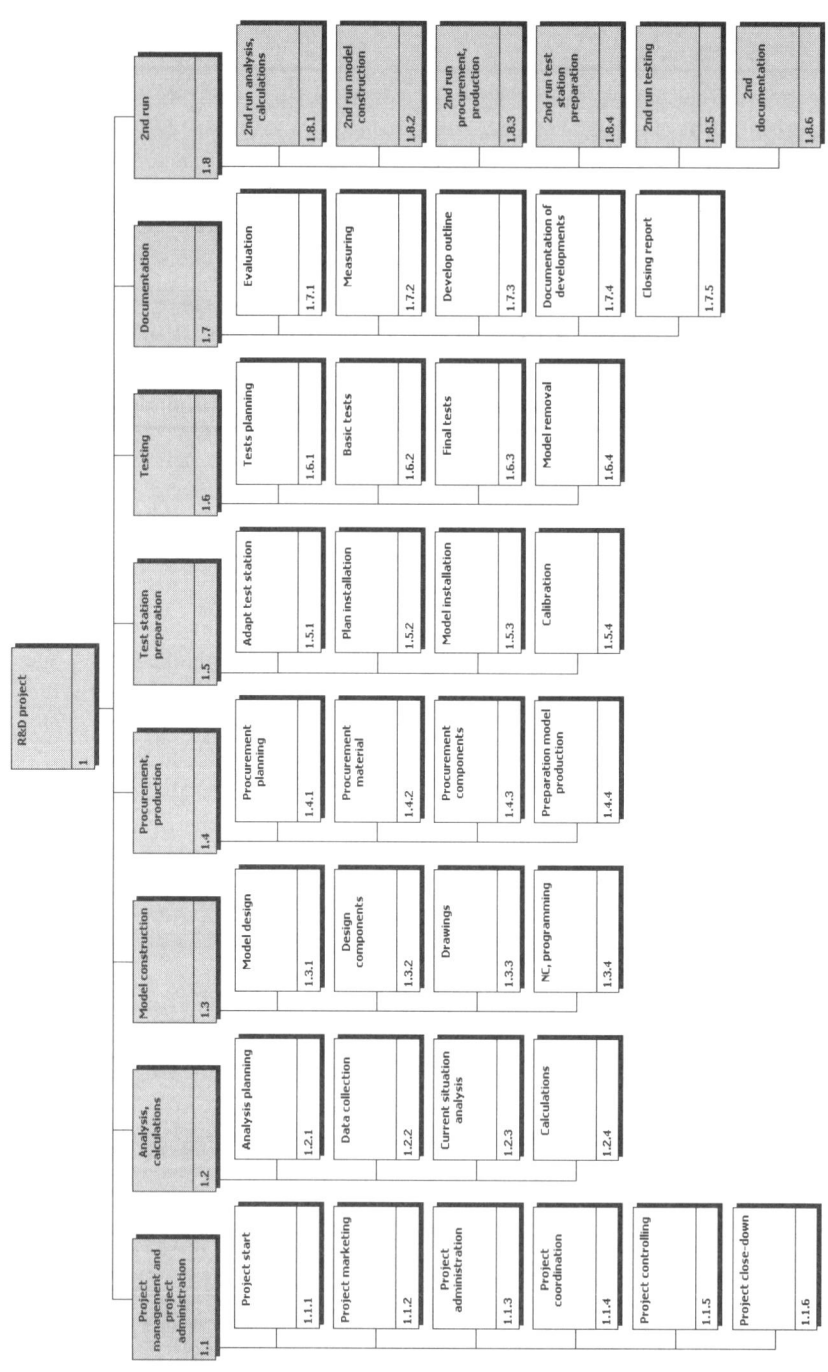

Fig. F1.91: Work breakdown structure with alternative scopes

An example of planning alternative project budgets is visualized in Figure F1.92. Three scenarios for the project budget (best case, planned case and worst case) were defined for the pm days '03 already during the project start process on the basis of assumptions about different developments of the project costs and the project income. This planning process increased the sensibility for influencing factors for the project success. The members of the project organization developed a common understanding about the possibilities for positively influencing these factors.

WBS code	Phase	Work package	Bad alternative	Good alternative	Medium alternative
Costs					
1.1.		Project management	-42.000,00	-37.000,00	-40.000,00
1.2.		Planning	-800,00	-500,00	-700,00
1.3.		Preparation first announcement	-12.000,00	-7.000,00	-9.000,00
1.4.		Preparation final announcement	-15.000,00	-9.000,00	-12.000,00
1.5.		Review and final marketing	-55.000,00	-48.000,00	-50.000,00
1.6.		Preparation for pm days '03	-7.000,00	-4.000,00	-5.000,00
1.7.		Performance and follow-up	-9.000,00	-8.000,00	-9.000,00
			-140.800,00	-113.500,00	-119.748,00
Income					
1.4.2		Sponsors	17.650,00	33.500,00	18.150,00
1.4.2		Exhibitors	17.000,00	30.000,00	18.500,00
1.4.2		Income other	3.000,00	3.400,00	3.200,00
1.7.1		Research conference	6.300,00	12.750,00	9.450,00
		Practice conference	41.150,00	97.550,00	87.850,00
		Expert seminars	12.750,00	25.500,00	11.000,00
			97.850,00	202.700,00	148.150,00
FC					

Fig. F1.92: Alternative project budgets for the project "pm days '03"

Developing standards for resolving project discontinuities

For selected situations standards for resolving project discontinuities can be developed. For example, an engineering company has defined the management of project discontinuities as a temporary management task in the job description of the profit center manager.

An international IT company has developed an "escalation" model for resolving technical crises in IT projects. This model stipulates which organizational unit of the company and which resources are involved during each type of situation.

In preparing to resolve acute project discontinuities in training situations project crises or project chances can be simulated.

CHAPTER

F

In social networks relationships for resolving discontinuities can be developed. Through "loose coupling" with other projects in networks of projects possibilities for acquiring information and for short-term recruiting of experts can be created.

F1.11.5 Use of methods for avoiding and/or promoting and providing for project discontinuities

Strategies and measures for avoiding and/or promoting and for providing for project discontinuities are to be planned and implemented as early as possible in the project. In case it does not seem emotionally sensible to perform this project management task in the project start process, possible project discontinuities should be discussed before the first project controlling.

The strategies and measures should be planned and implemented in the project team – with the consent of the project owner team.

The requirement for measures for avoiding and/or promoting and providing for project discontinuities is dependent on the type of project. For external contracting projects crisis avoidance measures will be set during the pre-project phase. During bidding, on the one hand, detailed object planning (development of contract specifications), project planning (estimating, scheduling, designing the project organization) and the contract itself will be developed. On the other hand, often risk management methods will be used which enable risk management measures (insurance policies, risk contingencies, etc.). Since the risk and the complexity of bidding are high, engineering companies define projects for the development of large bids. This is also to be considered as a measure of discontinuity management.

Internal projects, such as research and development projects or organizational development projects, are subject to more discontinuities than external projects because of their higher social complexity and dynamics. Project stoppings are more common in internal projects than in external projects. The use of the methods for discontinuities management appears, therefore, to be especially important for internal projects.

References

Briner, W., *Hastings,* C., *Geedes,* M., (Project leadership) Project leadership, 2nd ed., Gower, Aldershop, 2001

Jann, B., (Statistik) Einführung in die Statistik, R. Oldenbourg, München Wien, 2002

Konrad, L., (Strategische Früherkennung) Strategische Früherkennung – eine kritische Analyse des „Weak Signal"-Konzeptes, Brockmeyer, Bochum, 1991

Simon, D., (Schwache Signale) Schwache Signale – die Früherkennung von strategischen Diskontinuitäten durch Erfassung von „weak signals", Service Fachverlag an der Wirtschaftsuniversität, Wien, 1986

Reibnitz, U. von, (Szenario-Technik) Szenario-Technik: Instrumente für die unternehmerische und persönliche Erfolgsplanung, Gabler, Wiesbaden, 1992

F 2 Methods for coordinating projects and programmes

F 2.1 Overview of methods of coordination

The objective of project coordination is to inform the members of the project organization and representatives of relevant environments about the developments in the project. Relationships between work packages are reviewed with project team members. In doing so, the project progress and the quality of the work package results are ensured.

Project coordination tasks are performed continuously by the project manager. They include continuous communication with the project owner team, the project team members and the project contributors, as well as continuous project marketing toward the relevant project environments.

Project progress and quality of the work packages are ensured by coordinating the project resources, controlling the progress of work packages and accepting (intermediary) results of work packages.

These project coordination tasks can be performed in meetings between the project manager and a project team member or in sub-team meetings. For structuring and supporting these communications, above all, the project management documents developed in the project start process are to be used, such as the project objectives plan, the work breakdown structure, the project bar chart, the project cost plan and the project environment analysis. In addition, specific coordination methods, i.e. meeting protocols, to-do lists and acceptance certificates for work packages can be used.

The project management methods to be used in the project coordination process can be selected with the help of the list depicted in Figure F2.1:

methods for project coordination	small project	project
project management documents of the project start	must	must
to-do list	must	must
minutes of meetings	must	must
acceptance certificate for work packages	must	must

Fig. F2.1: Checklist: Use of project management methods

F 2.2 Meeting minutes

To ensure commitment and understanding in the project the project manager is to create meeting minutes after each project meeting. This formalism is justifiable because in projects sometimes persons not knowing each other work in different locations on relatively unique tasks. The documenting of agreements is, therefore, an important coordination method in projects.

Meeting minutes should be as short as possible and formulated in key words only. As an appendix to the minutes the measures agreed on are to be documented in a to-do list (see Chapter F2.3).

Forms should be used for developing meeting minutes to standardize the project communication. Meeting minutes should contain the objectives of the meeting and its content, but also formal information, such as the date and duration of the meeting, the participants and the location.

The content of the meeting minutes can be structured according to the agenda of the project meeting and should be oriented to the project phases and the work packages of the project. An example of meeting minutes for a sub-team meeting in the project "Realization e-application" is given in Figure F2.2.

✪ Realisation e Application	MEETING MINUTES		
meeting:	sub-team meeting organization, personnel and marketing	date:	13. 02. 2003
participants:	Müller, Berger, Gruber	location:	RGC
agenda			
• information about current status • planning: Organization, personnel • planning: Marketing • planning of further actions			
results			
planning: organization, personnel • drafts of role descriptions prepared (attached) • drafts of business process descriptions prepared (attached) • definition of administration process completed: Develop a new seminar, delete a seminar, edit a seminar planning: marketing • ideas collected for marketing activities, more detailed work agreed • coordination with re-launch of homepage completed, no apparent conflicts • idea: Marketing activities at the "pm platform" event planning of further actions • see to-do list			
appendix			
• to-do list • marketing plan			
Version: 1.0 Date: 13. 02. 2003 Author: GS			

Fig. F2.2: Meeting minutes of a sub-team meeting for the project "Realization e-application"

F 2.3   To-do list

The to-do list is an important method for project coordination. It is to be used as a supplement to the work breakdown structure, as it contains detailed activities (to-dos) for the fulfillment of the work packages[1] For each project the project manager should have only one to-do list.

The work breakdown structure is to be detailed to a point where the project manager can make agreements about the fulfillment of the work packages with the responsible project team members. On the basis of the work package specifications the responsible project team member can further structure the fulfillment of the work package. Necessary agreements between the project team member (or sub-team) and the project manager about these details can be documented in a to-do list.

The to-do list should be structured by project phases and work packages. In addition to the definition of the individual activities and their relationships to the work packages, it should also contain the responsibilities, the corresponding agreement and due dates, the status of performance and interpretations of the status. The status of an activity can be defined either by indicating percents or by verbal progress descriptions, such as "not started", "being performed", "completed". The agreement date is the date on which the agreement was made. The documentation of this date is also useful for project coordination.

The to-do list is to be created progressively. As part of project coordination the completed activities are to be noted with the status "completed", additionally agreed activities are added to the list and labeled with the agreement date.

The completed activities should remain in the to-do list in order to document the project progress. This gives the to-do list the character of a "project diary". In case the document with the completed activities becomes too large, parts of the to-do list can be discarded.

If the to-do list is adapted during a meeting, it is to be added to the meeting minutes as an appendix. An example of a to-do list of the project "Realization e-application" is depicted in Figure F2.3.

---

1) In project practice sometimes only a to-do list is used, without the "big project picture" that the work breakdown structure provides.

✪ Realisation e Application		TO-DO LIST				
WBS code	activity	responsi-ble	agreed on	due date	status	interpreta-tion
1.2.3.	review role and process descriptions with manage-ment	Müller	13. 2. 2003	03. 03. 2003	being per-formed	due date agreed
	detailed description of ad-ministration processes	Müller, Berger	13. 2. 2003	03. 03. 2003	being per-formed	
	hand over results to sub-team SW + HW	Müller	13. 2. 2003	03. 03. 2003	not started	
1.2.4.	detailed description of marketing activities	Gruber	13. 2. 2003	10. 03. 2003	not started	
	review marketing activi-ties with the management of the "pm platform"	Gruber	13. 2. 2003	10. 03. 2003	not started	delayed
Version: 1.0		Date: 13. 02. 2003		Author: GS		

Fig. F2.3: To-do list of the project "Realization e-application"

F 2.4 Acceptance certificate for a work package

The quality of the fulfillment of individual work packages and their progress are to be continuously controlled. This task is performed as part of project coordination.

After the completion of a work package by a project team member or a sub-team the work package is to be accepted by the project manager. This formal conclusion of the work in a work package is performed by the project manager and the project team mem-ber issuing an acceptance certificate. The project team member is thus formally dis-charged for this work package.

An example of an acceptance certificate for a work package of the project "Realization e-application" is shown in Figure F2.4.

⊛ Realisation ⊛ e Application		**WORK PACKAGE** **ACCEPTANCE CERTIFICATE**	
WBS code:	1.2.2	WP label:	detailed planning application functions

WP content
• final definition of user groups • description of user and administrator functions • description of the user interface • create "script" for important functions

WP results
• "Script" of important application functions, considering user groups • 3 layouts for the user interface

acceptance	
date of completion: 23. 02. 2003	actual costs: EUR 4,300

comment
• deviation from schedule by 1 week because of unclear user groups • WP results are of good quality

Project manager	WP responsible

Version: 1.0	Date: 23. 02. 2003	Author: GS

Fig. F2.4: Acceptance certificate of a work package of the project "Realization e-application"

F 3   Methods for controlling projects and programmes

F 3.1   Overview of methods for controlling projects and programmes

Objects of project controlling are, on the one hand, "hard facts", i.e. the project objectives, the objects of consideration of the project, the project scope, the project costs and the project income, the project resources, the project schedule and the project risks, and on the other hand, "soft facts". Objects of "social" project controlling are the project organization, the project culture and the relationships of the project to relevant project environments. The business case of the investment initialized by a project is also to be controlled.

Project controlling tasks are the project control, the project direction, the adaptation of project plans and the development of project reports (see Chapter E1.5). The control of the above defined objects includes the following tasks:

- determining the status of each object,
- performing planned versus actual analyses and
- performing deviation analyses.[1]

In addition to these control tasks, the preparation of project controlling meetings includes developing suggestions regarding directive measures, developing adapted project plans and developing drafts of project controlling reports.

Project controlling requires a project team meeting and a project owner team meeting. In these meetings the drafts of adapted project plans and controlling reports will be discussed and decided upon. The adapted project plans and the project controlling reports are completed as follow-up to the meetings.

To fulfill the project controlling process project controlling methods are to be used. The methods and the communication structures to be used are to be selected in the project start process. A checklist of project controlling methods which can be used is shown in Figure F3.1.

methods for project controlling	small project	project
project control and adapting the project plans		
project objectives plan	must	must
project objects of consideration plan	can	must
work breakdown structure	must	must

---

1) Project control is to be differentiated from overall project controlling, which also includes directive measures, etc.

methods for project controlling	small project	project
work package specifications	can	must
project milestone plan	must	must
project bar chart	can	must
CPM schedule	can	can
project resource plan	can	can
project cost plan	must	must
project income plan	can	can
project environment analysis	must	must
business case analysis	can	must
project other projects analysis	can	must
pre- and post-project phase analysis	can	must
project organization chart	must	must
project role descriptions	must	must
project risk analysis	must	must
methods for reporting		
project progress report	must	must
earned value analysis	can	can
project trend analysis	can	can
project score card (plus interpretations)	must	must
methods for project direction		
to-do list	must	must

Fig. F3.1: Selection of project controlling methods

Project controlling is to be performed for each object of consideration of project management (project objectives, project progress, project schedule, project costs, etc.). In the earned value analysis an integrated consideration of the progress, costs and schedule is performed. Directive measures are to be planned considering the consequences on the project scope, the project schedule, the project costs, the project organization, etc.

The control and adapting can be made for individual work packages or for the entire project. Project reports are developed for the entire project.

To present the application of project controlling methods again the case study of the project "Realization e-application" is used. The results of the project controlling at March 3rd 2003 are shown.

F 3.2 Methods for the control and the adaptation of project plans

The project control is based on the project plans developed in the project start process. Since the project plans are to be adapted, the methods for project control are outlined together with the adaptation of project plans.

F3.2.1 Objectives and objects of consideration

Control and adaptation of the project objectives

The realization of the project objectives is controlled for the entire project. It is to be determined, how far the work packages fulfilled until the controlling date have led to the realization of the project objectives. It is difficult to evaluate the realization of the project objectives up to the controlling date because the project objectives are not formulated relative to a schedule, but as a result at the end of the project. It must, therefore, be assessed whether the adherence to the objectives is still realistic.

The project objectives can, however, be adapted in a detailed manner. No more relevant objectives are to be deleted. New objectives are to be formulated as operationally as possible. Reasons for adapting the project objectives are mostly changes in the objects of consideration. Changes of the project objectives can relate to quantitative and to qualitative changes. Qualitative or quantitative changes in the objectives caused by customers in contracting projects are also referred to as "change management".

A change in the project objectives requires adaptation of most of the project plans. Not only the scope is to be adapted in the work breakdown structure, but also the project schedule, project costs and probably also the project organization and the project environment analysis must be adjusted.

The changes in the project objectives of the project "Realization e-application" are shown in Figure F3.2. In addition to the project objectives agreed to in the project start process, the main objective "Concept for all RGC seminars prepared" was defined.

CHAPTER
F

Realisation e Application	PROJECT OBJECTIVES PLAN
main objectives	adaptation as at 03. 03. 2003
• e-application tested in pilot operation introduced for 3 RGC seminars as per specification (use, administration and maintenance processes) • adequate organization, 3 e-application tutors and e-application administrators trained • marketing implemented (presentation at events, hints in brochures) • financing for the application of the e-application secured	• concept for all RGC seminars prepared

• therefore reductions in the durations of the seminars and increase in the teaching quality (more information beforehand, tests) • prerequisite for increase in customer loyalty developed	
additional objectives	adaptation as at 03. 03. 2003
• gaining experiences for further e-projects • further developing cooperative relationship with supplier	
non-objectives	adaptation as at 03. 03. 2003
• establishing new seminars • continuous operation and maintenance activities	
Version: 1.0	Date: 03. 03. 2003	Author: GS

Fig. F3.2: Adapting the project objectives for the project "Realization e-application"

Control and adaptation of the objects of consideration of the project

During control of the objects of consideration it should be checked whether or not all the defined objects of consideration have been considered in the fulfillment of the work packages.

In adapting the objects of consideration possible additional objects of consideration are to be defined (e.g. "Concept for all RGC seminars"). A further breakdown of already defined objects of consideration can also be necessary (such as a breakdown of the administration functions into "Developing a new seminar", "Deleting a seminar" and "Editing a seminar").

✪ Realisation e Application	PROJECT OBJECTS OF CONSIDERATION	
objects of consideration	object of consideration	adaptation as at 03. 03. 2003
hardware	server	
	PCs	
	IT-network	
software	e-application user	
	e-application administrator	
	database	
	office software	
	IT interfaces	
content	content seminar 1	
seminars	content seminar 2	
	content seminar 3	

seminar participants	regular seminar customers	
	new seminar customers	
organization: roles	customer/user	
	tutor	
	administrator	
	internet provider	
organization: business processes	seminar application process	
	seminar administration process	• develop a new seminar • delete a seminar • edit a seminar
	technical maintenance process	
documents	business process documentation	
	role descriptions	
	manuals (user, administrator)	
	training manuals	
	source code documentation	
	maintenance plan	
		concept for further RGC seminars
personnel	tutors	
	administratord	
financing	own financing	
	external financing	
marketing	brochures	
	events	
pilot operation	pilot operation	
Version: 1.0	Date: 03. 03. 2003	Author: GS

Fig. F3.3: Adapting the objects of consideration for the project "Realization e-application"

F3.2.2   Scope

The objective of the quantitative control of the project scope is to determine the progress of individual work packages, of the project phases and of the project. The actual progress up to the controlling date is to be compared with the planned progress.

In the "empowered" project organization the qualitative control of the work package results lies with the individual project team members and with the project team. The results can be reviewed in presentations and discussions during project team meetings as well as during project coordination meetings.

Methods for measuring progress

The following methods can be used for measuring the progress of a work package:

- 0% or 100% assumption,
- intuitive estimation by a project team member,
- output measurement,
- definition of progress milestones.

In the "0% or 100% assumption" method the work packages in progress are evaluated as either not yet started (progress = 0%) or as completed (progress = 100%). This method presupposes a highly detailed work breakdown structure. It is the most inexact of the methods.

In the "Intuitive estimation by a project team member" method the project team member charged with performing a work package gives an intuitive estimate of the progress.

In the "Output measurement" method progress is determined on the basis of the output achieved. This method is suitable for relatively continuous work, such as excavation, pouring concrete and laying pipes.

CHAPTER

F

work	measured quantity	
excavation	m³	earth excavated
pouring concrete	m³	concrete poured
pipe laying	lm	pipes laid
equipment assembly	t	equipment assembled

Fig. F3.4: Examples of measures for "Output measurement"

For services, such as designing, training, documenting, testing, etc., progress is not continuous. To measure the progress of such services "progress milestones" can be defined. These can be assigned a progress percentage rate by the project manager and the respective project team member. The manner in which the progress of the work packages is to be measured should be defined in the course of developing the work package specifications.

progress milestones of the work package: Developing construction plans		
number of the progress milestone	label of the progress milestone	progress (cumulative)
1	completion of the construction plans for internal review	40%
2	completion of the construction plans for discussion with the customer	50%
3	completion of the construction plans for tendering	80%
4	delivery of the construction plans	100%

Fig. F3.5: Progress milestones of the work package "Developing construction plans"

At the task level of a detailed schedule (such as a CPM schedule) a proportional relationship between progress and time can be assumed.

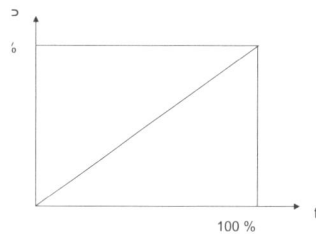

Fig. F3.6: Progress over time at the task level

At the work package level the assumption of a proportional relationship between progress and time cannot be upheld. As can be seen in the example in Figure F3.7, the result is a function with different slopes.

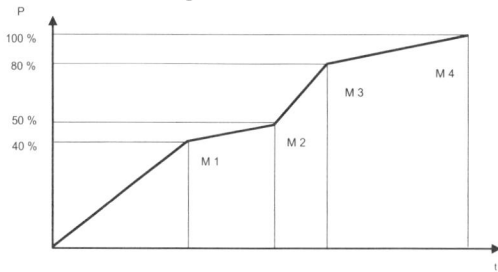

Fig. F3.7: Progress over time for the work package "Developing construction plans"

Progress control in the work breakdown structure and in the CPM schedule

The work breakdown structure is the most important method for visualizing project progress. For individual work packages the progress can be shown in the work breakdown structure. In Figure F3.8 the progress of the project "Realization e-application" is visualized at the first controlling date with the symbols "check" (for completed) and "waved line" (for in progress).

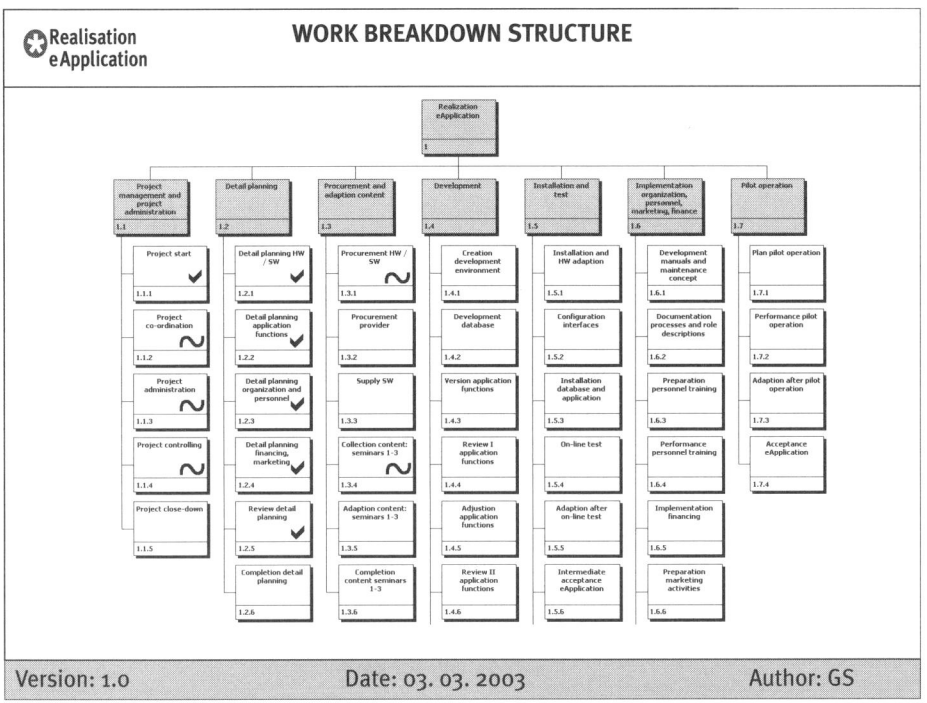

Fig. F3.8: Progress in the project "Realization e-application"

In the CPM schedule the progress of activities can also be depicted graphically. The symbols usually used for this are shown in Figure F3.9.

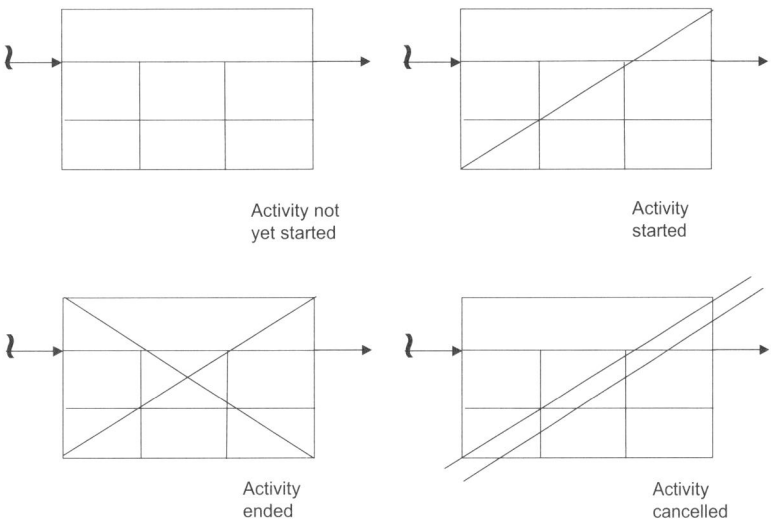

Activity not
yet started

Activity
started

Activity
ended

Activity
cancelled

Fig. F3.9: Documentation of the progress of activities in a CPM schedule

Weighting work packages with the relevance tree method

With the help of the relevance tree method the progress of the project phases and the overall project can be calculated on the basis of the progress of the work packages.

By assigning weights to the work packages and the work package groups in the work breakdown structure a relevance tree can be developed. The weights determine the relevance of work packages in regard to the work package groups to which they belong ("relative relevance") and in regard to the project ("absolute relevance"). The relevance of a work package for the entire project ("absolute relevance") can be calculated by multiplying the relative relevance of the work package with the absolute relevance of the work package group to which it belongs.

The sum of the weights of the work packages belonging to a work package group is equal to 100%. For all the work packages belonging to a work package group a uniform weighting criterion (e.g. the number of person-days) is to be used. If different resources are used in a project then the costs as a uniform weighting criterion are to be applied.

A general example of a relevance tree is shown in Figure F3.10. The relevance tree for the project "Realization e-application" is shown in Figure F3.11.

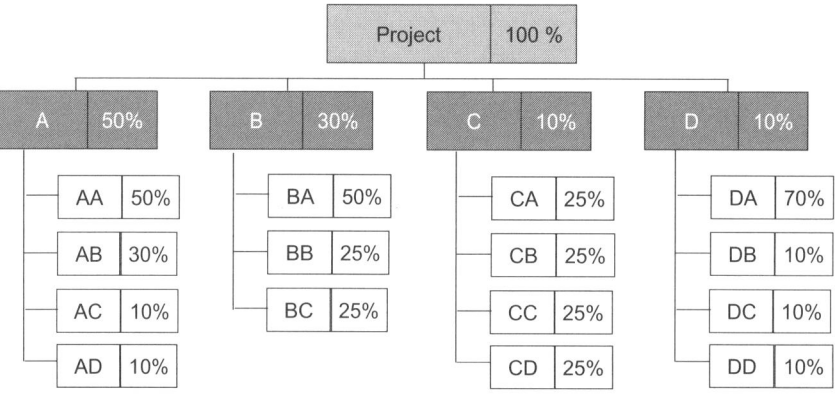

Fig. F3.10: Relevance tree (general)

The relevance of the individual work packages of the project "Realization e-application" was weighted based on the work package costs.

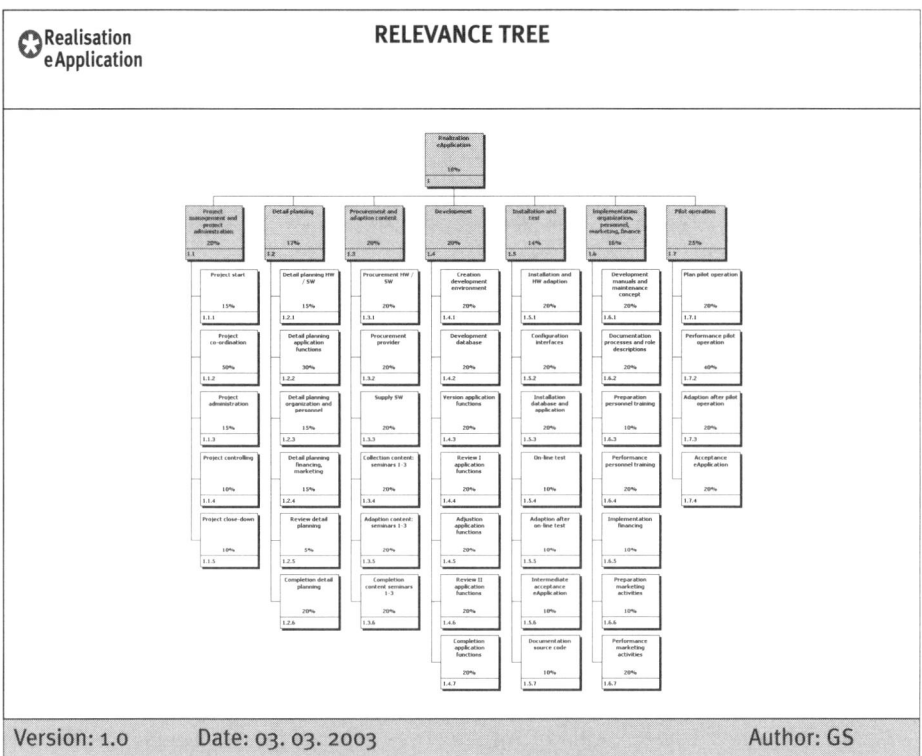

Fig. F3.11: Relevance tree for the project "Realization e-application"

Documentation of the changed scope in a work breakdown structure

Changes in the project objectives and in the objects of consideration require changes in the scope. The documentation of the changes in the scope of a project can be made in the work breakdown structure. Possibly, additional work packages to be performed must be added, work packages which are no longer necessary must deleted. These changes in the work breakdown structure should be visualized, e.g. by highlighting the changed work packages in grey (see Figure F3.12).

On the basis of the definition of the additional project objective "Concept for all RGC seminars prepared" an additional work package "Preparation concept for all RGC seminars" was to be added to the work breakdown structure of the project "Realization e-application".

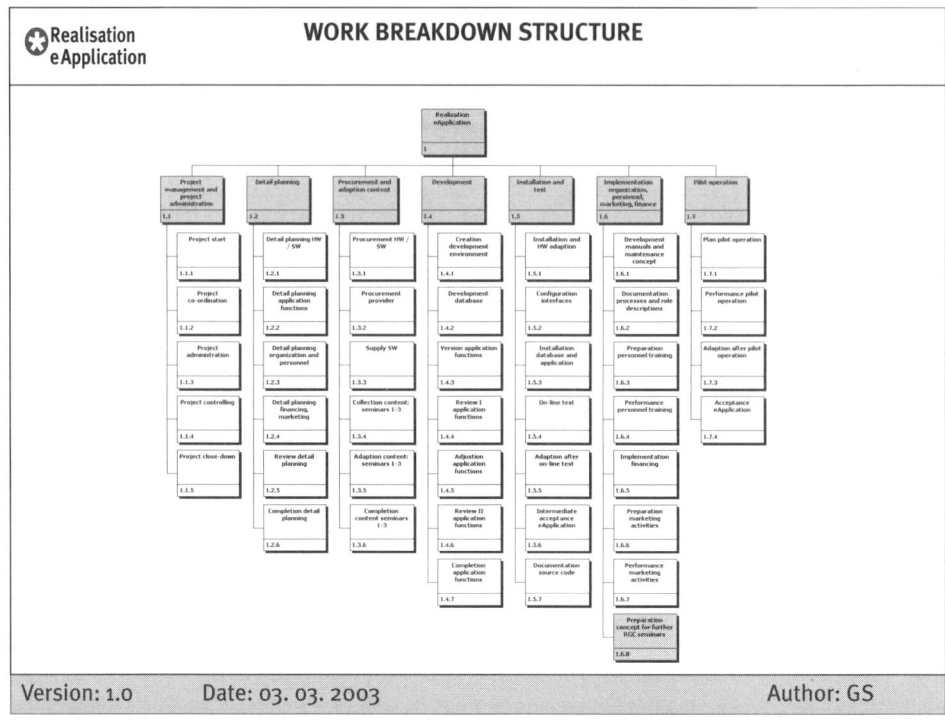

Fig. F3.12: Adapted work breakdown structure of the project "Realization e-application"

F3.2.3 Schedules

The objective of the control of the project schedule is to determine the scheduling status of the individual work packages of the project. In schedule control not all work packages of a project need to be considered, but only those performed in the period between two project controlling dates. Progress control forms the basis for schedule control because statements about schedule deviations can only be made in relation to progress. The schedule can be controlled in milestone plans, date lists, bar charts or CPM schedules.

An adaptation of the project schedules is necessary when schedule changes are made during the project performance and/or when the scope of the project has changed.

Schedule control in milestone plans and date lists

In milestone plans the actual dates of the milestones achieved are recorded and compared with the planned dates.

In date lists the actual dates of completed work packages and of those which are in progress are recorded and compared with the planned dates (see Figure F3.13). Possible deviations between the planned and actual dates must be interpreted in order to provide information for the planning of directive measures.

In the project "Realization e-application" there were delays and an increased demand of resources in the work packages 1.2.2 "Detailed planning application functions" and 1.2.6 "Completing planning". The cause for the delays was a knowledge deficit in the

CHAPTER

F

sub-team with regard to the interface between the application functions and the underlying database.

⊛ Realisation e Application		PROJECT DATE LIST				
work package		planned dates		actual dates and adapted dates as at 03. 03. 2003		
WBS code	label	start date	end date	start date	end date	
1.1.1	project start	01.02.03	10.02.03	01.02.03	10.02.03	
1.2.1	detailed planning HW + SW	11.02.03	15.02.03	11.02.03	15.02.03	
1.2.2	detailed planning application functions	11.02.03	20.02.03	11.02.03	23.02.03	
1.2.3	detailed planning organization and personnel	11.02.03	15.02.03	11.02.03	15.02.03	
1.2.4	detailed planning financing, marketing	11.02.03	15.02.03	11.02.03	15.02.03	
1.2.5	review detailed plans	21.02.03	22.02.03	24.02.03	24.02.03	
1.2.6	completion detailed planning	23.02.03	27.02.03	25.02.03	03.03.03	
1.3.1	procure HW + SW	26.02.03	20.03.03	28.02.03	24.03.03	
1.3.2	procure provider	28.02.03	28.02.03	04.03.03	04.03.03	
1.3.4	collect content: Seminars 1-3	21.02.03	02.03.03	24.02.03	04.03.03	
Version: 1.0		Date: 03. 03. 2003		Author: GS		

Fig. F3.13: Schedule controlling in the date list of the project "Realization e-application"

Schedule control in the bar chart

The schedule of a completed work package can be visualized in the bar chart through the positioning of the "actual" bar. For work packages being performed only the beginning date and the duration until the controlling date can be depicted.

Schedule control in the CPM schedule

In CPM schedules the logic as well as the dates can be controlled. The following data are to be recorded:

- actual structure of the project logic,
- start and end dates of the started and completed activities and
- actual durations of the activities (and of the relationships).

The structure of the logic is to be controlled directly in the CPM schedule.

Deviations in duration of activities in progress can be analyzed. By comparing the actual duration with the planned duration of an activity the deviation can be determined. The planned duration until the controlling date of an activity in progress is calculated by multiplying the progress by the planned duration of the activity.

✪ Realisation e Application		DEVATIONS IN THE ACTIVITY DURATIONS							
	activity	planned performance up to controlling date	actual performance up to controlling date	planned duration total (in days)	planned duration up to controlling date	"earned" duration until controlling date	actual duration up to controlling date	△ duration up to controlling date
WBS code	label
(1)	(2)	(3)	(4)	(5)	(6)=(3)x(5)=	(7)=(4)x(3)	(8)	(9)=(8)-(7)
1.2.2	detailed planning application functions	60%	50%	10	6	5	7	2
Version:			Date:			Author:		

Fig. F3.14: Analysis of deviations in activity durations for the project "Realization e-application"

Adapting the project schedules

Directive measures in project controlling may require adaptations of the milestone plan, the project date list, the bar chart and the CPM schedule necessary.

In the milestone plan the dates of the basis plan, the adapted planned dates and the actual dates can be depicted. This depiction visualizes the trend of change in the milestone plan. In the bar chart the basis plan and the adapted bars can be drawn one on top of the other.

The necessary adaptations of the milestone plan and the bar chart for the project "Realization e-application" are shown in Figures F3.15 and F3.16. The end date has shifted from August 30th to September 3rd 2003.

✪ Realisation e Application	PROJECT MILESTONE PLAN			
WBS code	label	basis plan	Planned plan as at 03. 03. 2003	actual
1.1.1	project assigned	01. 02. 2003		01. 02. 2003
1.2.6	detailed planning completed	27. 02. 2003	03. 03. 2003	03. 03. 2003
1.3.3	SW delivered	21. 03. 2003	31. 03. 2003	
1.4.3	draft application functions developed	04. 04. 2003	13. 04. 2003	
1.4.7	application completed	29. 04. 2003	10. 05. 2003	

1.5.6	intermediary approval of application made	20. 05. 2003	30. 05. 2003	
1.6.4	training performed	26.06. 2003	05. 07. 2003	
1.1.5	project approved	30. 08. 2003	10. 09. 2003	
Version: 1.0	Date: 03. 03. 2003	Author: GS		

Fig. F3.15: Adapted milestone plan for the project "Realization e-application"

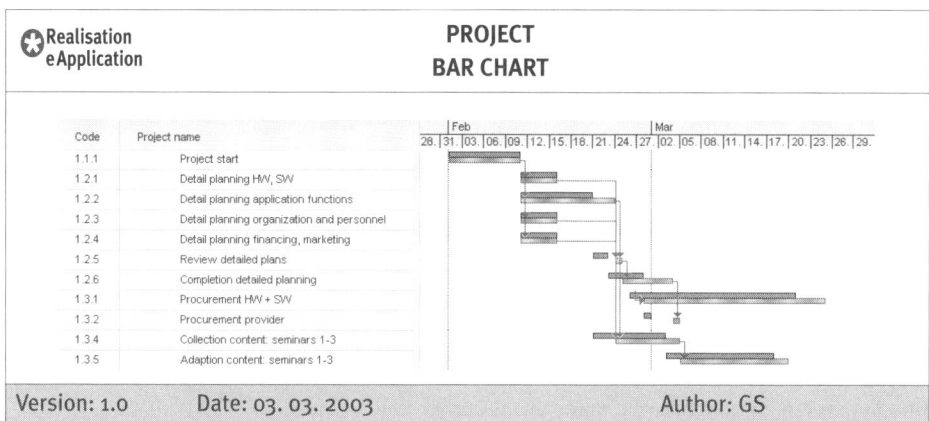

Fig. F3.16: Adapted bar chart for the project "Realization e-application"

F3.2.4 Costs and income

The objectives of the control the project costs and the project income are to record the costs and income up to the controlling date and to determine deviations in the project costs, the project income and the project success up to the controlling date.

The project resources are controlled by controlling the quantities within the framework of project cost control.

Adapting the project costs and the project income on the basis of changes in scope

If the scope of a project is changed during project performance the planned costs must be updated. By comparing the planned costs of the basis plan with the updated planned costs the effect of the scope changes on the project costs can be determined.

updated planned costs (after changes in scope)

– planned costs (basis plan)

cost deviation (on the basis of changes in scope)

If in a contracting project a scope change can be charged to a customer then the project income must also be updated.

changes made by customer		own changes	
to be charged	• supplements on the basis of customer requirements	to be charged	• chargeable improvements
not to be charged	• goodwill • sales promotion	not to be charged	• contract defect • non-chargeable improvements

Fig. F3.17. Charges for changes in the scope of contracting projects

Planned versus actual work package costs

The analysis of planned versus actual work package costs reveals how project costs and project success can be influenced by planning mistakes, changes in method, quantity changes, as well as changed prices, salaries, tariffs and calculation rates. The planned versus actual analysis also serves to identify the potentials for cost savings and income increases.

The prerequisite for an efficient control of the project costs are adequate project structures. The structure of the actual costs recording must correspond to the progress measurements. The actual costs should, therefore, be recorded relative to the work packages.

The actual costs for work packages are recorded on the basis of

- invoices from suppliers,
- material invoices,
- records of hours worked by the employees of the project organization,
- travel expense reports,
- etc.

A form for recording the working hours of an employee in a project-oriented company is shown in Figure F3.18.

record of hours worked by...														
project or cost center	code	hours per month												Sum
		1	2	3	4	5	6	7	8	9	10	11	12	
hours for contracting projects														
project 1	220	45												45
project 2	230	10												10
project 3	240	25												25
hours for internal projects														
product development project	410	40												40
marketing project	420	10												10

quality management project	430	25												25
hours for other cost centers														
department 1	520	5												5
department 2	550	10												10
sum hours per month		180												180

Fig. F3.18: Record of hours worked by an employee

Possible problems in recording actual costs are:

- no uniform structures for planned and actual costs,
- no clear difference between costs and cash outflows,
- documents to be reviewed remain lying on the project manager's desk due to lack of priority,
- documents to be reviewed are often passed back and forth between different cost centers,
- internal personnel costs are not defined as part of the project costs,
- records of hours worked are turned in late,
- processing of data from accounting not online but only once a month,
- no recording of the costs from suppliers before invoicing.

A special problem in the recording of actual costs is the recording of supplier costs. The recording of the actual supplier costs must correspond to the progress. This requires that supplier performance is also structured into work packages and integrated into the work breakdown structure. "Disposed costs" which are possibly considered, are to be divided according to their proportion of actual costs and remaining costs.

The performance of a planned versus actual analysis for the project costs requires the recording of the progress and a calculation of the earned value (see Chapter F3.2.5).

Adapting the project costs

The project costs are adapted by determining the remaining costs for the work packages in progress and by possibly adapting the rest costs of the work packages still to be fulfilled. The determination of the remaining costs is not an extrapolation of work package costs already incurred, but a systematic adaptation of the costs considering the improved level of information and the directive measures which have been set (changed use of resources, consideration of new price information, etc.).

A relatively inexact adaptation of the work package costs can be made on the basis of an estimation of the remaining costs still to be expected, without defining the already achieved progress. Planning the remaining costs of a work package requires information about the progress, as it is presented in the earned value analysis (see Chapter F3.2.5).

The project costs are to be adapted by the project team members and the project manager. Each project team member determines the remaining and/or total costs of the work pack ages. In the process, possible cost deviations of work packages to be performed in the future are to be pointed out as early as possible.

The adapted cost plan of the project "Realization e-application" is shown in Figure F3.19. The adapted planned costs of Phase 1.2 "Detailed planning" have increased by € 4,300. The additional costs were caused by an external software expert and by an increased number of person-days for internal personnel.

Work package		Cost type	Planned costs as at 01.02.	Planned costs as at 03.03.	Actual costs as at 03.03.
WBS	Label				
I.1.1	Project start	Personel	€ 2.000,00	€ 2.000,00	€ 2.000,00
I.1.2	Project coordination	Personel	€ 7.200,00	€ 7.200,00	€ 1.200,00
I.1.3	Project administration	Personel	€ 1.440,00	€ 1.440,00	€ 266,00
		Material	€ 1.000,00	€ 1.000,00	€ 100,00
I.1.4	Project controlling	Personal	€ 2.400,00	€ 2.400,00	€ 360,00
I.1.5	Project close down	Personel	€ 1.600,00	€ 1.600,00	
Phase 1.1	Project management and project administration		€ 15.640,00	€ 15.640,00	€ 3.926,00
I.2.1	Detailed planning HW, SW	Personel	€ 400,00	€ 400,00	€ 400,00
		Services	€ 1.000,00	€ 1.000,00	€ 1.000,00
I.2.2	Detailed planning application functions	Personel	€ 800,00	€ 1.500,00	€ 1.500,00
		Services	€ 2.000,00	€ 2.800,00	€ 2.800,00
I.2.3	Detailed planning organization and personel	Personel	€ 1.200,00	€ 1.200,00	€ 1.200,00
I.2.4	Detailed planning financing, merketing	Personel	€ 1.200,00	€ 1.200,00	€ 1.200,00
I.2.5	Review detailded plans	Personel	€ 800,00	€ 800,00	€ 800,00
		Services	€ 500,00	€ 500,00	€ 500,00
I.2.6	Completion detailed planning	Personel	€ 400,00	€ 1.200,00	€ 1.000,00
		Services	€ 1.000,00	€ 3.000,00	€ 700,00
Phase 1.2	Detailed planning		€ 9.300,00	€ 13.600,00	€ 11.100,00
I.3.1	Procure HW + SW	Personel	€ 800,00	€ 800,00	€ 400,00
I.3.2	Procure provider	Personel	€ 400,00	€ 400,00	
I.3.3	SW delivery	Material	€ 3.000,00	€ 3.000,00	
..	... etc.				

Realisation e Application — PROJECT COST PLAN

Version: 1.0   Date: 03. 03. 2003   Author: GS

Fig. F3.19: Adaptation of the costs of the project "Realization e-application"

F3.2.5   Earned value

By calculating the "earned value" in the earned value analysis it is possible to give the progress of a work package, a project phase or a project a monetary value. This creates a basis for the estimation of remaining costs.

In the earned value analysis the relationship between progress, costs and schedule is considered. An integrated analysis of these objects can be made for the entire project, but also for project phases and individual work packages. The earned value analysis is, above all, a controlling method. On the basis of the earned value analysis the planned costs can be adapted.

First, the earned value approach for a global analysis at the project level is described. Following that, the operational use of the earned value analysis at the work package level is shown.

Earned value analysis at the project level

A planned versus actual analysis is only reasonable if the costs to be compared are based on the same performance basis. A comparison of planned costs (for the planned performance) with the actual costs (for the actual performance) expresses very little. As a uniform basis for the comparison of planned and actual costs the actual performance is to be used.

The planned costs as determined during cost planning assume a planned performance (up to a controlling date). On the controlling date the actual work completed ("actual performance"), which leads to the actual costs, is measured. Knowledge of the actual performance enables calculating the originally planned costs for this actual performance. These planned costs of the actual performance are referred to as "earned value". In order to detect a cost deviation the earned value is to be compared with the actual costs.

The actual performance is recorded at the lowest level of the work breakdown structure. The earned value analysis presupposes the possibility of the aggregation of data through the use of the relevance tree. In regard to the relationship between performance, costs and schedule the following assumptions are made:

- at all levels of consideration (project, project phase, work package, activity) a proportionality between progress and costs exists,
- proportionality between progress and time is not assumed.

The data of the planning and controlling of performance, costs and schedule enable the development of the following cumulative curves:

- curve of the planned costs,
- curve of the planned performance (this corresponds to the curve of the planned costs),
- curve of the actual costs,
- curve of the actual performance and
- curve of the earned value (this corresponds to the curve of the actual performance).

These curves can be depicted with the project duration on the x-axis and the progress and/or the project costs on the y-axis of an earned value graph. On the basis of the assumption of proportionality between performance and costs, both the progress as well as the costs can be shown on the y-axis.

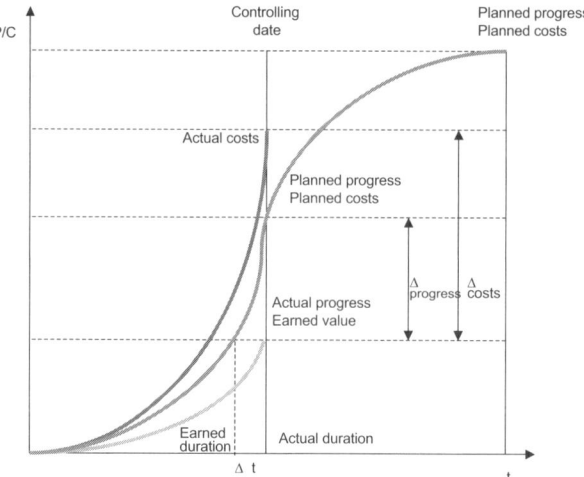

Fig. F3.20: Graphic depiction of the earned value analysis

On a controlling date

- the actual performance of the project and the planned performance (results in delta performance at the controlling date) and

- the actual costs of the project and the earned value (results in delta project costs at the controlling date)

can be compared.

By projecting the actual performance onto the planned performance curve the planned duration (for the actual performance of the project) can be graphically determined. By comparing the planned duration with the actual duration a (theoretical) deviation in the project duration ($\triangle$t) can be determined.

Earned value analysis at the work package level

The earned value of a work package can be calculated by multiplying the planned costs of the work package by the actual performance (in percent).

status of the work package	earned value of the work package
completed (P = 100%)	planned costs of the work package x 100%
in progress (0% < P < 100%)	planned costs of the work package x actual performance in%
not yet started (P = 0%)	costs of the work package are not yet considered

Fig. F3.21: Determining the earned value per work package

The application of the earned value analysis at the work package level is shown in Figure F3.22 for several work packages of the project "Realization e-application".

Realisation e Application — EARNED VALUE ANALYSIS

Work package WBS Code	Work package label	Planned progress as at 03.03.	Actual progress as at 03.03.	Planned costs as at 01.02.	Earned value as at 03.03.	Actual costs as at 03.03.	Cost deviation as at 03.03.	Remaining costs as at 03.03.	Total planned costs as at 03.03.	Total cost deviation as at 03.03.
1.1.1	Project start	100	100	€ 2.000,00	€ 2.000,00	€ 2.000,00			€ 2.000,00	
1.1.2	Project co-ordination	15	15	€ 7.200,00	€ 1.080,00	€ 1.200,00	€ 120,00	€ 6.000,00	€ 7.200,00	
1.1.3	Project administration	15	15	€ 2.440,00	€ 366,00	€ 366,00		€ 2.074,00	€ 2.440,00	
1.1.4	Project controlling	15	15	€ 2.400,00	€ 360,00	€ 360,00		€ 2.040,00	€ 2.400,00	
1.1.5	Project close-down			€ 1.600,00				€ 1.600,00	€ 1.600,00	
Phase 1.1	Project management and project administration			€ 15.640,00	€ 3.806,00	€ 3.926,00	€ 120,00		€ 15.640,00	
1.2.1	Detail planning HW / SW	100	100	€ 1.400,00	€ 1.400,00	€ 1.400,00			€ 1.400,00	
1.2.2	Detail planning application functions	100	100	€ 2.800,00	€ 2.800,00	€ 4.300,00	€ 1.500,00		€ 4.300,00	€ 1.500,00
1.2.3	Detail planning organization and personnel	100	100	€ 1.200,00	€ 1.200,00	€ 1.200,00			€ 1.200,00	
1.2.4	Detail planning financing, marketing	100	100	€ 1.200,00	€ 1.200,00	€ 1.200,00			€ 1.200,00	
1.2.5	Review detail planning	100	100	€ 1.300,00	€ 1.300,00	€ 1.300,00			€ 1.300,00	
1.2.6	Completion detail planning	100	100	€ 1.400,00	€ 980,00	€ 1.700,00	€ 720,00	€ 2.500,00	€ 4.200,00	€ 2.800,00
Phase 1.2	Detail planning			€ 9.300,00	€ 8.860,00	€ 11.100,00	€ 2.220,00		€ 13.600,00	€ 4.300,00
1.3.1	Procurement HW / SW	25	15	€ 800,00	€ 120,00	€ 400,00	€ 280,00	€ 400,00	€ 800,00	€ 2,00
1.3.2	Procurement provider	100		€ 400,00				€ 400,00	€ 400,00	
1.3.3	Supply SW	5		€ 3.000,00				€ 3.000,00	€ 3.000,00	

Version: 1.0   Date: 03. 03. 2003   Author: GS

F3.22: Earned value analysis at the work package level for the project "Realization e-application"

The earned value analysis enables the determination of cost deviations at a controlling date. It also offers an adequate basis for the determination of remaining costs for work packages in progress. This cost determination requires clarity about the remaining performance to be fulfilled.

Analysis of cost deviations at the work package level

By comparing the actual costs to the actual prices, actual costs to planned prices and the earned value (planned costs to planned prices) the (total) cost deviation of a work package can be differentiated into a price and quantity deviation. The price deviation is that part of the cost deviation which results from changes in the prices actually charged. The quantity deviation is that part of the cost deviation which results from changes in quantity.

Only the differentiating into price and quantity deviations enables an analysis of the causes of the cost deviation and a cost responsibility accounting.

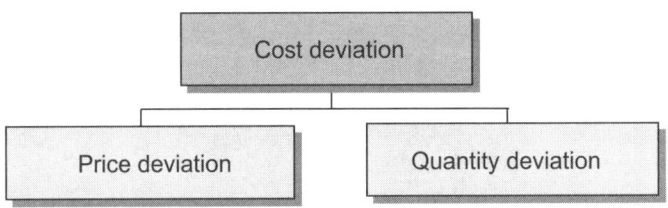

Fig. F3.23: Differentiating the cost deviations into price and quantity derivations

The cost deviation of a work package is determined based on the earned value analysis.

	actual costs (at actual prices)
–	earned value (planned costs at planned prices)
	cost deviation ($\triangle C$)

On the basis of this information the price deviation can be determined by comparing the actual costs at actual prices with the actual costs at planned prices:

	actual costs at actual prices (actual quantity x actual price)
–	actual costs at planned price (actual quantity x planned price)
	quantity deviation ($\triangle Q$)

F3.2.6   Risks and discontinuities

The controlling of project risks comprises the controlling of the set risk management measures, identification of new risks, evaluation of the newly identified risks, adaptation of the evaluation of the older, but still active risks and the planning and performing of new risk management measures.

For the fulfillment of these tasks, basically the methods employed in the project start process are to be used. A special method for controlling risks in projects is the risk trend analysis (see Figure F3.24). In the risk trend analysis the changes in the expected values of risks are documented over time (for one controlling date to the next). The depiction shows whether the risk management measures were successful.

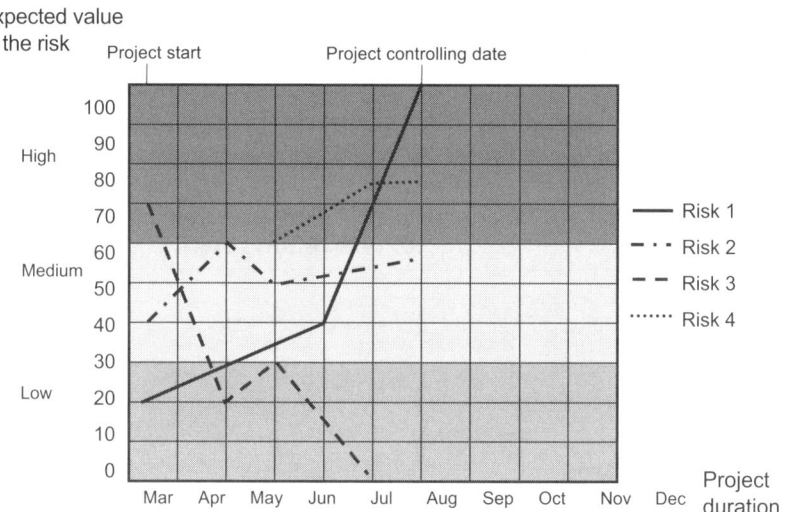

Fig. F3.24: Risk trend analysis

The controlling of potentials for project discontinuities comprises the identification of new weak signals and possibly the adaptation of the plans developed to provide for the resolving of a project discontinuity. The newly identified weak signals may necessitate the development of further project scenarios or an adaptation of the defined project scenarios. Usually, however, it is not necessary to apply the scenario technique during each controlling cycle.

F3.2.7 "Social" project controlling

"Social" project controlling comprises the controlling of the project organization, the project culture and the relationships of the project to relevant environments. The functionality of the organizational structures is to be checked, the developed project culture is to be reflected and the appropriateness of the measures for designing the project environment relationships are to be controlled. Changes in the project objectives and the objects of consideration or simply the improved level of information may require an adaptation of the project organization, the project culture and the relationships to the relevant environments.

Controlling the project organization

The controlling of the project organization relates to the design of the project organization chart, the description of the project roles, the composition of the project team and the project owner team, the communication structures and the project rules. The relationships within the project team can also be subjected to a "social" controlling through reflection and mutual feedback among the project team members (see Chapter D2: Leadership in projects).

Examples of adaptations of the project organization chart and the project communication structures of the project "Realization e-application" are shown in Figures F3.25 and F3.26. In the adapted project organization chart an additional database expert is depicted, in the figure showing the adapted communication structures the partial presence of the project owner team at the project start and project close-down workshop is documented.

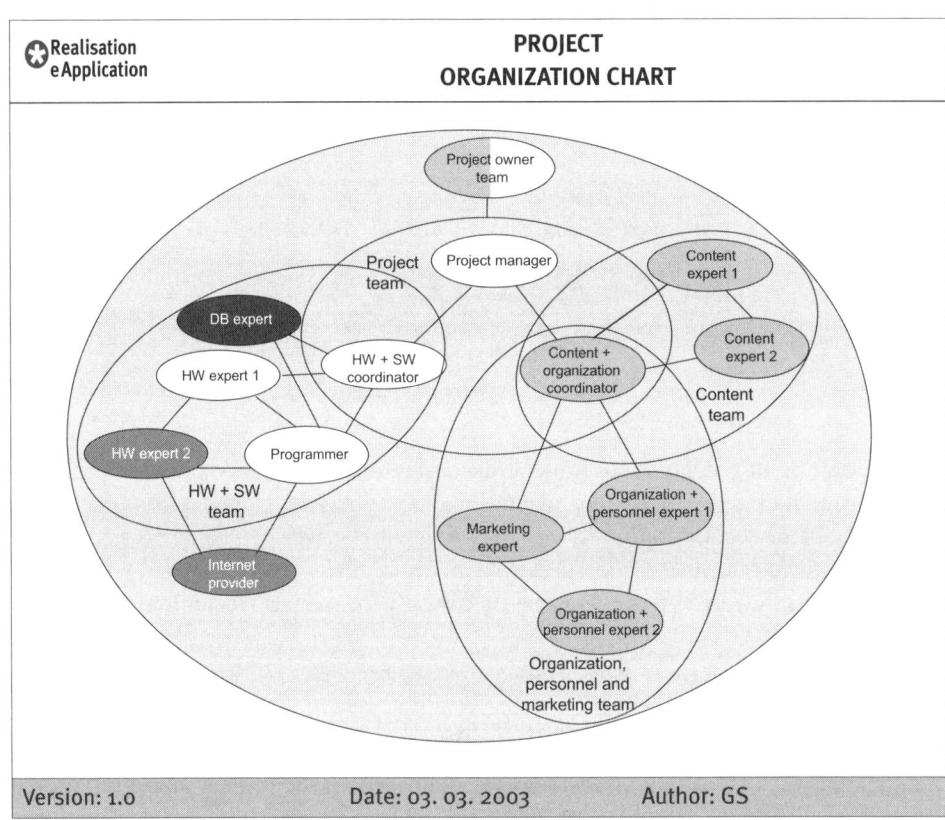

Fig. F3.25: Adapted project organization chart of the project "Realization e-application"

✪ Realisation e Application	PROJECT COMMUNICATION STRUCTURES			
meeting	content	participants	frequency	adaptation as at 03. 03. 2003
project start workshop	information on the project objectives, the project structures, strategies, organization, developing the "big project picture"	project team	1x in the project start process	project start workshop with project owner team
project owner team meeting	project status and project preview, context information, current problems, strategic decisions	project owner, project manager, selected project team members	every 2-3 weeks	
project team meeting	project status and project preview, review, definition of project strategies, operational decisions	all project team members, possibly selected representatives of environments	every 2-3 weeks	
sub-team meeting	problem-solving	sub-team members, possibly project manager and selected representatives of environments	1x per week	
poject close-down workshop	reflection, feedback, dissolving the social system, clarifying remaining tasks	project team and project owner	1x in the project close-down process	project close-down workshop with project owner team
Version: 1.0	Date: 03. 03. 2003		Author: GS	

Fig. F3.26: Adapted communication structures of the project "Realization e-application"

Controlling the project culture

In projects the elements of the project culture (project name, project mission statement, project values, project slogans) should be relatively stable. Once established, a project name, for example, should not be changed during a project controlling, even if a more attractive name comes up. In case the adaptation of the project culture elements appears to be necessary, then this must not be shied at. Changes in the project are to be promoted!

Controlling project environment relationships

Controlling of the relationships to the relevant project environments includes the following tasks:

- analysis of the status of the relationships of the project to individual project environments (strengths and weaknesses in the relationship, changes in the importance of the environment and the intensity of the communication with the environment, demand for an adaptation of the relationship),

- planning of strategies and measures for the new design of existing relationships,

- identification of environments which are no longer relevant,

- identification of additional relevant environments and

- planning of strategies and measures for the design of relationships to additional relevant environments.

CHAPTER

F

In Figure F3.27 the changes in the relationships of the project "Realization e-application" to existing environments and the consideration of the additional environment "DB expert" are shown.

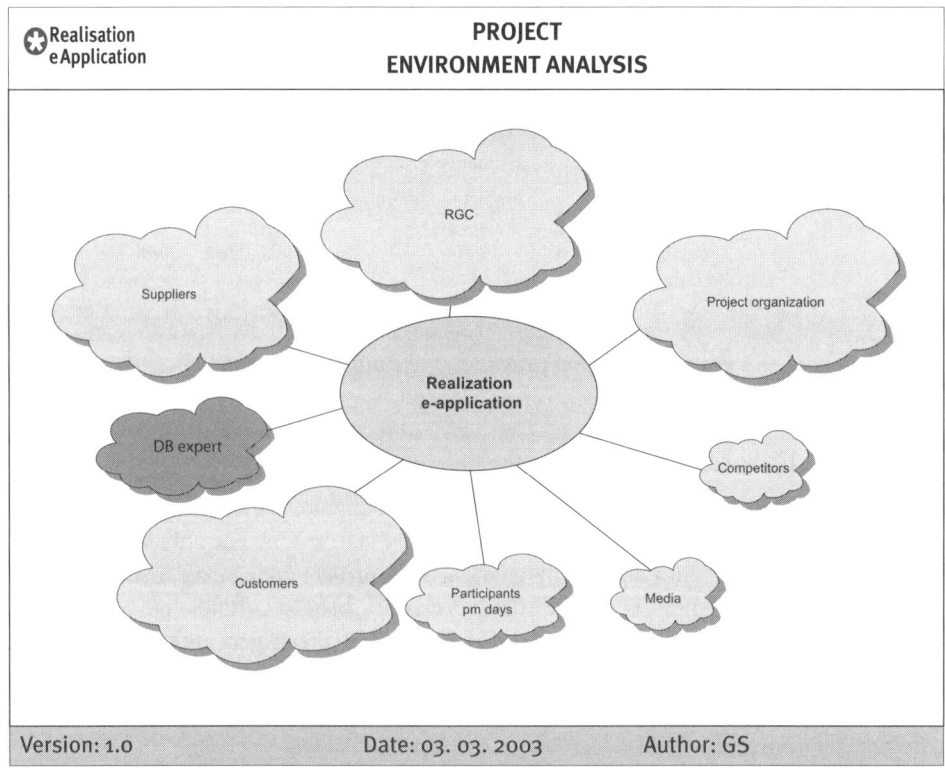

Fig. F3.27: Controlling of the project environment relationships of the project "Realization e-application"

F3.2.8 Controlling the business case analysis

In the project controlling process the business case analysis is also to be controlled and adapted if necessary. Assumptions made for the post-project phase are to be questioned and possibly changed, the realization of the company strategies by the investment and its benefits are to be verified.

The project organization also has a responsibility for the controlling of the business case during the performance of the project.

F 3.3 Methods for planning directive measures

Directive measures are to be planned and implemented on the basis of the project control. Interpretations of changes and deviations in the forms of project control provide the foundation.

Directive measures can lead to changes in the project objectives and the objects of consideration, changes in the project costs or the project duration, the redesign of the project organization as well as to changes in the project environment relationships.

Possible directive measures are, for example, working overtime to shorten the project duration, using qualified employees to improve the quality of the work packages, or using new technologies in order to comply with changed project objectives. Since the shortening of the project duration by the use of overtime, outside services, etc., usually leads to increased costs, changes in the sequencing of activities should also be considered. One such possibility is the overlapping of activities on the critical path, for example. If additional services are to be fulfilled in contracting projects, it should be checked whether these costs can be passed on to the customer.

In planning measures for project direction to-do lists, as described in Chapter F2, can be used. The to-do list which was the result of the planning of directive measures in the project "Realization e-application" is shown in F3.28.

✪ Realisation e Application	TO DO-LIST			
WP#	What?	Who?	Until when?	status
project management	new WP "1.6.8": Prepare concept for further RGC seminars			
project management	reinforced controlling of the performance in the sub-team "HW + SW"			
project management	adapt project plans			
project management	develop know-how: New sub-team member: DB expert (budget increase)			
Version: 1.0	Date: 03. 03. 2003	Author: GS		

Fig. F3.28: To-do list for project direction (project "Realization e-application")

F 3.4   Controlling reports

project score card

The project score card is a project management method for describing and communicating the project status on a controlling date. The project score card provides a holistic view of the project status. The status of the individual objects of consideration of project management is scored in an integrated form. Relationships between the criteria can be considered. (For example, the relationship to the customer cannot be very good when the progress of the project is bad.)

In an analogy to the balanced score card model by Kaplan and Norton (see Excursus F1) several quantitative and qualitative criteria for scoring the project status are brought together in the project score card. The criteria to be considered are dependant on the project management approach used. The project score card of the project "Realization e-application" shown in Figure F3.29 is based on *ROLAND GAREIS Project and Programme Management*®. Depending on the type of project and the specific project requirements some criteria of the project score card can be dropped or added. For example, for contracting projects the consideration of the business case analysis can be dropped. The relevant project environments to be considered must always be defined in a project-specific manner.

The individual criteria on the project score card can be evaluated by means of the traffic signal colors red, yellow and green or with a 5-color scale (as in Figure F3.29). The 5-color scale enables a better differentiation of the "scores". The entire status of a project can be evaluated either globally or with the help of an algorithm. An example of an algorithm could be: "If 2 or more criteria are red, then the total score of the project is red". If the project score card is not part of a comprehensive project progress report, the individual scores should be briefly interpreted. The causes for the scoring should be given.

To observe the development of the criteria on the project score card the scores from several controlling dates can be displayed next to each other.

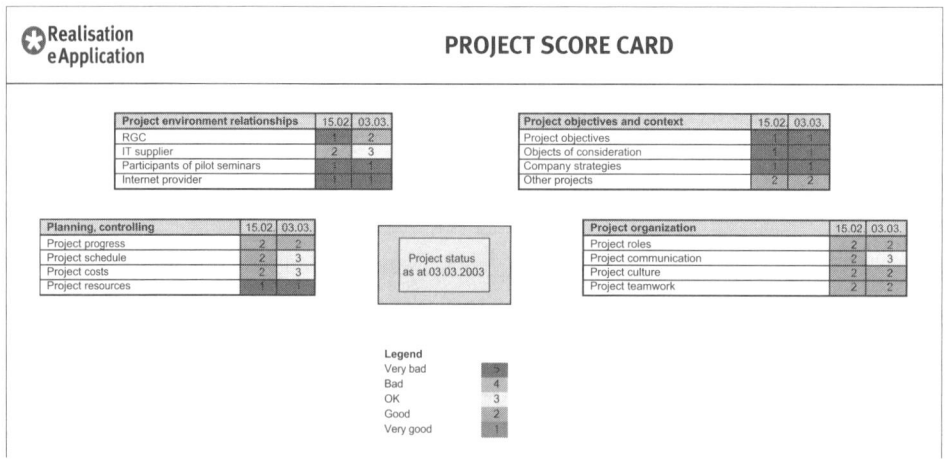

Interpretation of the project score card for the project "Realization e-application" as at March 3rd 2003 compared to February 15th 2003:	
planning, controlling: • project progress good, • project schedule and project costs were exceeded; additional controlling necessary project environment relationships: • relationship to IT supplier is OK; inclusion of a DB expert in the project team necessary	project objectives and context: • relationship to other projects is good project organization: • project communication is OK, inclusion of the project owner team in the project start workshop and in the project close-down workshop necessary
Version: 1.0 Date: 03. 03. 2003 Author: GS	

Fig. F3.29: Project score card of the project "Realization e-application"

The visualization of the project status in a project score card, taking into account different criteria, poses a new, important communication instrument in project management. A draft version of the project score card can be developed by the project manager. This draft version is to be adapted in the project team and then presented to the project owner team.

CHAPTER

F

⋯⋗ EXCURSUS F1: The balanced score card model by *Kaplan* and *Norton*

The idea for the balanced score card (BSC) came into being in the beginning of the 1990s. *Kaplan* and *Norton* recognized the necessity to provide a company not only with unilateral analyses by means of financial ratios (e.g. turnover, financial contribution and cash flow), but that additional factors which determine the success of a company, such as qualifications of the employees, customer satisfaction, innovation, delivery times, complaints, etc., also need to be considered.

The objective of the balanced score card model is not to create a new system for analyzing numbers, but to provide an integrated management approach from which balanced control units can be derived. For this, material as well as immaterial factors are considered. A balance between monetary and non-monetary, between short- and long-term and between internal and external factors should be developed. The balanced score card considers companies from the financial perspective, the customer perspective, the process perspective and the employee and/or potential perspective.

The **financial perspective** deals with the economic and financial objectives of a company and considers various characteristics of profitability. Profitability ratios and income ratios are the focus of attention.

The **customer perspective** considers factors such as customer loyalty, market share, customer satisfaction, cooperation with partners, etc.

In the **process perspective** only those processes are considered which have an influence on the achievement of the company's financial objectives or on customer satisfaction. The cost structure, delivery time or the manner of dealing with complaints in relation to these processes is considered.

The **employee and/or potential perspective** is of special importance because of its long-term effect on the company's success. The key to success of every company is investing in knowledge and using the abilities of each employee. Employee satisfaction, motivation and continuous education, and employee development are the main factors in this perspective.

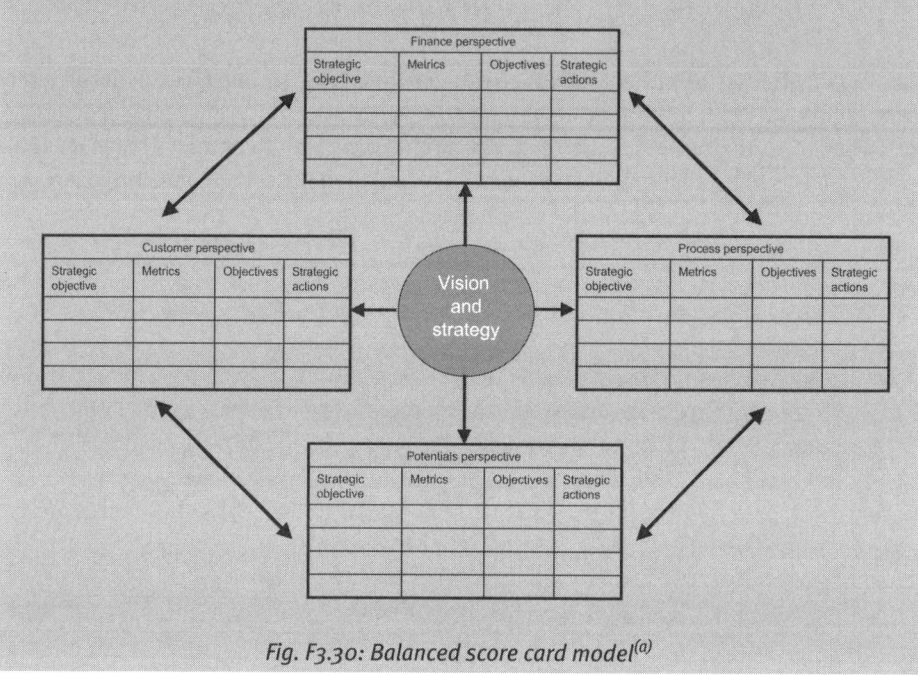

Fig. F3.30: Balanced score card model[a]

[a] *Kaplan, R., Norton, P.,* (Balanced Scorecard), 1997, p. 46.

Project progress report

A project progress report is a description of the project status at a controlling date. For each object of consideration of project management (project objectives, project scope, project schedule, etc.) possible deviations are to be interpreted and the agreed directive measures documented. The appendix of the project progress report also contains adapted project plans.

Apart from the project team members, the main target group for the periodical project progress report is the project owner team. The project progress report represents an important project-internal communication and marketing instrument. For contracting projects the project progress report also serves to inform the customer.

The project progress report is a formal, relatively extensive report based on project controlling. Therefore, it should only be developed in longer intervals (about once a month). To satisfy information requirements additional project score cards with short interpretations can developed at short notice (about every 2 weeks). With this combination of different project controlling reports the reporting effort can be minimized.

The structure of a project progress report should be oriented on the objects of consideration of project management which are being reported. After a short verbal description of the overall project status, the status of the project objectives and the objects of consideration, the project progress, the project schedule, the project costs, the project resources and project income, the project organization and the project culture as well as the project environment relationships are to be described. The assessment of the entire project status should correspond to the project score card.

The project progress report should be brief. Thus, it should be formulated in key words. To ease communication punctuation should be used. To supplement the verbal descriptions the project score card, the adapted project plans and, possibly, the deviation trend analysis should be included in the appendix of the project progress report.

An extract of a project progress report from the project "Realization e-application" is given in Figure F3.31.

CHAPTER

F

✪ Realisation e Application	**PROJECT PROGRESS REPORT** **as at 03. 03. 2003**			
❑ very good	❑ good	x OK	❑ bad	❑ very bad
overall status of the project				

- additional objectives: Preparation of a concept for all RGC seminars
- problems in planning because of lack of know-how
- schedule delay approx. 10 days and cost increase of EUR 4,000
- cooperation between RGC & IT supplier somewhat strained

status: Project objectives and objects of consideration

- additional objectives: Preparation of a concept for further RGC seminars agreed
- additional object of consideration: Concept for further RGC seminars

status: Project progress

- 1.1.1: completed
- 1.1.3-1.1.5: started
- 1.2.1-1.2.5: completed
- 1.2.2 and 1.2.6: difficulties because of know-how deficit about the new DB solution
- 1.2.6, 1.3.1, 1.3.4: started

status: Project schedule
• schedule delay of approx. 10 days • reason: Difficulties in WPs 1.2.2 and 1.2.6 • all other WPs are running as planned
status: Project costs, project resources, project income
• increased resource requirement in WP 1.2.2 and WP 1.2.6 • estimation: Further increased resource requirement in 1.2.6 • cost increase (actual + estimation) of EUR 4,000
status: Project environments, relationships to other projects
• RGC: Further course of the project unclear because of difficulties at the beginning • IT supplier: Conflict of interests on cost reduction – know-how deficit • internet provider: Configuration is no problem
status: Project organization and project culture
• organization basically OK • communication between RGC and IT supplier unclear at the beginning, but OK now
directive measures
• development of know-how: New sub-team member: DB expert (budget increase) • new WP "1.6.8": Prepare concept for further RGC seminars • increased controlling of the performance in sub-team "HW + SW" • adaptation of project plans
appendix
• project score card • adapted project plans
Version: 1.0   Date: 03. 03. 2003   Author: GS

Fig. F3.31: Extract of the project progress report of the project "Realization e-application"

Deviation trend analysis

Deviation trend analyses are graphic representations of deviations expected at controlling dates. Trend analyses can be made for different objects. Usually, they are made for the project costs, the project income or the project schedule. They represent the results of the adaptations at the controlling dates.

In Figure F3.32 a deviation trend analysis for the project end date is depicted. It is apparent that for each controlling date a later project end date is forecast.

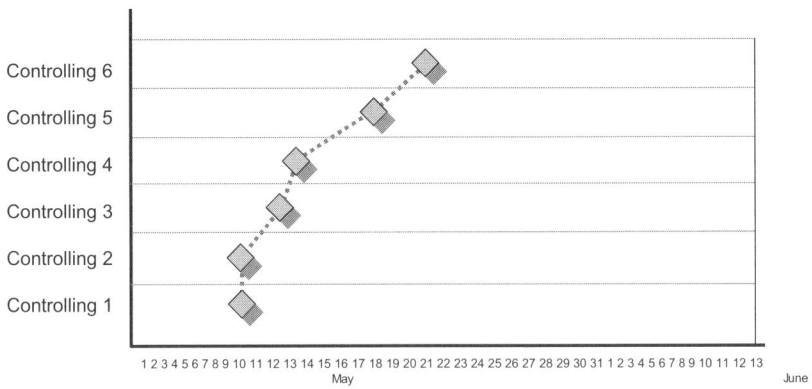

Fig. F3.32: Deviation trend analysis for the project end date

References

Kaplan, R., Norton, P., (Balanced Scorecard) Balanced Scorecard, Strategien erfolgreich umsetzen, Schäffer-Poeschel Verlag, Stuttgart, 1997

F 4 Methods for resolving a discontinuity of a project or a programme

F 4.1 Overview of methods for resolving a discontinuity of a project or a programme

In contrast to avoiding and/or promoting and providing for project discontinuities, which is to be performed in the project start process and the project controlling process, the resolution of a project discontinuity is a project management sub-process in its own right. This sub-process is to be performed in case an acute project discontinuity arises.

To resolve a project discontinuity the following tasks must be fulfilled:

- defining the project discontinuity,
- planning and performing immediate measures for resolving the project discontinuity,
- cause analysis,
- planning alternative strategies for resolving the project discontinuity,
- planning and performing additional measures for resolving the project discontinuity and
- ending the project discontinuity (see Chapter E1.6).

Methods which can be used for resolving project discontinuities are the definition the project discontinuity, the cause analysis and the planning of alternative resolution strategies, the planning and the controlling of immediate and additional measures, and ending the project discontinuity. These methods are described below.

The methods for resolving a project crisis do not differ from those of resolving a project chance. The process of resolving a project chance proceeds analogously to the process of resolving a project crisis. The two processes only differ with regard to the causes – existential threat in contrast to new potentials – and in their objectives.

A checklist for selecting methods to resolve a project discontinuity is depicted in Figure F4.1.

methods for resolving a project discontinuity	small project	project
defining the project discontinuity	must	must
planning immediate measures	must	must
cause analysis	must	must
planning alternative resolution strategies	must	must
planning of additional measures	must	must
ending the project discontinuity	must	must

Fig. F4.1: Selection of methods to be used for resolving a project discontinuity

In the following, methods for resolving a project chance are illustrated, taking the project "Realization e-application" as an example. The resolution of a crisis of the project "Smoke signals" was already described in Chapter E1.6 as a case study.

F 4.2   Defining a project discontinuity

The definition of a project discontinuity is a central task in the process of resolving a project discontinuity. The existence of a project discontinuity cannot be measured on the basis of objective criteria but is to be defined by means of a communication process in the project.

The resolution of a project discontinuity requires the conscious construction of the discontinuity as a new project reality. By defining a discontinuity the "crisis" or the "chance" is differentiated from the normality in the project. (Radical) measures for resolving the project discontinuity are legitimized only once they have been defined.

Even though the definition of a project crisis or a project chance is a situational construction, the communication process can be supported by using company-specific checklists, as depicted in Figure F4.2, for example.

CHAPTER

F

project crisis:
threat of occurrence of one or several of the following events:
• substantial deterioration of the results of the business case analysis (e.g. negative net present value) • substantial increase in the project scope (approx. by 50%) • substantial exceeding of the project costs (approx. by 50%) • substantial exceeding of the project duration (approx. by 50%) • absence of a substantial project partner • illiquidity of the customer (in contracting projects) • etc.
project chance
potential of occurrence of one or several of the following events:
• substantial improvement of the results of the business case analysis (e.g. 50% of the positive net present value) • learning potentials or reference potentials due to the use of a new technology • substantial extension of the scope of a customer contract (approx. 50% in contracting projects) • etc.

Fig. F4.2: Checklist for defining project crises and project chances

Whether a project situation is defined as a discontinuity or not cannot be decided by the project manager or the project team. It must be prepared by the project team and decided by the project owner team. The definition of a project discontinuity is a power-influenced negotiation process, in which the decision-makers must be involved.

In an Austrian engineering construction company projects with a contract value of € 30 million are considered a "flop" when a loss of approx. € 1 million is to be expected. The term "crisis" is not explicitly used. In the company practice the use of the term "crisis" is treated with reserve. However, terms like "flop", for example, are used, which communicate the same idea in a softer way.

✪ Realisation e Application	DEFINITION OF A PROJECT CHANCE
• In the phase "Procurement and adaptation content" of the project "Realization e-application" it was detected that synergies could be created by a joint development of e-applications for seminars of RGC, for courses at the PROJECT MANAGEMENT GROUP (PMG) from the University of Economics and Business Administration, Vienna, and for the university programme "International project management".	
• The potentials of a uniform solution for the three organizations cooperating in training programmes, i. e. sharing the development costs and improving the results of the business cases for all partners, were analyzed.	
• Since ensuring these synergies required a fundamental restructuring of the project, the situation was jointly defined by the project team and by the project owner team as a project chance.	
Version: 1.0 Date: 03. 04. 2003 Author: GS	

Fig. F4.3: Definition of a project chance

F 4.3   Planning of immediate measures and additional measures for resolving a project discontinuity

In planning and performing measures for resolving project discontinuities immediate measures are to be differentiated from additional measures.

Immediate measures of short- and medium-term effectiveness are to be planned and performed on the basis of a brief analysis of the situation. Additional measures are subsequently agreed on the basis of a more extensive cause analysis and the planning of alternative resolution strategies. Measures for resolving project discontinuities are to be recorded in the work breakdown structure. Some measures refer to project environments and are therefore to be considered in the project environment analysis.

In the following, extracts of TO DO-lists for planning immediate measures and additional measures for resolving the project chance of the project "Realization e-application" are depicted.

❋ Realisation e Application	TO-DO LIST: IMMEDIATE MEASURES			
WP#	What?	Who?	Until when?	status
3000	continuation content work			
1000	addition: Sub-process "Resolving project chance" (WP 1400)	Schubert	08. 04. 2003	
1400	definition benefit of extension of scope	Schubert	08. 04. 2003	
1400	definition and description of alternatives	Meier	08. 04. 2003	
1400	definition of the demand PMG	Meier	08. 04. 2003	
1400	definition of the demand university pro-gramme "International project manage-ment"	Meier	08. 04. 2003	
1400	analyzing capacity IT supplier	Hofer	08. 04. 2003	
Version: 1.0	Date: 03. 04. 2003	Author: GS		

Fig. F4.4: TO DO-list: Immediate measures project chance "Realization e-application"

❋ Realisation e Application	TO-DO LIST: ADDITONAL MEASURES			
WP#	What?	Who?	Until when?	status
1400	adaption of project plans	Schubert	08. 04. 2003	
1400	financial agreement RGC/PMG/university programme	Schubert	15. 04. 2003	
	additional offer IT supplier	Hofer	15. 04. 2003	
1400	definition new interfaces	Hofer	15. 04. 2003	
1400	specification for PMG and university programme	Meier	15. 04. 2003	
1400	establishing new project organization, new project values	Schubert	15. 04. 2003	
1400	communication to relevant project environ-ments	Schubert	15. 04. 2003	
1400	development of a new project assignment	Schubert	15. 04. 2003	
Version: 1.0	Date: 03. 04. 2003	Author: GS		

Fig. F4.5: To-do list: Additional measures project chance "Realization e-application"

F 4.4   Cause analysis of a project discontinuity

The basis for the planning of alternative resolution strategies and of additional measures for resolving a project discontinuity is as an analysis of the causes that led to the project discontinuity. Changes in the strengths and weaknesses of the project as well as in relevant environment relationships are to be identified. A common view of the members of the project organization regarding the assessment of the project situation is to be established.

Causes for project discontinuities can be, on the one hand, the internal dynamics of projects and, on the other hand, changes in the project environment relationships. Thus, the cause analysis can be performed differentiated by relationships to project-internal and project-external environments.

In Figure F4.6 the cause analysis for the project chance of the project "Realization e-application" is depicted.

✷ Realisation e Application	CAUSE ANALYSIS OF THE PROJECT CHANCE
relevant project environment	object of consideration
project	• identification of relationships between RGC – PMG – university programme • expectation of possible synergies
RGC	• interested in cooperation
PMG	• need for the e-application • interested in cooperation
university training course	• need for the e-application • interested in cooperation
IT supplier	• interested in extension of the contract
Version: 1.0	Date: 03. 04. 2003 Author: GS

Fig. F4.6: Extract of the cause analysis of the chance of the project "Realization e-application"

F 4.5   Planning of alternative resolution strategies

Apart from the "do nothing" alternative basic strategies for resolving a project discontinuity are:

- redesigning the project,
- stopping the project and
- interrupting the project.

Redesigning the project may require the redefinition of the project objectives and the project content, may necessitate the adaptation of project environment relationships, may lead to appointing a new project owner team, project manager or individual project team members and may include the development of a new project culture. Redesigning the project creates a new project identity. This further development serves to form the basis for a successful continuation of the project.

The stopping of a project poses a catastrophe in the development of the social system "project", its survival is no longer guaranteed.

A further strategic alternative is the interruption of the project. This presupposes that the project can be continued successfully after the period of interruption.

Not only a project crisis but also a project chance can lead to the stopping or to the interruption a project. When the existing project structures do not form an adequate basis for the change in identity, a new start of the project becomes necessary. In defining strategies for resolving the chance of the project "Realization e-application" (see Figure F4.7) it did not seem necessary to consider a new start as an alternative.

The planning of strategies for resolving a project discontinuity comprises, on the one hand, the definition of strategies to be considered and, on the other hand, their description. The description of the consequences of alternative resolution strategies should include the necessary changes in the business case, in the project objectives, in the objects of consideration, in the project scope, in the project costs, in the project schedule, in the project environment relationships, in the project organization and in the project culture (see Figure F4.7).

One of the alternative resolution strategies is to be selected. This decision is to be taken by the project owner team.

✪ Realisation e Application	PLANNING OF ALTERNATIVE RESOLUTION STRATEGIES		
	alternative resolution strategies		
criteria	no extension	perceived as a continual change	perceived as a project chance
objectives	as planned: Solution for RGC	• extension by solution for PMG and university programme "International project management"	• extension by solution for PMG and university programme "International project management"
objects of consideration	as planned: RGC seminars	• additionally: PMG courses and university programme "International project management"	• additionally: PMG courses and university programme "International project management"

environments	as planned: RGC employees, seminar participants, etc.	• PMG: Employees, students, WU • university programme: Employees, students, WU, TU	• PMG: Employees, students, WU • University programme: Employees, students, WU, TU
organization	as planned: Cooperation RGC and IT supplier integrated project organization	• no explicit adaptation of the organizational structures • no new contracts	• adaptation of the organizational structures • new contract structures
culture	as planned: Cooperation RGC and IT supplier	• no awareness of higher complexity • project values taken over	• adequate handling of higher complexity • redefinition of project values
business case	as planned: Benefit for RGC	• synergies due to common solutions and common financing only partially used	• synergies due to common solutions and common financing thoroughly used
Version: 1.0	Date: 03. 04. 2003	Author: GS	

Fig. F4.7: Planning alternative resolution strategies, chance of the project "Realization e-application"

F 4.6   Ending a project discontinuity

The transition of resolving the project discontinuity to "normal" project management must be made consciously. Just like defining a project discontinuity, ending the project discontinuity presents an act of symbolic management.

The ending of the discontinuity should be performed as early as possible and as late as required to establish a new identity. In the process of ending the project discontinuity it is to be agreed which new project structures apply after the discontinuity (see Figure F4.8). Achieved changes must be visualized and appreciated in ending the project discontinuity.

✪ Realisation e Application	**PHASE AFTER THE PROJECT DISCONTINUITY**
agreements for the phase after the project discontinuity	
• completion of the project on the basis of new project plans • implementation of the newly designed project organization: Integrated project organization with representatives of RGC, PMG, of the university programme and the IT supplier	

- living the new project values: Cooperation of RGC, PMG, the university programme and the IT supplier
- project marketing toward old and new relevant project environments
- short-term reflection and possible adaptation of the new project structures
- etc.

| Version: 1.0 | Date: 03. 04. 2003 | Author: GS |

Fig. F4.8: Agreements for the phase after the project discontinuity, chance "Realization e-application"

The ending of a project discontinuity represents an important learning chance for the project and its environments. The experiences gained in resolving the project discontinuity must also be secured for the project-oriented company.

F 4.7 Organization and communication for resolving a project discontinuity

To resolve project discontinuities resources of the project team usually do not suffice. The contribution of members of the project owner team and the involvement of additional experts (legal experts, PR experts, etc.) in the project team become necessary. If the members of the project owner team and the project team are not themselves the cause for the project discontinuity, they should retain their roles during the resolution the project discontinuity.

On the one hand, the competencies for the cause analysis, the strategic planning and the resolution measures need to be available. On the other hand, the formal authority, to be able to react on short notice, and social relationships, to ensure the acceptance of the resolution measures, are required.

The communication structures of the project must be newly designed. The extent and the intensity of the project communications increase during the resolving of the project discontinuity. The objectives and the participants of meetings must be newly defined. Project controlling reports to be developed on short notice serve as a basis for meetings for resolving the project discontinuity.

Communicating the causes of the project discontinuity, the resolution strategies and the respective status of the resolution measures to the members the project organization and to representatives of relevant project environments is a challenge in the process of resolving a project discontinuity. The success of resolving a project discontinuity is dependent on the handling of project information. The basic communication policy – broad, open communication or selective, restricted communication – must be defined. The forms and the frequency of the transfer of information and the target group-specific information design must be determined.

For resolving the project chance in the project "Realization e-application" the organizational rules described in Figure F4.9 were determined:

✪ Realisation e Application	ORGANIZATIONAL RULES FOR RESOLVING THE PROJECT CHANCE

- informing selected employees of RGC, PMG and university programme about the definition of the project chance
- extending the project team for resolving the project chance by employees of PMG and the university programme "International project management"
- performing 2 half-day workshops with the extended project team for planning the project strategies, for planning immediate and additional measures, and for restructuring the project
- informing all employees of RGC, PMG and the university programme about the cooperation in the project and the new project structures

Version: 1.0	Date: 03. 04. 2003	Author: GS

Fig. F4.9: Organizational rules for resolving the chance of the project "Realization e-application"

F 5 Methods for closing-down a project or a programm

F 5.1 Overview of methods for closing-down a project or a programme

Objectives of the project close-down process are the content-related close-down and the emotional close-down of a project.

To achieve these objectives, the remaining work is to be planned and performed, the post-project phase is to be planned, the know-how transfer into other projects and into the contributing project-oriented organizations is to be organized, the environment relationships are to be designed, the performance of the members of the project organization is to be assessed, the project success is to be evaluated, and the project is to be formally approved.

These tasks of the project close-down process are to be supported by symbolic actions, such as sending out thank-you letters or handing over presents, organizing a closing "social" event, etc.

For the fulfillment of these tasks specific project management methods, such as project close-down reports, exchange of experience workshops, or assessments of performance, can be used in the project close-down process. The project management methods used in the project start and in the project controlling process can be used as a basis. Thus, e.g. the project success can be evaluated by comparing the project objectives planned in the project objectives plan with the actually realized project objectives, and the project environment analysis can be used for dissolving the project environment relationships.

Professional project management is characterized by the continuity of the project management approach. The relatively great effort in the project start process pays off when the project management methods and the project communication forms are utilized in the following sub-processes of project management.

The project management methods to be used in the project close-down process can be selected with the help of the checklist depicted in Figure F5.1.

methods for the project close-down	small project	project
planning of measures		
• TO DO-list: Remaining work	must	must
• designing the environment relationships	must	must
• TO DO-list: Post-project phase	must	must
• adaptation business case analysis	can	must
know-how transfer		
• project close-down report	must	must

• special reports		can	can
• actual project management documentation		must	must
• project presentation		can	can
• articles in newsletters, on homepage, in journals		can	can
• exchange of experience workshop		can	can
assessment of performance			
• evaluation of project success		must	must
• assessment of the members of the project organization		can	must
symbolic actions in the project close-down			
• "social" end event		can	must
• closing project cost center		must	must
• project acceptance certificate		must	must

Fig. F5.1: Checklist for selecting the project management methods for the project close-down

F 5.2   Planning of measures

In the project close-down process the remaining work to be fulfilled, the dissolution of the project environment relationships and the design of new environment relationships as well as the post-project phase must be planned.

The remaining work still to be fulfilled in a project can be planned by means of the TO DO-list "Remaining work". The fulfillment of the planned remaining work can also be controlled with the TO DO-list "Remaining work".

As an example the TO DO-list "Remaining work" of the project "Realization e-application" is depicted in Figure F5.2.

✹ Realisation e Application		TO-DO LIST: REMAINING WORK					
WBS code	activity	responsible	agreed on:	due date	status	interpre- tation
	close-down meeting project owner team	Schubert	01. 10. 2003	15. 10. 2003		
	completion application	Sommer	01. 10. 2003	12. 10. 2003	being per- formed	
	formal approval	Huber, Fischer/ Schubert	01. 10. 2003	15. 10. 2003		
	operating contract	Schubert	01. 10. 2003	15. 10. 2003		
	development of project management documen- tation, reflection	Schubert	01. 10. 2003	15. 10. 2003		
Version: 1.0		Date: 01. 10. 2003			Author: GS		

Fig. F5.2: TO DO-list "Remaining work" of the project "Realization e-application"

In the design of the environment relationships two tasks must be fulfilled in the project close-down process. Since the project is dissolved as a temporary organization, firstly, the relationships of the project to its relevant environments must be dissolved. Secondly, new relationships between relevant environments and permanent departments of the organizations, which were performing the project, must be established for the post-project phase. For example, the project-performing organization may be interested in winning a supplier, it has cooperated with in the project for the first time, as regular supplier. The responsibility for managing the relationship with the supplier in the post-project phase lies with the purchasing department.

The project environment analysis is to be used for planning the dissolution of existing project environment relationships and for establishing new relationships. The planned measures are to be added to the TO DO-list "Remaining work".

An example for the use of the project environment analysis in the project close-down process is depicted in Figure F5.3 for the project "Realization e-application".

CHAPTER

F

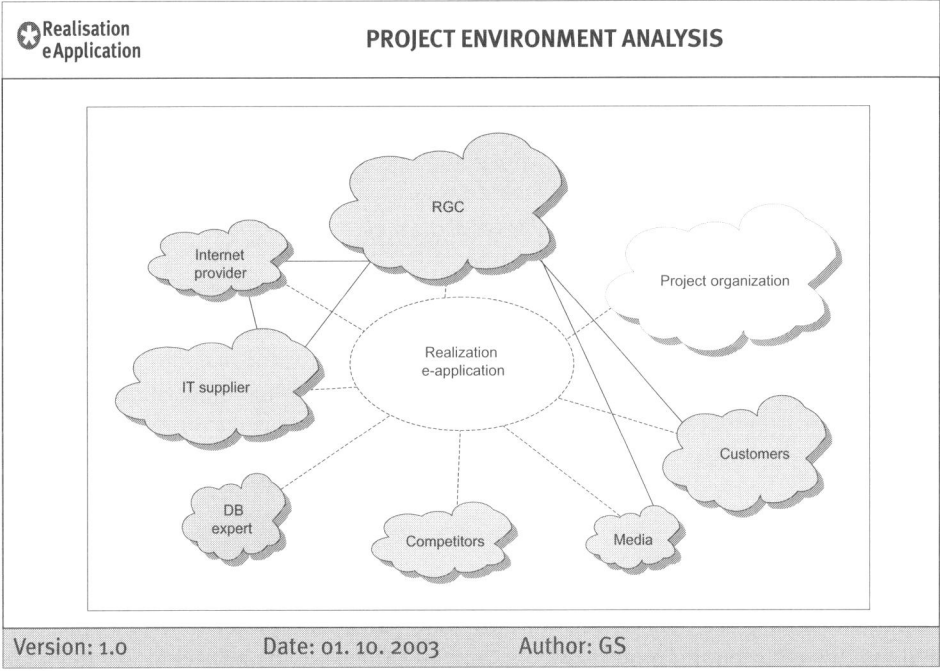

Fig. F5.3: Planning the environment relationships in the project close-down process of the project "Realization e-application"

Instruments for the dissolution of project environment relationships are thank-you letters, presents and a "social" project event. Thank-you letters and small presents to customers, partners and suppliers serve to express, that the good cooperation in the project is appreciated. A closing evening together at a restaurant enables emotional close-down of a project and supports the dissolution of the project team.

The planning of the post-project phase comprises the development of a TO DO-list "Post-project phase" and the final adaptation of the business case analysis. In the TO DO-list "Post-project phase", above all, tasks to be fulfilled immediately after the end of the project and the responsibilities in relation to these are to be planned. It has to be ensured that even after the dissolution of the project organization the quality in fulfilling the tasks initialized by the project is guaranteed. Typical tasks in the post-project phase are e.g. maintenance tasks. The TO DO-list "Post-project phase" should also display whether an investment evaluation is planned after the project end.

As an example the TO DO-list "Post-project phase" of the project "Realization e-application" is depicted in Figure F5.4.

✪ Realisation e Application	TO DO-LIST: POST-PROJECT PHASE		
WP#	What?	Who?	Until when?
	adapt all seminars RGC, PMG, university programme	all	12/2003
	internal seminars: Marketing plan, implementation plan	Meier, Huber	9/2003
	marketing to seminar participants	Gruber, all trainers	7/2003
Version: 1.0	Date: 15. 10. 2003	Author: GS	

Fig. F5.4: TO DO-list "Post-project phase" of the project "Realization e-application"

The final adaptation of the business case analysis aims at incorporating the definite actual data of the project and the most up-to-date assumptions regarding the post-project phase of the investment initialized by the project, into the business case analysis. These results of the business case analysis at the end of the project are an important criterion for evaluating the project success. The business case analysis adapted in the project close-down process forms the basis for an investment evaluation.

F 5.3 Know-how transfer

Important methods for transferring the know-how gained in a project into other projects and into the contributing project-oriented organizations are the project close-down report, special reports, the actual project management-documentation and exchange of experience workshops.

The project close-down report summarizes the basic projects results and the know-how gained in the project. It serves as final project information for all members of the project organization and selected representatives of relevant environments. Generally, only one project close-down report is to be developed. Target group-specific adaptations are possible, however. In case specific focal points of larger scope are of interest only to individual target groups, the project close-down report is to be supplemented by special reports for these target groups. The project close-down report of the project "Bhumibol" (see Chapter E1.7), for example, was supplemented by a special report for the design department on experiences with the use of the "digital control".

The project close-down report can be structured according to project phases and relevant project environments. It should be as concise as possible and be written in note form.

An extract of the project close-down report of the project "Realization e-application" is depicted in Figure F5.5.

 **Realisation e Application**

PROJECT
CLOSE-DOWN REPORT
as at October 15th 2003

overall impression

- project objectives achieved
  - e-application tested in pilot operations for RGC seminars introduced according to specification;
  - adequate organization implemented, personnel trained, marketing for pilot operations performed, financing for the application of the e-application secured;
  - experiences for further e-projects gained;
  - cooperative relationship to suppliers further developed;
  - concept for all further RGC seminars developed;
  - adaptation of the e-application for PMG and university programme completed;
  - joint contract and financing solution.
- project discontinuity (chance) successfully resolved. Measures taken:
  - extension e-application for use by PMG, university programme;
  - joint contract and financing solution.
- cost and schedule deviations due to increased efforts in detailed planning and due to extension for PMG and university programme
  - cost deviation of 14,800,
  - schedule deviation of 1.5 months.

reflection: Fulfillment of the planned tasks, adherence to schedule

- Performance generally good;
- difficulties during phase 1.2 "Detailed planning" due to lack of know-how in the project organization, solved by involving an external DB expert;
- high quality and good feedback of participants in performing the pilot operations (1.7);
- schedule delays of approx. 1.5 months due to difficulties during phase 1.2, as well as resolving the project chance. Since pilot seminars could be performed adequately, the schedule delay poses no problem.

reflection: Adherence to resource and cost planning

- total cost deviation from basis plan of + € 14,800;
- increase through increased (internal, external) demand of resources in phase 1.2, deviation of + € 4,300;
- increased demand of resources due to redefinition of the objectives (extension to include PMG and university programme), deviation of the project costs of + € 10,500.

reflection: Project environment relationships, relationships to other projects

CHAPTER

F

347

- Relationship to RGC: Relationship to RGC was generally positive, initial uncertainty due to difficulties during 1.2 "Detailed planning" could be resolved. Relationship RGC – IT supplier challenging at the beginning, good cooperation by clearly defining the contact persons and closer content-related cooperation.
- Relationship to IT supplier: Difficulties at the beginning due to conflicting interests as well as increased effort (involvement of external DB expert), continual improvement in the course of the project.
- Relationship to participants: Experiences from the pilot seminar were swiftly adapted for the other 2 pilot seminars. Positive feedback on new seminar didactics.
- Relationship PMG, university programme: Positive due to inclusion of new impulses for the project.
- Relationship to other environments: Nothing striking.

reflection: Team work, project management

- Project team: After repeated adjustments of the roles and communication structures good cooperation in the project team.
- Sub-teams: Work in the sub-team "HW + SW" challenging since many persons involved came form different organizations and because of repeated adaptations (RGC, IT supplier, external DB expert, later also PMG/university programme, IT expert). Sub-teams "Content" and "Organization, personnel and marketing" trouble-free.
- Project management adequate, "model project".

Summarizing experiences for other projects

- Integrated project organization proved to be efficient despite initial difficulties.
- Attention to detailed planning is a success factor, no making false economies. Know-how deficits during detailed planning should be eliminated immediately by involving additional, even external, experts.
- Timely marketing and ongoing, active communication with the participants necessary to ensure acceptance of the e-application.
- Definition of the project discontinuity (chance) served to increase the overall benefit of the project. Resolution process adequate.

| Version: 1.0 | Date: 15. 10. 2003 | Author: GS |

Fig. F5.5: Extract of the project close-down report of the project "Realization e-application"

The actual project management documentation is to be added to the project close-down report as an appendix. In the actual project management documentation final versions of the project plans are summarized. The actual project management documentation is a knowledge management instrument of the project-oriented organization. The realized project plans can serve as a basis for the planning of similar projects.

If, for example, in the course of a pilot project extensive know-how was gained, it may be reasonable to organize an exchange of experience workshop for all those interested in the gathered experiences. The costs of such a workshop must not be charged on the project, however, but are overhead costs of the project-oriented organization.

A final project marketing can be performed by means of a project presentation and possibly by articles on the project results in a newsletter and on the homepage of the project-oriented organization, or in trade journals.

F 5.4   Assessment of performance

In the project close-down process the project success is to be evaluated and the performance of the members of the project organization is to be assessed.

The evaluation of the project success is based on the project objectives, possibly adapted during project performance. Those results should be delivered, which are expected by the project owner team (and relevant project environments) at the end of the project, and not those formulated at the project start. Sometimes, it only becomes apparent in the project close-down process that additional objectives have been realized in the project. This clarification serves as a "retrograde" sense-making of the project.

A central project objective is to contribute to the optimization of the results of the investment initialized by a project. The fulfillment of this objective therefore poses an important criterion for evaluating the project success. In the project "Realization e-application" a rough cost-benefit analysis for documenting the results of the investment in the e-application was used (see Figure F5.6).

The (not discounted) investment success in an analysis period of 4 years results in a benefit-cost difference of € 146,000.

CHAPTER

F

✪ Realisation eApplication	COST-BENEFIT ANALYSIS						
	costs				benefits		
	personnel	material	services	total	description	assessment	
conception e-application	30 PD at € 400	-	€ 5,000	€ 17,000	learning for the management of e-projects	profit of 1 additional contract	€ 5,000
realization e-application	30 PD at € 400	€ 2,000	€ 8,000	€ 22,000	learning for the management of e-projects Performance of 3 pilot seminars	5 additional participants in pilot seminars	€ 5,000
utilization year 1 • internet provider • maintenance • additional administration • per year	10 PD at € 400		€ 500 € 500	€ 5,000	additional customers in external seminars additional contracts	20 additional participants 4 contracts	€ 20,000 € 20,000
utilization year 2				€ 4,000			€ 40,000
utilization year 3				€ 4,000			€ 40,000
utilization year 4				€ 4,000			€ 40,000
adaptation				€ 8,000			€ 40,000
total				€ 64,000			€ 210,000
Version: 1.0	Date: 15. 10. 2003			Author: GS			

Fig. F5.6: Extract of the cost-benefit analysis of the investment "e-application"

The project success can be evaluated by the members of the project organization and by the users of the results. In contracting projects the external view of the customer is of interest. Obtaining feedback from customer representatives is part of the quality management system in many organizations.

For the evaluation of the project success various methods can be used. Figure F5.7 depicts a project success matrix. In a project close-down workshop the members of the project organization enter their initials in this matrix to assess the axes "Project results achieved" and "Process of cooperation in the Project". The result serves as a basis for a joint interpretation and discussion.

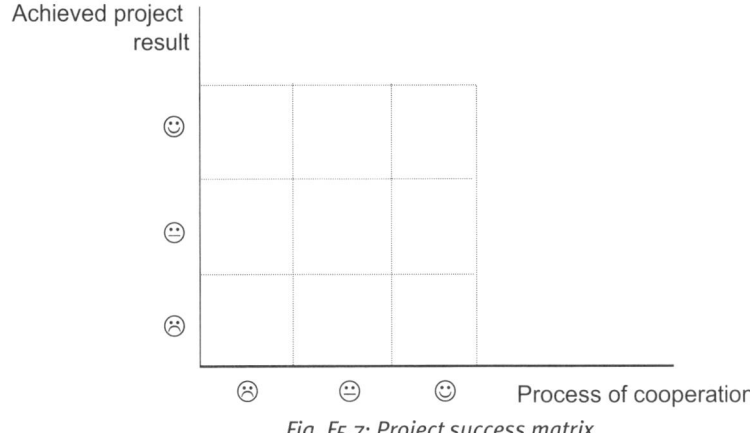

Fig. F5.7: Project success matrix

A more formal method of reflection is a questionnaire for the "Evaluation of project success". This questionnaire should be filled in by each member of the project organization. The results of the analysis of the questionnaires are to be interpreted and jointly discussed.

In the project close-down process the project performance of the members of the project organization should be explicitly assessed. Objectives of these assessments of performance are the appreciation of the achievements and the feedback to individual members of the project organization. These personal learning chances should be utilized by the project manager, the individual project team members, but also by the members of the project owner team.

Criteria for assessing the project performance of individual members of the project organization are their contributions to realizing the project success. Thereby a direct relationship between the project success and the project-related achievements of employees is established.

For assessing the performance of individual members of the project organization personal feedback and questionnaires can be used. Personal feedback can be organized in meetings between the project manager and individual project team members or in the project team in the course of a project close-down workshop.

A questionnaire for a 360° feedback to the project manager and the analysis of the results are depicted in Figures F5.8 and F5.9. The assessment of the project performance of individuals requires appropriate social competencies of the persons giving and taking the feedback.

Assessment of the project manager of the project	by ☐ the project owner team ☐ the project team ☐ others				
Assessment of the performance	**very poor**	**poor**	**average**	**good**	**very good**
• Achieving the project objectives					
• Considering the project context					
• Planning/controlling of schedule, resources					
• Designing the project organizati-on, the project culture					
• Product and industry competences					
• Designing the project management process					

Fig. F5.8: Assessment of the project manager

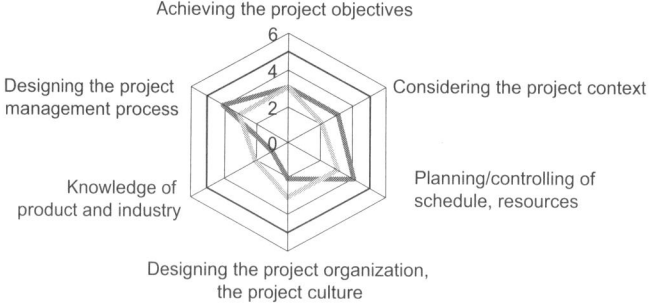

Fig. F5.9: Results of the assessment of the project manager by different environments

The assessments of the performances of the members of the project organization can serve as a basis for project-related bonuses. The results of the assessments of perform-ance can be used with regard to the project but should also be available as information for the planning of personnel development and personnel disposition measures by the respective superior of an individual in the permanent organization of the project-oriented organization (e. g. Expert Pool manager).

F 5.5 Symbolic actions in the close-down

Important symbols for closing down projects are closing the project cost center, organ-izing a closing "social" event and the formal project approval by the project owner team.

The project end can be communicated to the "project-internal" environments by closing the project cost center. Thereby it becomes clear to all members of the project organization that no costs can be accounted on the closed-down project anymore.

A closing "social" event, such as a common visit to a restaurant, promotes closing reflections on the project in an informal atmosphere and enables taking leave of the members the project organization. An example for the design of the closing "social" event of the project "pm days" is described in Case Study F3.

···⟩ CASE STUDY F3: Closing "social" project event of the project "pm days"

project:
- organizing the "pm days", the annual international project management event
- see Case Studies D1 in Chapter D and E7 in Chapter E

organizer:
- ROLAND GAREIS CONSULTING (RGC) in cooperation with the
- PROJEKTMANAGEMENT GROUP (PMG) from the University of Economics and Business Administration, Vienna

structures of the project "pm days '03":
- pm days: Closing event,
- location: "Mozart und Meisel" (pub in the 19th district in Vienna),
- participants: All members of the project organization, approx. 15 people,
- duration: Approx. 3 hours, 8 p.m. to 11 p.m.
- Objectives:
  Emotional close-down of the project; reflection by the members of the project organization on the event "pm days" (not on the project "pm days")[a]; information on the attitude and the satisfaction of the exhibitors, speakers, participants, cooperation partners.
- Agenda:
  Common meal in a separate room; recounting of anecdotes about the event "pm days" by each member of the project organization, laughs were permitted; other members asked questions about the anecdotes; thereby a common picture of the event "pm days" was created
- Timing:
  Immediately after the event "pm days", not at the end of the project (after performing the remaining work, such as evaluating the event, settling of bills, the documentation work, conception of the follow-up event).
- Interpretation of the timing:
  After the performance of the event "pm days" the project organization was reduced from 15 to 3 people for the fulfillment of the remaining work. A common

> close-down with all members was therefore only possible immediately after the event. From an emotional point of view, too, this was when the need to share impressions regarding the event was greatest.

(a) The closing reflection of the project is done in the project close-down process by means of a questionnaire which is filled in by all members of the project organization. The analysis is made available to the members of the project organization and serve mainly to develop a conception of the follow-up event.

The project can be formally approved by the project owner team either by means of a project acceptance certificate or by an approval remark on the project assignment. The project approval is the final event in the project management process. The project approval discharges the project manager and the other project team members formally of the project responsibility.

An example of a project approval in the form of a remark on the project assignment is depicted in Figure F5.10 for the project "Realization e-application".

✪ Realisation e Application	PROJECT APPROVAL	
achieved project objectives		
<ul><li>e-application tested in pilot operations introduced for 3 RGC seminars according to specification (to reduce attendance times and improve didactics at RGC seminars);</li><li>adequate organization implemented, personnel trained, marketing for pilot operations performed, financing for the application of the e-application secured;</li><li>experiences for further e-projects gained;</li><li>cooperative relationship to suppliers further developed;</li><li>concept for all further RGC seminars developed;</li><li>adaptation of the e-application for PMG and university programme completed;</li><li>joint contract and financing solution for RGC/PMG/university training course exists.</li></ul>		
end date: 15. 10. 2003	actual costs: Externals EUR 9,800	actual benefits: -
approval		
After a substantial extension of the scope (project chance) the project has been successfully completed.		
15. 10. 2003 date of approval		project owner team
Version: 1.0	Date: 15. 10. 2003	Author: GS

Fig. F5.10: Project approval by means of an acceptance protocol (project "Realization e-application")

G

Programme management

A programme is a temporary organization for the fulfillment of a complex and unique business process. The projects which are part of a programme serve to realize programme objectives.

Programme management is a business process of the project-oriented company, which includes the sub-processes programme start, programme coordination, programme controlling, possibly the resolution of a programme discontinuity and programme close-down.

G Programme management

Contents

G 1 Construct "Programme" and the programme management process

G 1.1 Construct "Programme"

A programme is a social construct. It is used as a temporary organization to fulfill a unique and complex business process of large scope.

Description of business processes	Characteristics		
Frequency	often	once	once
Importance	small	medium-large	large
Scope	little	medium-high	high
Duration	short	short-medium	medium-long
Resources required	few	some	many
Costs	low-medium	medium-high	high
Organization	few	some-many	many
	↓	↓	↓
Organization form	Permanent organization or working group	Project	Programme

Fig. G1.1: Characteristics of business processes which are organized as programmes

It is of great strategic importance for the company performing the programme, it is unique and it is limited in time. The projects which are part of the programme serve to realize programme objectives.

The objective of a programme – in the organizational meaning – can be, for example, the performance of a service (contracting programme), the establishment of a new infrastructure (a construction programme or an IT programme) or the development of a new organization (reorganization programme). Examples of programmes are shown in Figure G1.2.

programme name	programme objective	company
railway stations initiative	revitalization of 40 Austrian railway stations	Austrian Railway Company
LKS 2000	reorganization of the Landeskliniken Salzburg	Landeskliniken Salzburg
Statoil SAP	implementation of SAP in the Statoil concern	Statoil
programm I austria	the Austrian project management initiative	PROJEKTMANAGEMENT GROUP and Projekt Management Austria
STAR 2000	migration of the SW systems from Lufthansa and AUA	Austrian Airlines
EBanking & eBrokerage	establishment of eBanking and eBrokerage companies	ABN AMRO
Symphonie	merger of the banks Bawag and PSK	Bawag/PSK

Fig. G1.2: Examples of programmes

The yearly investment programme or the strategic priorities of a company are not programmes in an organizational meaning. The same is true for a television programme, a software program, etc. The term "programme" is used every day in different ways. The organizational meaning of the word must still make its mark.

Programmes are a new possibility for differentiating organizations of the project-oriented company. In practice, the term project is often used for temporary organizations which should be managed as programmes. In order to visualize the difference in scope and complexity of such organizations, some companies refer to these as "total projects" or "large projects". But in doing so, the following organizational potentials are lost, which result from the differentiation between projects and programmes:

- differentiation between the programme owner team and the owner teams of the projects,
- use of different project owner teams for different projects in the programme,
- promotion of the autonomy of the individual projects in a programme, such as the development of a project-specific culture and the design of project-specific environment relationships,
- substitution of the hierarchical organizational structure of a complex project with several sub-projects (see Figure G1.3) with a flat programme organization (see Figure G1.4),
- ensuring clear terminology, discarding the unpopular name "sub"-project manager,
- reduction in the complexity of "large projects" through the establishment of smaller, easy-to-survey project organizations,

CHAPTER

G

- development of several, easy-to-read project documents and one lean, integrated programme documentation instead of a "large project" documentation.

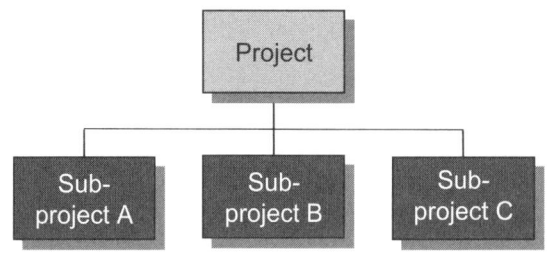

Fig. G1.3: Hierarchical organizational structure of "large projects" (Project and sub-projects)

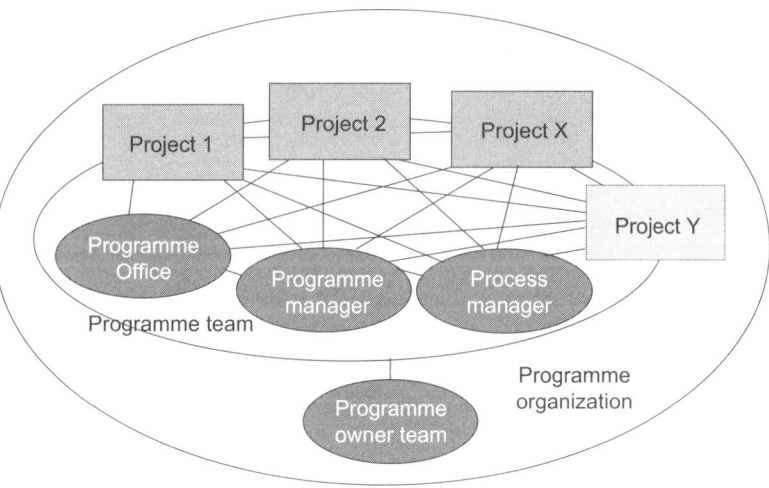

Fig. G1.4: Flat programme organization (standard organization chart)

The use of programme management in the project-oriented company ensures better quality, lower costs, shorter durations and lower risks of the programmes.

G 1.2   The programme management process

Programme management is a business process of the project-oriented company which includes the sub-processes programme start, programme coordination, programme controlling, possibly the resolution of a programme discontinuity, and programme closedown. The programme management process corresponds exactly to the project management process. Programmes must also be started on the basis of a programme assignment, controlled and closed down. Because of its great importance, programme marketing should be seen as its own sub-process of programme management and not – as in project management – as a part of the start, controlling and close-down processes.

The assignment of the programme to the programme manager by the programme owner team is the starting event of a programme. The programme approval by the programme owner team is the formal programme close-down.

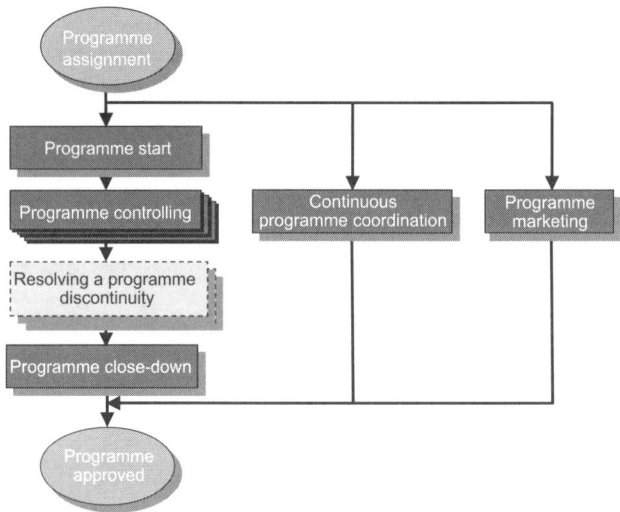

Fig. G1.5: The programme management process

Programme management is to be performed in addition to the management of the individual projects which make up the programme. The instruments for integrating projects into a programme are: The planning and controlling of programme objectives, the programme schedule, the programme budget and the programme risk, the design of the programme environment relationships, the design of the programme organization and of the programme culture. It is to be ensured that the projects follow the programme standards. The projects of a programme are closely coupled by the programme management.

G 1.3   Programme management sub-processes

The programme start and the programme close-down processes are limited in time and are performed only once in a programme, the programme controlling process in a programme is performed several times. The programme coordination and the programme marketing are continuous processes.

The sub-processes of programme management are described in Figures G1.6-G1.11.

The programme start process

Objectives of the programme start process

- Information transfer from the pre-programme phase into the programme,
- definition of expectations regarding the post-programme phase,
- developing adequate programme plans for managing the programme objectives, deliverables, schedule, resources, costs, income, risks and financing,
- designing the programme organization, adequate integration of the programme into the permanent organization,
- developing the programme culture,
- establishing communication relationships between the programme and other programmes and relevant programme environments, initial programme marketing,
- communicating the "big programme picture" to all members of the programme organization,
- planning of measures for discontinuity management,
- definition of the structures for the programme management-sub-processes to follow,
- developing the documentation "Programme start" and
- efficient design of the programme start process.

CHAPTER

G

Time boundaries of the programme start process

- Start: programme assigned
- End: documentation "Programme start" filed
- Duration: 2 – 4 weeks

Tasks and responsibilities in the programme start process

Tasks \ Responsibilities	Programme owner team	Programme manager	Programme office	Programme team member	Programme team	Representatives of relevant programme environments	External	Documents
Planing the programme start								
• Checking the programme assignment and the results of the preprogramme phase		P						
• Selecting start communication form		P	C					
• Selecting programme team members (and a PM consultant)		C				P		
• Selecting PM methods and PM templates to be used		P	C					
• Agreeing with programme owner team	D	P						1)
Preparing the programme start communication								
• Hiring of a PM consultant (possible)	D	P	C					
• Preparing start communications I, II, etc.		P	C					
• Inviting participants	I	P	C			C		2)
• Documenting the results of the pre-programme phase		C	P	C		C		
• Developing drafts for planning, organizing and marketing the programme		C	P	C		C		
• Developing information material for start communication		C	P	C		C		3)
Performing the programme start communication								
• Distributing information material to participants	I	C	P					
• Performing start communication I	C	C1			P		C	
• Developing draft of PM documentation "Programme start"		C	P					
• Performing start communication II, etc.	C	C1			P		C	
Follow-up to the programme start communication								
• Completing draft of PM documentation "Programme start"		C	P					
• Agreeing with programme owner team	D	P	C					4)
• Programme marketing: Initial information	C	P	C	C			C	
• Distributing PM documentation "Programme start"		C	P			I		
• Filing of PM documentation "Programme start"	C	C1			P		C	
Performing first work packages (parallel)		C1			P		C	

Legend:

P ... Performance
I ... Contribution
I ... Information
C1 ... Coordination
D ... Decision

Results/documents:

list of programme management methods to be used
invitation of participants to programme start workshop
information material for programme start workshop
programme management documentation „Programme start"

Fig. G1.6: Description of the programme start process

The programme coordination process

Objectives of the programme coordination process

- continuous information of the members of the programme organization and of representatives of the relevant environments,
- continuous coordination of the programme resources,
- ensuring continuous progress in the programme by controlling the progress of the projects and the work packages and the acceptance of the projects and work packages, and
- coordination of the relationships between the projects of the programme.

Time boundaries of the programme coordination process

- Start: programme assigned
- End: programme approved

Tasks and responsibilities in the programme coordination process

Tasks / Responsibilities	Programme owner team	Programme manager	Programme office	Programme team members	Sub-team	Expertpool manager	Representatives of relevant programme environments	Documents
Communicating with programme owner team	C	P		C				1)
Communicating with programme team members and programme contributors		P	C	C		I		1)
Communicating with representatives of relevant environments	C	P	C	C			C	1)
Participating in sub-team meetings		C	C		P			3)
Approving projects, work packages	I	P	C	C				2)

Legend:
P ... Performance
I ... Contribution
I ... Information

Results/documents:
1) to-do lists
2) acceptance certificates
3) meeting minutes

Fig. G1.7: Description of the programme coordination process

Programme marketing process

Objectives of the programme marketing process

- Information of relevant programme environments about the structure, the objectives and the results of the programme,

- ensuring management attention and resources for the programme,
- efficient design of the programme marketing process.

Time boundaries of the programme marketing process
- Start: programme assigned
- End: programme approved
- Duration: programme duration

Tasks and responsibilities in the programme marketing process

Tasks \ Responsibilities	Programme owner team	Programme manager	Programme office	Programme team member	Programme team	Expertpool manager	Representatives of relevant programme environments	Documents
Planning the programme marketing								
• Planning communication forms	I	C	P					
Planning programme marketing instruments	I	C	P					1)
Preparing the programme marketing								
• Developing programme marketing instruments		C	P	C			C	2)
• Preparing events								
Performing the programme marketing								
• Applying programme marketing instruments	C	C	P				C	
• Performing events	C	C	P			I	C	
Follow-up to the programme marketing								
• Analyzing the application of programme marketing instruments	I	C	P					3)
• Follow-up to the events	I	C	P					4)

Legend:
P ... Performance
I ... Contribution
I ... Information
C1 ... Coordination

Results/documents:
1) programme marketing plan
2) marketing instruments
3) analysis report: Application marketing instruments
4) reflection report: Event

Fig. G1.8: Description of the programme marketing process

The programme controlling process

Objectives of the programme controlling process

*   Determining the programme status, design of a common programme reality,
*   agreeing on directive measures,
*   reviewing the programme objectives, further development of the programme culture and the programme organization,
*   updating the programme plans,
*   developing controlling reports (programme progress report, programme score card, financial plan),
*   organizing organizational learning in the programme and
*   efficient design of the programme controlling process.

Time boundaries of the programme controlling process

*   Start: initialization of the programme controlling
*   End: programme controlling reports filed
*   Duration: 1-2 weeks

Tasks and responsibilities in the programme controlling process

Tasks	Programme owner team	Programme manager	Programme office	Programme team member	Programme team	Expertpool manager	Representatives of relevant programme environments	Documents
Planning the programme controlling								
• Adapting established programme controlling structures		P	C					
• Reviewing programme controlling structures	D	P						1)
Preparing the programme controlling communications								
• Actual data collection and planned versus actual analysis		P	C	C				
• Deviation analyses, planning directive measures		P	C	C		C		
• Developing drafts of adapted programme plans and programme controlling reports		C	P	C				
• Preparing programme controlling communications		C	P					
Performing the programme controlling communications								
• Distributing information material to participants	I	C	P	I		I	I	2)
• Performing programme controlling communication I	C	C1				P	C	
• Performing programme controlling communication II	C	C1				P	C	
Follow-up to the programme controlling communications								
• Completing adapted PM documentation, programme controlling reports		C	P					3,4)
• Initializing updating of programme portfolio database		P	C					
• Programme marketing	C	P	C	C			C	
• Distributing programme controlling reports, programme score card	I	C	P	I		I	I	
Performing work packages (parallel)		C1			P			

Legend:
P ... Performance
I ... Contribution
I ... Information
C1 ... Coordination

Results/documents:
1) adapted programme controlling structures
2) invitation of participants to the programme team meeting
3) adapted programme management documentation
4) programme controlling report

Fig. G1.9: Description of the programme controlling process

Resolving a programme discontinuity

Objectives of resolving a programme discontinuity

- Resolving a programme discontinuity,
- limiting the damage to the programme,
- developing the basis for a successful continuation of the programme and
- efficient resolution of the programme discontinuity.

Time boundaries of resolving a programme discontinuity

- Start: definition of a programme discontinuity
- End: end of the programme discontinuity communicated
- Duration: several weeks

Tasks and responsibilities in resolving a programme discontinuity

Tasks	Programme owner team	Programme manager	Programme office	Programme team member	Programme team	Representatives of relevant programme environments	External	Experts	Documents
Defining the programme discontinuity									
• Decision to define the programme discontinuity	P	C			C	I			
• Communicating the programme discontinuity to relevant programme environments	C	C₁	C		P			C	1)
Planning and performing immediate measures									
• Planning immediate measures	C	C₁	C		P			C	
• Deciding on immediate measures	P	C₁			C	C			2)
• Performing immediate measures		C₁	C	P			C	C	
• Controlling the success of immediate measures	P	C₁	C		C				
• Communicating the results of immediate measures	C	P	C	I		I	I	I	
Cause analysis, planning strategies									
• Cause analysis	C	C₁	C		P			C	
• Planning alternative strategies for resolving the programme discontinuity		C₁	C		P	C		C	
• Deciding on an alternative strategy	P	C							3)
• Communicating the results of the planning	C	P	C	I		I	I	I	
Alternative: Redesign of the programme: Planning and performing additional measures									
• Planning additional measures	C	C₁	C		P	C		C	4)
• Performing additional measures		C₁	C	P			C	C	
• Controlling the success of additional measures	P	C₁	C		C				
• Communicating the results of additional measures	C	D		I		I	I	I	
Ending the programme discontinuity									
• Evaluating the resolution of the programme discontinuity	P	C₁	C	C	C				
• Defining the end of the discontinuity	P	C			C				
• Adapting the programme management documentation	I	C₁	C		P			C	5)
• Defining learning points	I	C₁	C		P				
• Communicating the ending of the programme discontinuity	C	P	C	I		I	I	I	
Performing work packages		C₁		P					
Alternative: Programme stopping (see programme close-down process)									
Alternative: Programme interruption (see programme close-down process)									

CHAPTER

G

Legend:

P ... Performance

I ... Contribution

I ... Information

C1 ... Coordination

Results/documents:

1) communication plan

2) immediate measures plan

3) documentation of the scenarios and strategies

4) additional measures plan

5) adapted programme management documentation

Fig. G1.10: Description of the process of resolving a programme discontinuity

Description of the programme close-down process

Objectives of the programme close-down process

- Content-related and emotional programme close-down,
- completion of remaining work,
- development of actual programme management documentation,
- make agreements for the post-programme phase,
- make agreements about a potential investment controlling,
- development of programme close-down reports,
- transfer of know-how gained in the programme into the permanent organization and to other programmes,
- dissolution of the programme environment relationships,
- final programme marketing,
- evaluation of the success of the programme, assessment of the contributions of the members of the programme organization to the success of the programme,
- appreciation for the contributions of the members of the programme organization and dissolution of the programme team and
- efficient design of the programme close-down process.

Time boundaries of the programme close-down process

- Start: initialization of the programme close-down
- End: programme approved
- Duration: 2-4 weeks

Tasks and responsibilities in the programme close-down process

Tasks / **Responsibilities**	Programme owner team	Programme manager	Programme office	Programme team member	Programme team	Representatives of relevant programme environments	External	Documents
Planning the programme close-down								
• Establishing the programme close-down structures	D	P						1)
Preparing the programme close-down								
• Documenting the remaining work to be fulfilled		P	C					
• Planning of tasks and responsibilities in the post-programme phase	D	P	C	C				
• Preparing the evaluation of the programme success		P	C	C			C	
• Preparing the dissolution and establishment of environment relationships	I	P	C	C				
• Developing drafts of the programme close-down reports		C₁	P	C			C	
• Developing drafts of the actual programme management documentations		C₁	P	C				
Performing the programme close-down communications								
• Distributing information material to participants of the close-down communications	I	C	P			I	I	2)+3)
• Performing close-down communication I, II, etc.	C	C₁	C		P		C	
• Performing exchange of experience workshop		C₁	C		P	C	C	
• Performing "social" end event	C	C₁	C		P	C	C	
Follow-up to the close-down communications								
• Dissolution and establishment of environment relationships	C	P	C	C			C	
• Completing programme close-down reports, special reports, actual documents		C₁	P	C			C	4)
• Initializing updating of project portfolio database		P						
• Distributing programme close-down reports	I	C₁	P	I		I	I	
• Closing the programme cost center	I	P	C					
• Final programme marketing	I	P	C					
Performing remaining work packages (parallel)		C₁			P			

Legend:
P ... Performance
I ... Contribution
I ... Information
C₁ ... Coordination
D ... Decision

Results/documents:
1) adapted programme close-down structures
2) invitation of participants to the programme close-down workshop
3) information material for programme close-down workshop
4) programme close-down reports, actual programme management documentation, special reports

Fig. G1.11: Description of the programme close-down process

G 1.4   Design of the programme management process

The design of the programme management process is about the use of programme management methods, standard programme plans, programme communication forms, ICT instruments, project management consultants and project management coaches, as well as programme management checklists (see Chapter E2: Design of the business process: "Project management").

The selection of programme management methods is described here. The use of programme-specific standards is described in Chapter G3.6.

The methods of programme management correspond to the project management methods, i.e. a programme objectives plan, an objects of consideration plan, a programme work breakdown structure, a programme bar chart, a programme environment analysis, etc. are all used (see Chapter F: Methods of project and programme management). The programme management methods to be used in a programme can be selected by using a checklist (see Figure G1.12).

methods for the programme start	must/can
planning the programme objectives, programme scope, programme schedule, programme resources and programme costs	
programme scope planning	
• programme objectives plan	must
• programme objects of consideration plan	must
• work breakdown structure	must
• work package specifications	must
programme scheduling	
• programme milestone plan	must
• programme bar chart	must
• programme CPM schedule	can
programme resource and programme cost planning	
• programme resource plan	can
• programme cost plan	must
• programme finance plan	can
designing the programme context relationships	
• programme environment analysis	must
• cost-Benefit or Business case analysis	must
• analysis: Relationships to other programmes and projects	must
• analysis: Pre- and post-programme phase	must
• presentations, vernissages, publications	must

designing the programme organization	
• programme assignment	must
• programme organization chart	must
• programme role descriptions	must
• programme responsibility matrix	can
• programme communication structures	must
• programme rules	must
developing the programme culture	
• programme name	must
• programme logo	can
• specific "social" events	must
programme risk management and discontinuity management	
• programme risk analysis	must
• programme scenario analysis and alternative planning	can
methods for programme coordination	
• to-do list	must
• meeting minutes	must
• project and work package acceptance certificates	must
methods for programme marketing	
• programme presentations	must
• programme homepage	must
• programme folder	must
• programme events	can
methods for programme controlling	
programme control and adaptation of the programme	
• adaptation of programme plans	must
programme controlling reports	
• programme progress report	must
• earned value analysis	can
• programme trend analyses	can
• programme score card (plus interpretations)	must
programme controlling	
• to-do lists	must

methods for resolving a programme discontinuity	
• definition of the programme discontinuity	must
• planning immediate measures	must
• cause analysis	must
• scenario analysis and alternative planning	must
• planning additional measures	must
• ending the programme discontinuity	must
methods for the programme close-down	
planning of measures	
• to-do list: Remaining work	must
• designing the environment relationships	must
• to-do list: Post-programme phase	must
• adaptation business case analysis	must
know-how transfer	
• programme close-down report	must
• special reports	can
• actual programme management documentation	must
• programme presentation	must
• articles in newsletters, on homepage, in journals	can
• exchange of experience workshop	can
assessment of performance	
• evaluation of programme success	must
• assessment of the members of the programme organization	can
symbolic actions in programme close-down	
• "social" end event	can
• closing programme cost center	must
• programme acceptance certificate	can
designing the programme management process	
programme communication	
• kick-off meeting	can
• programme workshops	must
• programme team meetings	must
IT support	
• project management software	must
• video conferencing	can

Fig. G.1.12: Checklist: Programme management methods

The programme management methods listed in Figure G1.12 have already been described in Chapter F. Chapter G3 will deal with the specifics of applying individual programme management methods.

G 1.5 Competencies for programme management

For professional programme management individual and organizational competencies are required in the project-oriented company.

Specific requirements for the performance of programme management roles are: Dealing with the programme complexity and the fulfillment of integrative functions in the programme. Both require many years of experience in project management and a high level of social competence.

The EDS-defined maturity level for the development of programme management competencies in the project-oriented company are shown in Figure G1.13. Chapter J will deal with the specifics of programme management maturity according to *ROLAND GAREIS Management of the Project-oriented Company*® in detail.

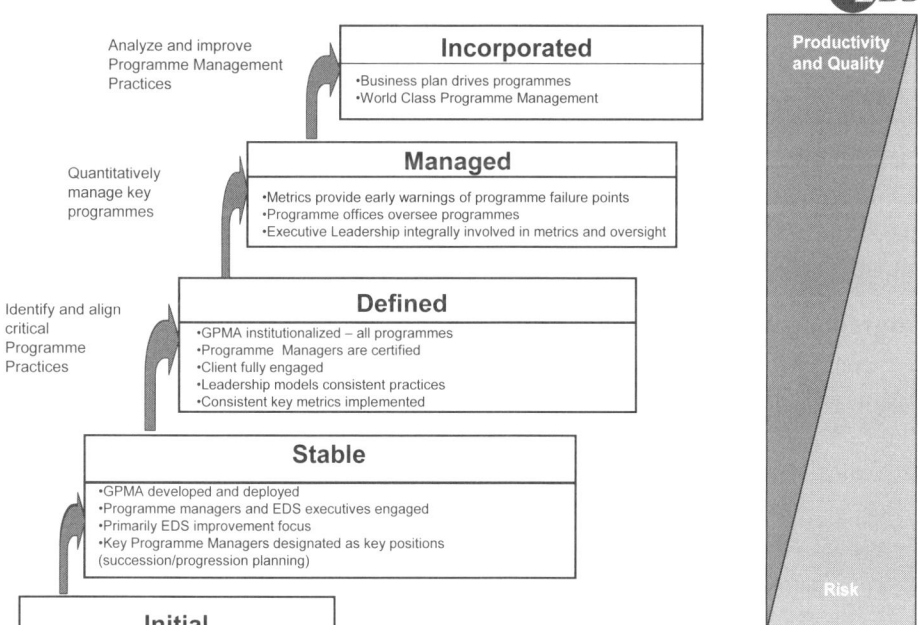

Fig. G1.13: Model of programme management maturity of EDS

G 2   Programme management case studies

The following case studies serve to clarify the difference between project and programme management and give examples of programme management documentations. In the analysis of the specifics of programme management in Chapter G3, reference will be made to, among others, the experiences of these two case studies. The author was a programme management consultant in both "LKS 2000" and "Railway stations initiative".

G 2.1   Programme "LKS 2000"

Objectives and context of the programme "LKS 2000"

In the 1990s it was first attempted to implement the concept "new public management" in public administration in Austria. To this end, the office of the Salzburg state government performed the programme "Landesdienst 2000" (State Services 2000). In order to account for the specific situation in the state hospitals of Salzburg, the directors of the state hospitals and the political department responsible for "new public management" planned the programme "LKS 2000" to be realized in the state-owned hospitals: State Hospital, State Mental Hospital and St. Veit Hospital.

On the basis of the project "Conception: LKS 2000" the Salzburg state government decided in February 1996 to perform the programme "LKS 2000". Objectives of the programme were to improve the quality of treatment for the patient, to intensify the cooperation between the 3 hospitals, to make the management of the hospitals more professional, to save costs and to increase employee orientation.

The programme end event was a presentation of the achieved results in December 1998. Since state elections were planned for the beginning of 1999, the main programme objectives were to be realized by this time.

Programme work breakdown structure of "LKS 2000"

The scope of the programme "LKS 2000" is shown in the work breakdown structure in Figure G2.1.

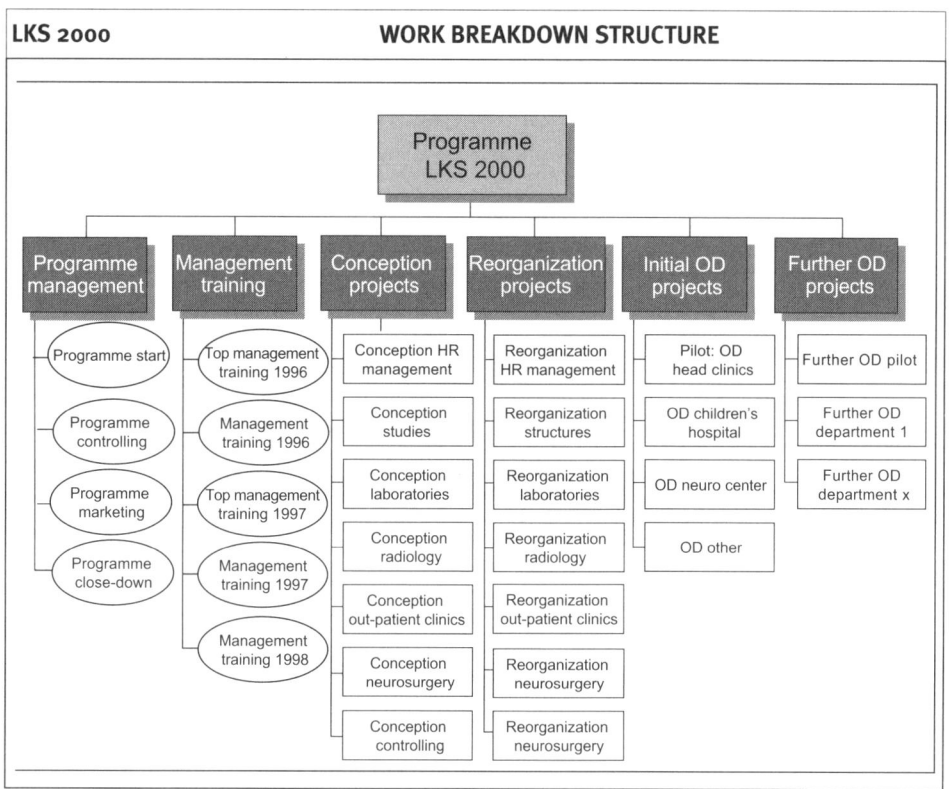

Fig. G2.1: Programme work breakdown structure of "LKS 2000"

The programme work breakdown structure shows that, in addition to projects, tasks which were not project-worthy were also included, i.e. the work packages for programme management and management training. The following types of projects were found in the programme: Conception projects, reorganization projects, basic as well as advanced organizational development projects, pilot projects and follow-up projects.

Objects of consideration of the programme which were used in the structuring of the work breakdown structure were: The three hospitals, the hospital departments, the affected departments of the Salzburg state government, the management training, the reorganizations and the organizational development (see Figure G2.2).

LKS 2000	OBJECTS OF CONSIDERATION PLAN
type of object	object of consideration
hospitals	LKA
	LKH
	St. Veit
hospital wards	ENT
	ophthamology
	oral surgery
	childern's hospital
	etc.
department of the Salzburg state government	hospital department
	personnel department
	finance department
	etc.
management training	management training 1996
	management training 1997
	management training 1998
reorganizations	reorganization laboratories
	reorganization radiology
	reorganization outpatient clinics etc.
organizational development (OD)	OD ENT
	OD ophthamology
	OD oral surgery
	OD children's hospital
	etc.

Fig. G2.2: Objects of consideration of the programme "LKS 2000"

Programme environment analysis of "LKS 2000"

The relevant environments of the programme "LKS 2000" are shown in the programme environment analysis in Figure G.2.3.

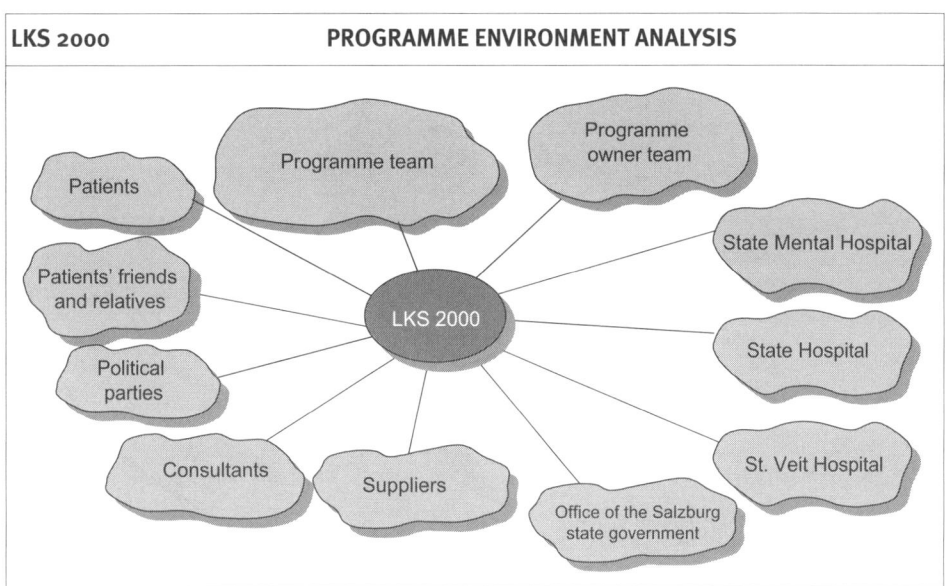

Fig. G2.3: Programme environment analysis of the programme "LKS 2000"

Apart from the three hospitals and the offices of the Salzburg state government, the hospital patients and their families were considered relevant environments. The relationships between the programme and the individual programme environments were analyzed and differentiated strategies and measures for the design of the programme environment relationships were planned.

The programme environment analysis defined the framework for the project environment analyses, which were developed for each project of the programme.

Programme organization chart of "LKS 2000"

The programme organization chart of "LKS 2000" shows that the programme manager was supported by a Programme Office. The Programme Office consisted of a programme secretary, a programme marketing expert and a programme controller. The consulting team of ROLAND GAREIS CONSULTING performed the consulting in the programme and in the individual projects, as well as the management training. This way, important integration functions could be performed, e. g. ensuring a uniform programme language, the use of documentation standards, and uniform information about the programme's progress.

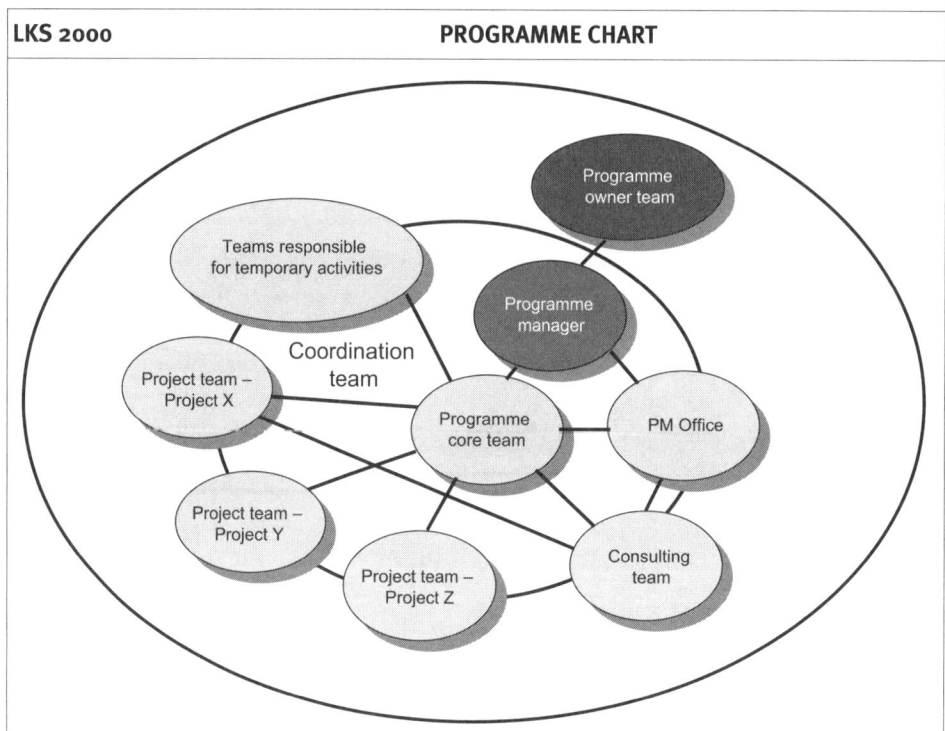

Fig. G2.4: Organization chart of the programme "LKS 2000"

G 2.2   Programme "Railway stations initiative"

Objectives and context of the programme "Railway stations initiative"

The objectives of the programme "Railway stations initiative" were the revitalization of 40 properties in Austria, optimization of the use of these properties, reorganization of the management of the properties, and ensuring financing of the revitalization work.

The programme was performed in the context of the new organization of the concern and the real estate department.

The most important objects of consideration of the programme are shown in Figure G2.5.

Railway stations initiative	OBJECTS OF CONSIDERATION PLAN
type of object	object of consideration
properties	property 1
	property 2
	property 3 etc.
businesses in the properties	lease and rent businesses
	commercial businesses
	offices
	tenements
	special facilities
organization and personnel management for the properties	organizational structure
	operational structure
	personnel management
programme standards	process management standards
	project management standards
training	management training
	project management training
financing	financing by federation
	financing by states
craft	construction
	heating/climate control/ventilation
	electric installations
	etc.

Fig. G2.5: Objects of consideration of the programme "Railway stations initiative"

Programme work breakdown structure of "Railway stations initiative"

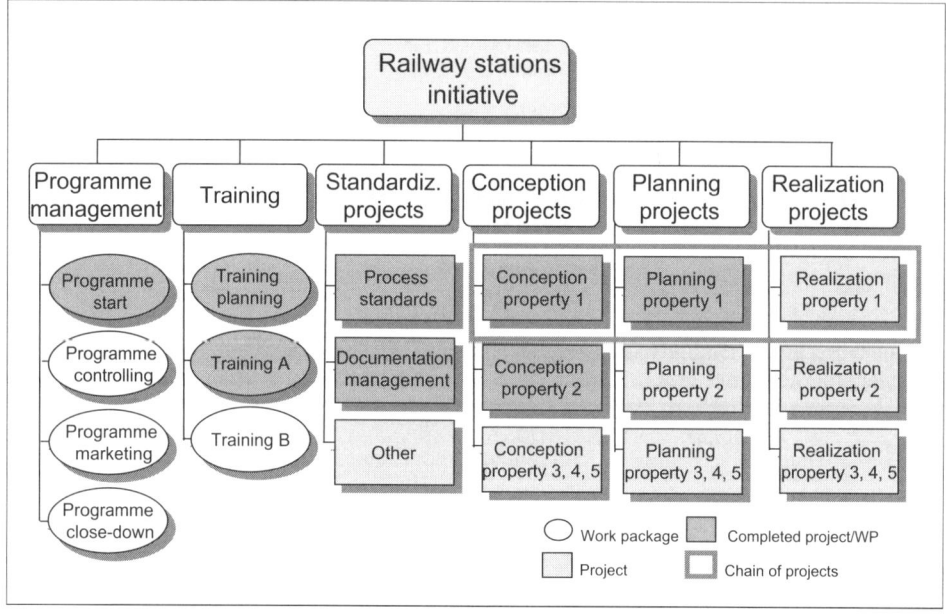

Fig. G2.6: Programme work breakdown structure of "Railway stations initiative"

The programme "Railway stations initiative" was mainly structured according to the phases of the property renovation. The conception projects had the objectives of analyzing the current situation, defining renovation alternatives and selecting the most appropriate renovation alternative. The focus of the planning projects was the architectural competition and the decision for an architectural solution. The realization projects had as objectives the detail planning, the tendering of the construction services, the construction services themselves and the commissioning of the property. One chain of projects was established per property when the property was large, and smaller properties were put together in groups within common geographical areas. In addition to the construction-related projects there were also standardization projects for the organization of learning within the programme.

The individual projects were each planned and documented in project handbooks. The work packages of the programme, such as the programme management or training, were described in work package specifications.

Programme environment analysis

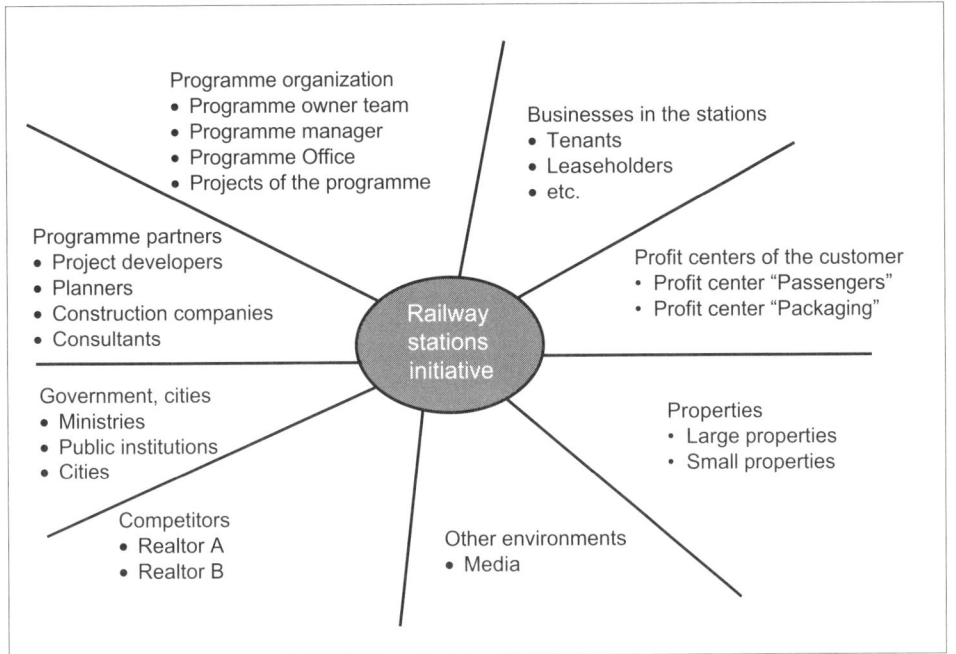

Fig. G.2.7: Environment analysis of the programme "Railway stations initiative"

The number of environments to be considered in the programme environment analysis was very high due to the complexity of the programme. A segment graph was therefore chosen for illustration purposes. Strategies and measures were developed for designing the relationships with the selected environments.

Programme organization chart

The organization chart of the programme "Railway stations initiative" was structured according to the chart shown in Figure G1.4.

The programme owner team was made up of 4 persons. All were members of the concern's board of directors. The programme manager was the manager of the real estate department. This means that the programme was placed very high in the hierarchy, but had the problem of very limited availability of time of the programme owner team. The programme manager was also not available full-time, which would have made sense in a programme of this scope.

The Programme Office was made up of 10 to 14 persons. The organization chart of the Programme Office is shown in Figure G2.8.

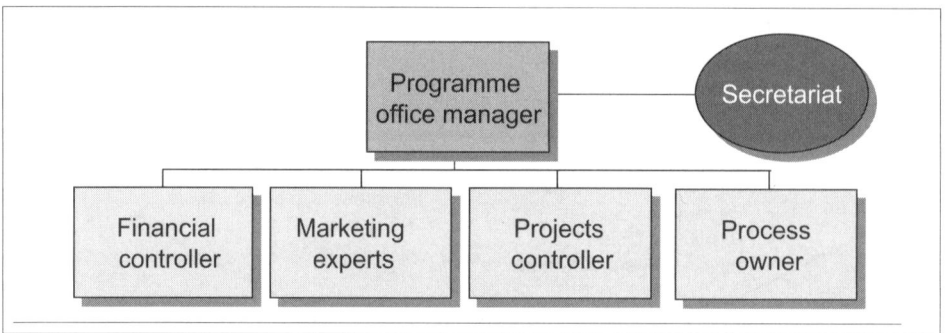

Fig. G2.8: Organization chart of the Programme Office for the programme "Railway stations initiative"

The objectives of the Programme Office were the support of the programme manager and of the projects in the programme. This should serve to make a positive contribution to the successful performance of the programme. The Programme Office fulfilled the following tasks:

- programme planning and controlling,
- definition of programme standards,
- programme marketing,
- promoting communications between the projects, and
- organization of training for the members of the programme organization.

⋯⟩ CASE STUDY G1: Development of programme standards in the project "Documentation management"

As part of the programme "Railway stations initiative" also organization projects were performed. An important project for organizing learning in the programme was the project "Documentation management". To describe the development of programme standards the structures and selected results of the project "Documentation management" are described below.

Structures of the project "Documentation management"

Project objectives:

- contribution to organizational learning of the programme "Railway stations initiative",
- developing project management and result standards,
- implementing an IT-supported documentation management system to enable access to the documentations of the "Railway stations initiative"

Project context:

- performing repetitive conception, planning and realization projects,
- possibility to benefit from experiences from pilot projects,
- know-how transfer between the projects and into the programme required.

Project scope:

- The scope of the project "Documentation management" is depicted in the work breakdown structure in Figure G 2.9. After the actual analysis and the development of the to-be concept SW and HW were implemented and standards developed. These solutions were communicated to members of the programme organization in information events and trainings, and were applied in a pilot phase.

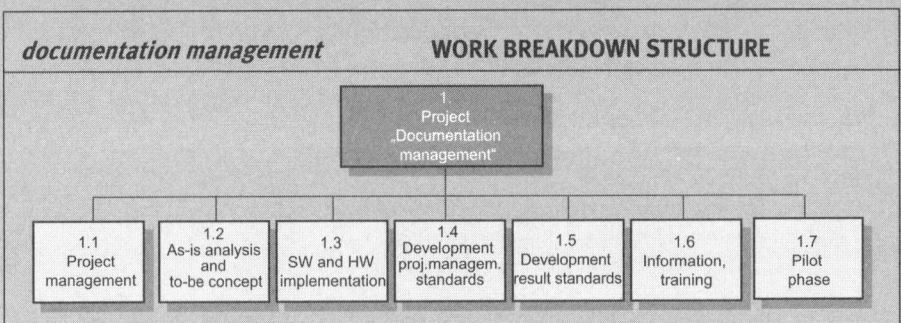

Fig. G2.9: Work breakdown structure of the project "Documentation management"

Project organization:

- The project "Documentation management" was performed in the form of an integrated project organization. The project owner team consisted of the Programme Office manager of the programme "Railway stations initiative" and the managing director of ROLAND GAREIS CONSULTING. To ensure a holistic solution the project team included representatives of the programme "Railway stations initiative", of the consulting company and of a SW-development company acting as sub-supplier to the consulting company (see project organization chart in Figure G2.10).

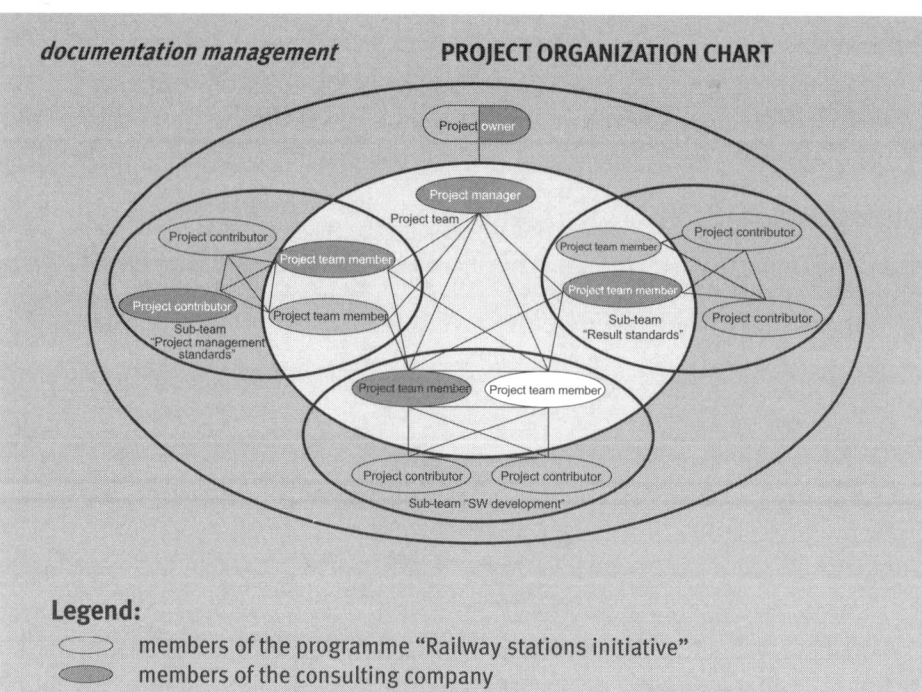

Legend:

⬭ members of the programme "Railway stations initiative"
⬬ members of the consulting company
⬭ member of the SW-development company

Fig. G2.10: Integrated organization of the project "Documentation management"

Results of the project "Documentation management"

- Results of the project were, on the one hand, project standards for conception, planning and realization projects of the programme (standard project plans and standardized result documents including forms), and on the other hand, the SW-solution "doku management", which enabled IT support in the administration of common access to the standard documents.

- Figure G2.11 depicts a sample standard work breakdown structure for a realization project. As examples for result standards Figures G2.12 and G2.13 depict a specification of a utilization plan of a property and the form "Utilization plan".

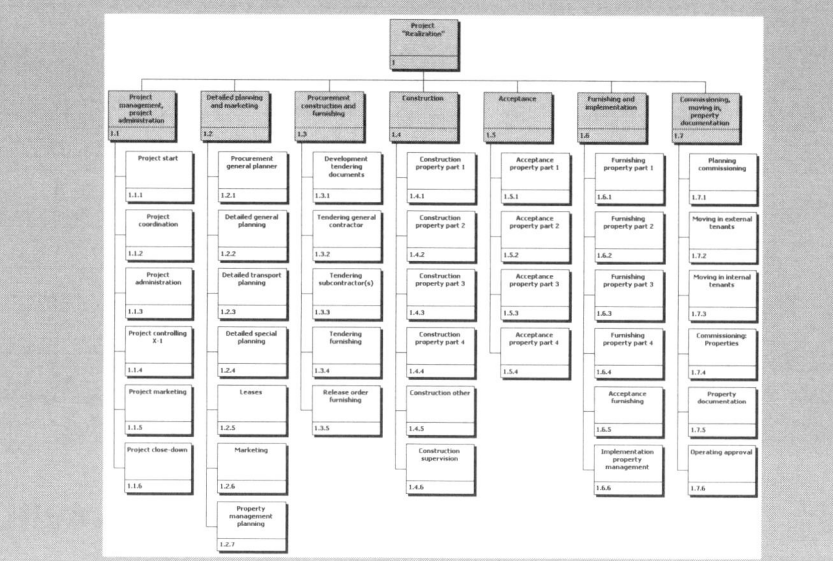

Fig. G2.11: Example of a standard work breakdown structure for a "Realization project"

Standard: Lease plan
(Result of WP 1.6.1 according to standard work
breakdown structure of the planning project)

Objectives
• Presentation of the lease plan for property XY

Contents
• List of business areas for lease

• List of deadlines for leases

Formal specifications
• Volume: 1-2 pages
• File type(s): Excel

Form
• Form: Lease plan

Interpretations
• Is a continuation of the lease concept from
  the conception project. The information on
  the lease plan is more concrete and detailed
  than in the concept.

Fig. G2.12: Sample result standard (specification "Utilization plan")

Lease plan									
Unit	Line of business	m² shop space	m² storage space	m² effective space	Total income potential	Income /m² per month shop space	Income/m² per month effective space	Notes	Preliminary contracts to be signed by
	Sum total								

Fig. G2.13: Sample result standard (form "Utilization plan")

G 3 Specifics of programme management

G 3.1 Conception projects as the basis for programmes

Because of the complexity of programmes it is advisable to perform a conception project as the basis for the investment decision and the programme assignment. The objectives of such a conception project are to describe the need for investment, to analyze the current situation, and to analyze the costs and benefits of the investment. Other objectives are planning the content, the organization, the budget and the schedule of the planned programme in order to initialize the investment.

For example, the investment decision of the Salzburg state government, which was based on the results of the project "Conception LKS 2000", served as a basis for performing the programme "LKS 2000" (see Chapter G2.1).

The sequence of a conception project and a programme for initializing an investment becomes a chain of a project and a programme.

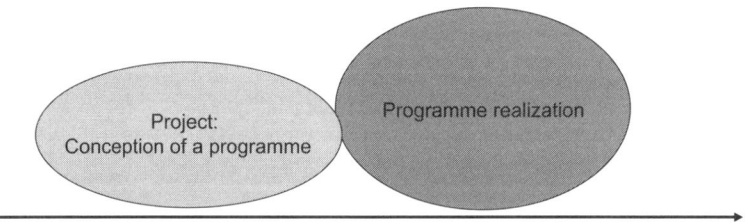

Fig. G3.1: Chain of a conception project and a programme

In the conception project the fundamental structures for realizing the programme to follow are planned. In the programme start process these structures are detailed and concretized. To ensure continuity in this chain, members of the project organization of the conception project can also take on roles in the programme organization.

G 3.2   Definition of programme objectives

Because of the large scope and the long duration of programmes the objectives of programmes are more dynamic than those of projects. During the programme assignment the global programme strategy and the programme objectives are defined, which will become adapted and more detailed during the performance of the programme.

In the programme "Railway stations initiative" (see Chapter G2.2), for example, the revitalization of 40 properties was formulated as the objective. The actual number of properties to be renovated was dependent upon the financial possibilities. These were, however, not yet clear at the time of the assignment. Securing financing for the individual projects was defined as a task of the programme "Railway stations initiative".

In the assignment of the programme "LKS 2000" fundamental savings objectives were defined. The actual solutions for realizing these objectives could only be established in the framework of the various conception projects in the programme "LKS 2000". Conception projects in programmes are, therefore, a central instrument for planning the objectives of potential follow-up projects and also for adapting the programme objectives.

CHAPTER

G

The medium- to long-term durations and the dynamics of programmes can make it necessary to plan alternative programme objectives with different end events and schedules. The boundaries of the programme should, however, always be defined. Programmes should not have an "open end".

G 3.3   Programme organization chart and programme roles

The main difference between projects and programmes lies in their organizational design. This differentiation requires that projects and programmes are perceived as temporary organizations.

Elements of the organizational structures of programmes are the projects of the programme as well as the programme-specific roles, i.e. programme owner team, programme manager, possibly programme process experts, programme team and Programme Office. The programme roles and their relationships to each other can be visualized in a programme organization chart. In Figure G3.2 a standard programme organization chart is depicted.

The programme owner team assigns the programme manager and the Programme Office with the performance of the programme. Strategic decisions of programme management, such as the selection of projects to be started, the changing of programme priorities, the definition of fundamental programme strategies toward relevant programme environments, are made by the programme owner team. The programme owner team also decides about project-related issues which are beyond the decision-making authority of the programme team or the relevant project owner and project manager.

The programme owner team should be made up of managers of those areas of the company which will be affected most by the programme. There should not be more than four persons in the programme owner team. A speaker for the programme owner team is to

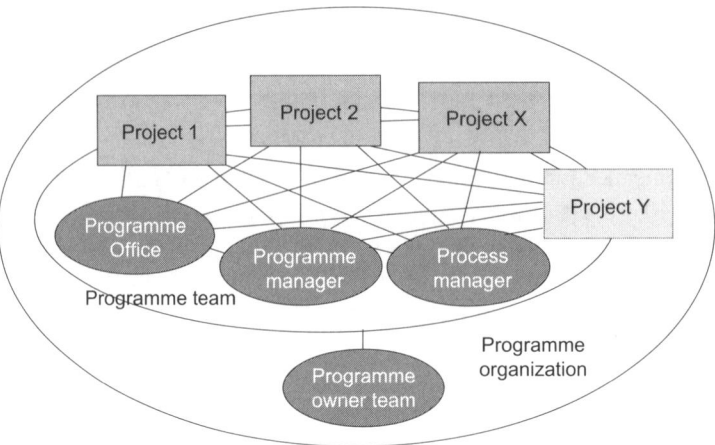

Fig. G3.2: Standard programme organization chart

be nominated who will act as the first contact person for the programme manager. An expanded programme owner team can also include as members representatives of partners and from important suppliers. The role of the programme owner team is described in Figure G3.3.

Role: Programme owner team
objectives
• coordinate programme and company interests • assign a programme manager to perform the programme • perform strategic programme controlling • lead the programme manager
position in the organization
• part of the programme organization • programme manager reports to the programme owner team
tasks
tasks in the programme start
• assign the programme • provide support in the selection of the programme team members • establish the programme context • ensure the availability of programme resources • contribute to the initial programme marketing • participate in the initial programme team meeting • hold programme owner team meeting

tasks in programme controlling
• hold programme owner team meetings
• coordinate the programme with other programmes, projects and company objectives
• analyze the programme progress report
• select the project owner team and project managers
• ensure the availability of programme resources
• continually contribute to programme marketing

tasks in resolving a programme discontinuity
• define a programme discontinuity
• collaborate in the development of and decisions about immediate measures
• collaborate in the cause analysis of the programme crisis
• decide about alternative strategies
• collaborate on the performance of corrective measures and checks on their success
• end a programme discontinuity

tasks in programme close-down
• hold programme owner team meeting
• participate in the closing programme team meeting
• formal programme approval

formal authority
• selecting the programme manager
• providing the programme budget
• changing the programme objectives
• definition of a programme discontinuity
• programme stopping
• programme close-down

Fig. G3.3: Role description: Programme owner team

The programme manager is responsible for the realization of the programme objectives together with the programme process experts and the Programme Office. The programme manager ensures the professional fulfillment of the programme management processes.

Because of the extensive management and marketing effort in programmes it is recommended that a Programme Office is established for the operational support of the programme managers. This is a major contribution to securing success in the programme. The integrative programme management tasks are institutionalized in the Programme Office. It establishes a "home base" for the programme. Programme management experts are available in the Programme Office for the individual projects and for representatives of the relevant environments.

In the programme "LKS 2000" the Programme Office was staffed by the programme manager (part-time: 50%), a programme secretary (100%), a programme marketing expert (50%), a programme controller (50%) and the programme consultant.

The roles of the programme managers, the Programme Office and the programme process experts are described in Figures G3.4-G3.6.

Role: Programme manager

objectives

- realize programme interests
- ensure the realization of the programme objectives
- lead the programme team
- represent the programme toward representatives of relevant environments

position in the organization

- member of the programme team
- reports to the programme owner team
- gets organizational support from the Programme Office
- leads the project owner team and the project managers

tasks

tasks in the programme start

- design the programme start process (together with the programme team)
- hold programme team meetings
- design adequate programme plans and the programme organization (together with the programme team)
- risk management and development of a specific programme culture (together with the programme team)
- design programme context relationships and perform programme marketing
- develop programme management documentation "Programme start" (with support from the Programme Office)

tasks in programme controlling

- design the programme controlling process (together with the programme team)
- hold programme team meetings and participate in programme owner team meetings
- coordinate all resources required in the programme
- agree on or perform controlling measures (together with the programme team)
- set project priorities within the programme
- adapt programme plans (with support from the Programme Office)
- develop programme progress reports (with support from the Programme Office)
- perform programme marketing tasks (with support from the Programme Office)
- start new projects (together with the relevant project owner team and project manager)
- perform strategic controlling of the current projects together with the respective project owner team; networks between current projects and the programme
- define a project discontinuity (together with the respective project owner team)
- contribute to project close-downs, transfer of know-how into other projects of the programme

tasks in resolving a programme discontinuity

- work out immediate measures (together with the programme team)
- perform cause analysis (together with the programme team)
- work out alternative strategies (together with the programme team)
- perform measures to resolve the crisis and check for success (together with the programme team)
- end the programme discontinuity (together with the programme owner team and the programme team)

tasks in programme close-down
• design the programme close-down process (together with the programme team)
• hold the programme close-down meetings (together with the programme team) and partic- ipate in the closing programme owner team meeting
• develop the programme close-down report
• transfer know-how into permanent organization and make agreements for the post-pro- gramme phase
• perform final programme marketing

formal authority
• decisions about coordinating the programme (together with the programme team)
• decisions on the project priorities within the programme
• responsibility for the programme budget
• calling programme owner team meetings
• coordination of resources required in the programme
• starting projects (together with the respective project owner team and project manager)
• changes to project objectives, definition of a programme discontinuity and closing down projects (together with the respective project owner team)

CHAPTER

G

Fig. G3.4: Role description: Programme manager

Role: Programme Office

objectives
• support the programme manager and the projects in the programme
• ensure the adherence to management standards in the programme
• administer the programme finances

position in the organization
• reports to the programme manager
• led by the Programme Office manager
• programme Office manager is a member of the programme team

tasks
• support the programme manager by adapting programme plans and developing programme progress reports
• prepare programme owner team and programme team meetings, prepare decisions for these meetings
• individual members of the Programme Office participate in programme owner team and pro- gramme team meetings
• perform essential programme marketing tasks
• support individual projects in project management and project marketing
• ensure the infrastructure for programme management and for project management (offices, computers, etc.)
• develop profitability analyses, financial analyses
• develop and ensure the use of programme management standards

formal authority
• joint decisions in the Programme Office

Fig. G3.5: Role description: Programme Office

Role: Process manager
objectives
• develop programme-specific standards for the fulfillment of content-related processes • ensure the quality of content-related processes in the programme • bring in expert knowledge to fulfill programme objectives • contribute to programme team
position in the organization
• assigned by the programme owner team • member of the programme team
tasks
• develop standards for content-related processes • ensure the adherence to standards, quality assurance • participate in programme team meetings
formal authority
• decisions regarding the design of content-related processes • agreements with programme owner team and programme manager

Fig. G3.6: Role description: Process manager

The programme team is made up of the programme manager, representatives of the Programme Office and, possibly, programme process experts. The active projects of a programme, or those which are due to start soon, are represented in the programme team by the respective project manager. The composition of the programme team changes over time since different projects are active at different times.

The tasks of the programme team are to ensure synergies in the programme and to set priorities between the projects of the programme. The role of the programme team is described in Figure G3.7.

Role: Programme team
objectives
• develop "added value" through the interaction of the programme team members in meetings and workshops • realize the programme interests • jointly responsible for programme success
position in the organization
• central communication structure in the programme • led by the programme manager • includes the programme manager, the Programme Office manager, members of the Programme Office, the process manager, project managers of current projects

tasks
• exchange information with the programme team members
• coordinate the programme
• make agreements
• construct a common point of view
• hold programme team meetings
formal authority
• agreements within the programme team
• agreements with the programme owner team

Fig. G3.7: Role description: Programme team

G 3.4   Programme work breakdown structure

The objectives to be realized in a programme require the performance of projects. The projects of the programme are, therefore, to be depicted in a programme work breakdown structure. Programmes also have other tasks to be fulfilled, such as training tasks or the programme management tasks, which do not require a project organization but can be performed as work packages of the programme (see Chapter G2.1, Programme "LKS 2000").

It is recommended that the objects of consideration of the programme are used as a basis for the development of the programme work breakdown structure. Objects of consideration can be, for example, services and products, regions and markets, objects (buildings, IT infrastructure), organizations and personnel groups.

When structuring a programme into projects a rough process orientation must be ensured. Due of the high degree of complexity of programmes there is a stronger object orientation (such as structuring by location) as in the structuring of projects. Programmes are to be structured so as to develop projects with holistic boundaries.

CHAPTER

G

⋯⟩ **CASE STUDY G2: Structuring the programme "eBanking & eBrokerage of the ABN AMRO Bank"**

Company, performing the Programme:

- ABN AMRO Bank is an international bank, originated in the Netherlands in 1824. It is the fourth largest bank in Europe and ranks eigth in the world. In 76 countries it operates some 3500 branches. It employs over 125 000 staff.
- Recently the bank created a European Division (see Figure G3.8). Within the Business Development organization the programme: eBanking & eBrokerage was performed.

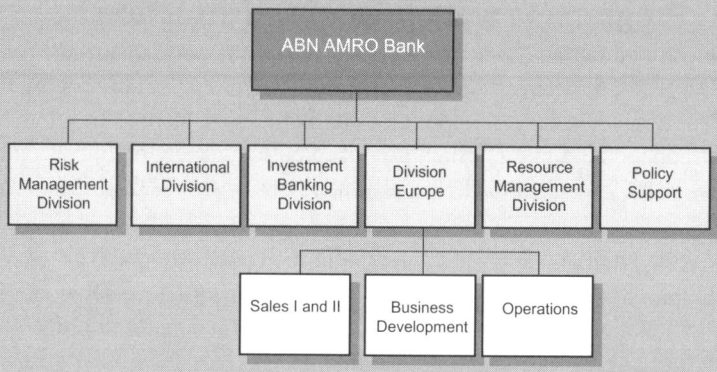

Fig. G3.8: Organization chart of ABN AMRO

- The realization of the single European market and technological developments, such as the Internet, force banks to reconsider their positions. New distribution channels such as call centers and the Internet are used to cater the needs of existing and new customers.
- ABN AMRO's main strategic challenge is to transfer from a traditional line organization into an innovative, project-oriented organization.

Programme Objectives:

- In March 2000 the business case for a Pan-European eBanking & eBrokerage infrastructure was accepted by the managing board of ABN AMRO. With the business case the development of a platform for Pan-European eBanking & eBrokerage and the launch of a pilot project in one country were decided.
- The programme started with the development of the platform. Also ABN AMRO Marketing and Sales started to acquire ventures. In July 2000 the first venture (Money Planet) was acquired.

Objects of Consideration of the Programme:
- The major groups of objects of consideration of the programme were: countries, products and services, market segments, the IT-architecture, the e-system, the organization, vendors and alliances, and the financing.
- The IT-architecture, the e-system, etc were fundamental for the programme. Their developments were defined as „conditional projects" of the programme.
- The objects, which could determine the structure of the programme, were the implementing countries (Netherlands, France, Germany, etc), the offered products and services (derivates, mortgages, pensions, etc) and the market segments (personal banking, private banking, small business).

Alternatives for Structuring the Programme:
- Based on the criteria, identified as basically influencing the overall structure of the programme, three alternative programme structures were developed and analyzed.
- The possible structures „Geographic Pioneer", „Products and Services", and „Market Segments" are presented in the Figures G3.9, G3.10, and G3.11.

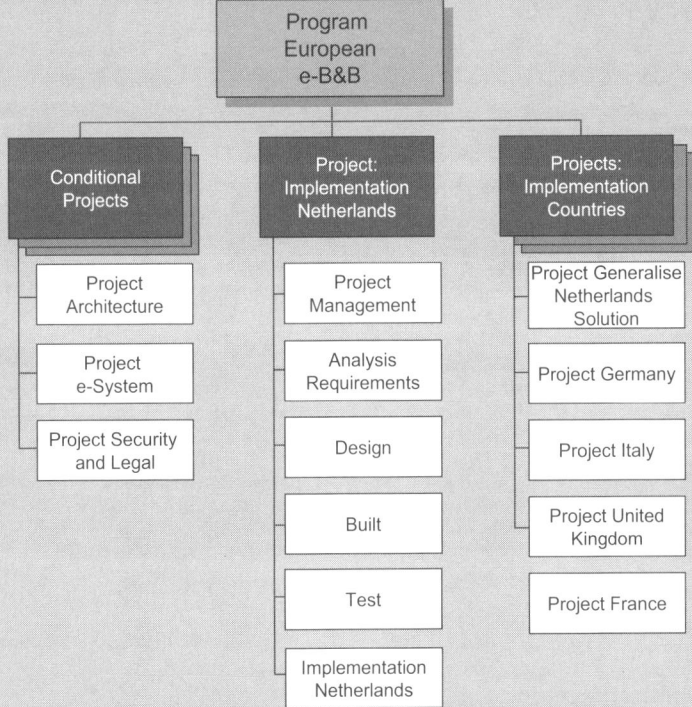

Fig. G3.9: Programme "eBanking & eBrokerage": Geographic pioneer

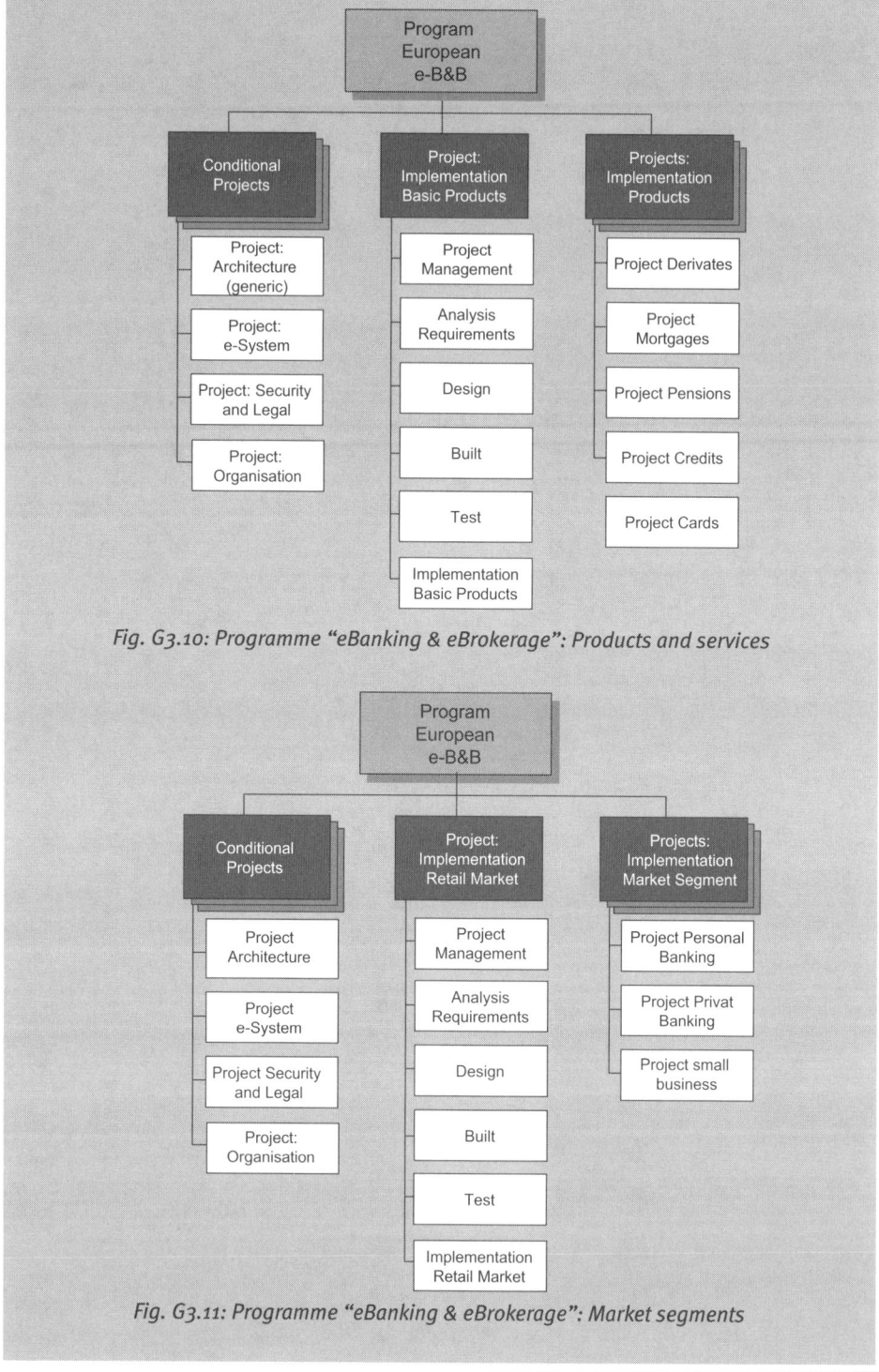

Fig. G3.10: Programme "eBanking & eBrokerage": Products and services

Fig. G3.11: Programme "eBanking & eBrokerage": Market segments

Interpretation:
- It was obvious for the programme team, that the structure considering the geographic criterion, was the most favourable. This was because
  - the marketing efforts are organized at a country level,
  - the Internet-Sites have to be in local languages,
  - possibilities to work with local partners,
  - legal barriers country by country,
  - the scalability of the IT-architecture and the e-system.

The projects of a programme are performed in parallel as well as sequentially. The sequential performance of projects in programmes leads to chains of projects. The design of chains of projects enables the differentiation between the projects in the chain. This is done by defining project-specific objectives, environments and organizations. Chains of projects also promote the integration of these projects, by planning the respective follow-on project in the previous project or by ensuring continuity of personnel in the projects of a chain (see e.g. the chain of projects "Conception laboratory" and "Realization laboratory" in the programme "LKS 2000").

Organizational learning in programmes can be ensured through the definition of pilot projects within the programme. Pilot projects should have available to them time, room for reflection and documentation of the experiences gained. The experiences gained in pilot projects are to be made available to follow-on projects (see e.g. the pilot project "OD Head Clinics" and the "OD follow-on" projects in the programme "LKS 2000").

During programme controlling it may become apparent that the dynamics of programmes require the splitting or merging of projects. Splitting projects into several projects or merging two or more project into one pose discontinuities for the affected projects. Their structures must be redesigned.

CHAPTER

G

G 3.5 Programme risk management

Due to the size, the complexity and the novelty of programmes, the element of risk is high. The danger of failure or of stopping before completion is especially high for programmes.

The object of consideration in programme risk management is the programme. This is in addition to the individual projects of the programme.

In programmes, too, risk analysis, measures to avoid or promote risk, as well as making provisions for risks and risk controlling are management tasks to be performed. In programmes a global as well as a detailed risk management can be performed.

In global risk management, the risks for the programme are analyzed without taking into account the risks of the individual projects and their relationships to one another. This method is the same as for projects. The programme is viewed as an individual project.

A detailed risk management for programmes accounts for the risks of the individual projects and the relationships between the risks of the projects of the programme. Because of the relationships between the projects, the programme risk is not equal to the sum of the project risks. The risks of several projects together can have positive or negative correlations or they can have a neutral effect on each other.

Only when there is no correlation between the risks of the projects of a programme is the programme risk equal to the risks of all the project risks. In the case of a positive correlation of project risks the programme risk is increased, in the case of a negative correlation the programme risk is decreased.

The basis of the programme risk analysis is the analysis of the risks of the individual projects of a programme. Then the correlation between the project risks can be analyzed. Positive correlation between project risks occurs, for example,

- in a chain of projects between the pilot and follow-on projects or a conception and a realization project,
- in cooperations with the same suppliers or partners in several projects of the programme,
- in the use of the same technologies in several projects of the programme,
- in the performance of several projects for the same customer and
- in the performance of several projects of the programme in the same country.

By deploying the same supplier or the same technology for several projects, "economies of scale" and learning potential can be utilized, which are reflected in lower project costs. On the other hand, a higher level of dependence, i.e. risks, on the chosen supplier or on the technology is created. With the loss of the supplier or if the technology is not yet ready for implementation, not only one project suffers, but several projects of the programme are damaged. Here one has to strike a balance between lower programme costs and the risk of an even higher loss for the programme.

A negative correlation between projects occurs when a risk in one project becomes acute and in doing so excludes a risk in another project. Such relationships between projects are seldom. Risk management in programmes must, therefore, concentrate on the reduction of positive correlation between the project risks.

Depending on the propensity for risk-taking (likes taking risks, neutral toward risks, adverse to risks) of the decision-makers of a programme (programme owner team, programme manager) different risk measures are possible. By adapting the correlations of the risks between projects the programme risk can be influenced.

G 3.6   Programme standards

In programmes there are often repetitive projects. In the programme "LKS 2000", for example, organizational development projects were performed for several hospital departments. In the programme "Railway stations initiative" a conception, a planning and a realization project were performed for each property.

For the performance of repetitive projects in a programme it is recommended that programme standards are developed. For the management of the projects standard project plans (such as standard work breakdown structures, standard work package specifications or standard milestone plans) can be developed and used.

For the fulfillment of content-related processes within the projects uniform structures, methods, rules and forms can be defined and provided for use in the projects. In the programme "Railway stations initiative", for example, the following standards for the fulfillment of content-related processes were developed and used:

- a standard frequency analysis in which the structure and the results of the analysis were specified, to analyze the number of people frequenting each property,

- a design manual, in which design standards, such as the use of typefaces and colors, the placement of vending machines, etc., were set down for the design of the properties and

CHAPTER

G

- standard leases and quality standards for potential tenants of the properties.

The objective of the use of programme standards is to develop a uniform way of working in the projects of a programme. Through the application of standards the projects are more closely coupled in the programme. The relative autonomy of the projects is, therefore, less than in projects which are not part of a programme.

The development and the use of programme standards ensure quality in the projects of programmes and promote organizational learning in programmes. The development of programme standards in programmes can be achieved in the form of a project (see Chapter G2.2). The programme manager and the process experts are responsible for the development and the use of programme standards.

G 3.7   Programme marketing

Because of the high strategic importance and the uniqueness of a programme the professional communication of the programme objectives and the programme structures play a large part in the success of the programme.

Professional marketing is, therefore, especially important in programmes. Only by professional marketing can an understanding of the meaning of the programme be communicated to the relevant programme environments, and the availability of management attention, of know-how and of resources for the performance of the programme ensured (see Chapter E1.8: Project marketing).

As a formal task of programme marketing in the programme "LKS 2000" a newsletter called "LKS 2000" was published roughly every three months (see Figure G3.12), There was a vernissage of the projects of "LKS 2000" for the employees of all three hospitals and the offices of the Salzburg state government, a symposium on "Project management in hospitals" was given and there were regular project presentations.

Fig. G3.12: Newsletter "LKS 2000" *Fig.G3.13: Symposium "Project management in hospitals"*

In the programme "Railway stations initiative" one member of the Programme Office was responsible for programme marketing. The Programme Office developed a marketing plan (see Figure G3.14) and had a corresponding marketing budget. Based on the intensive work on the programme culture (development of a programme logo, definition of programme colors, formulating of programme values and programme slogans) the programme-internal and the programme-external marketing measures were realized. Programme-internal measures were, for example, the configuration of the Programme Office according to the programme design, the production of notebooks, publishing a newsletter and performing project vernissages. Programme-external measures were advertising on the buildings during the renovation, advertising on Austrian television and in the print media, publication of programme brochures, etc.

	project team member	line manager	external partner	frequency
programme brochure	yes	yes	yes	every 2 years
programme newsletter	yes	yes	nein	every 6 months
project newsletter	yes	no	yes	every 6 months
programme vernissages	yes	yes	yes	as required

Fig. G3.14: Extract of the programme marketing plan "Railway stations initiative"

H

Assurance of the management quality in projects and programmes

Projects and programmes are relatively autonomous organizations, which can also be the object of consulting. Various consulting services, such as management coaching of members of the project organization, the moderation of project meetings, management consulting of projects, etc., can be fulfilled for projects and programmes.

The management of projects and programmes must be performed according to the guidelines for project and programme management of the respective project-oriented organization. Management consulting and management auditing of projects and programmes serve to ensure the application of these guidelines and with it the management quality of projects and programmes.

H Assurance of the management quality in projects and programmes

Contents

H 1 Management consulting of projects and programmes

Projects and programmes present new objects of consulting. In projects and programmes it can be differentiated between various types of consulting: Training and management coaching of individuals and teams, moderation of communication situations, consulting and auditing of content-related processes, management consulting and management auditing as well as external project or programme management.

Objectives of the consulting of projects and programmes are to ensure their content quality and their management quality. Quality management tasks in project-oriented organizations must be performed for their permanent and temporary organizations.

> ···⇢ **EXCURSUS H1: Qualitätsmanagement**
>
> Quality management originated in the beginning of the 20th century. The division of labor and mass production introduced by *Frederick W. Taylor* led to the requirement to control the quality of products.
>
> Quality management has since developed from a product-related quality control to an organization-related total quality management approach with the objective of continuous improvement of the business processes[a]. The individual development stages of quality management, according to *Krcza*[b] are briefly outlined below.
>
> The objective of quality control at the beginning of the 20th century was to check individual parts and products and eliminate defects. Products and individual parts were inspected. Since a 100% quality control was not possible, random samples were drawn – based on statistical parameters. It was *Walter Shewhart* who realized that it was necessary not only to control the product, but also the process. He developed control charts to visualize process output variation.
>
> *Deming*[c] is one of the founders of quality management. He showed that quality and productivity do not contradict each other, but correlate with each other positively, as long as production is perceived as a process which accounts for customer relations and customer feedback. *Deming* describes the following chain reaction: Quality improvement leads to lower production costs, since less defect correction measures are necessary. Fewer defects lead to a better utilization of the machines and the material. Lower production costs lead to higher productivity and enable better quality at lower prices. This leads to a higher market share, which in turn secures the existence of the company and thus that of the jobs. *Deming* also introduced the *Deming* cycle, which accounts for defect correction as well as defect prevention. The *Deming* cycle consists of the following steps: "Plan, Do, Check and Act – PDCA" (see Figure H1.1). This is the basis for the continuous improvement of business processes.

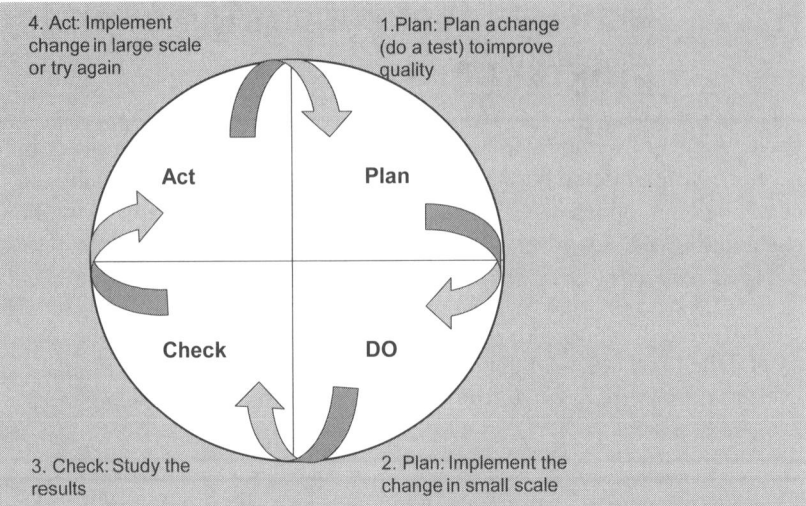

4. Act: Implement change in large scale or try again

1.Plan: Plan a change (do a test) to improve quality

Act

Plan

Check

DO

3. Check: Study the results

2. Plan: Implement the change in small scale

Fig. H1.1: Deming's PDCA cycle

Deming's approach was further developed by *Ishikawa* and *Taguchi* in Japan. *Ishikawa* postulated that all company units and all staff members are responsible for quality and that quality is defined by the customer. Customer does not only mean the end customer who pays for the product, but always the next person in the process. Therefore, every staff member is both a customer and a supplier at the same time[d].

From *Deming*'s approach and the further developments of the Japanese the total quality management approach emerged, which can be summarized by seven principles[e].

- customer orientation,
- continuous improvement of systems and processes,
- process management,
- search for the true reason of a problem: The problem is only a symptom – the true reason lies in the system or in the process,
- data collection and statistical methods,
- people orientation,
- team orientation.

In modern companies many different quality management-methods can be applied. These include certification (e.g. ISO certification, www.iso.ch), accreditation, excellence models (e.g. European Quality Award, www.efqm.org/International Project Management Award, www.gpm-ipma.de), benchmarking, auditing and reviewing, evaluation, coaching and consulting[f].

[a] Cp. *Seaver, M.,* (Quality Management) 2003.
[b] Cp. *Krczal, A.,* (Von Qualitätskontrolle zu Qualitätsverbesserung) 1999.
[c] Cp. *Deming, W. E.,* (Crisis) 1992.
[d] Cp. *Ishikawa, K., Lu, D.,* (Quality Control) 1985
[e] *Bounds, G. M., Dobbins, G. H., Fowler, O. S.,* (Quality Perspective) 1995; *Krczal, A.,* (Von Qualitätskontrolle zu Qualitätsverbesserung) 1999, p. 399 ff.
[f] Cp. *Huemann, M.,* (Improving Quality), 2004.

H 1.1   Overview of consulting types in projects and programmes

Traditional objects of consulting are permanent organizations, for example companies, profit centers or divisions. Due to the complexity and dynamics of projects and programmes a relatively high degree of support is required for the performance of content processes and management processes. Defining projects and programmes as temporary organizations enables their perception as objects of consulting. Projects and programmes can constitute "client" systems in consulting.

It is the objective of the consulting of projects and programmes to ensure their quality. The performance of the consulting of projects and programmes presupposes the definition of the consulting process and the definition of roles and methods for consulting.

The following consulting types for projects and programmes can be differentiated analogous to the general types of consulting according to *Titscher*[1].

- management training of persons and of teams in projects and programmes,
- management coaching of persons in projects and programmes,
- management coaching of teams in projects and programmes,
- moderation of communication situations in projects and programmes,
- content-related consulting and management consulting of projects and programmes,
- content-related auditing and management auditing of projects and programmes,
- external project or programme management.

Management training as well as management coaching of persons and teams are instruments of personnel and team development in projects and programmes, whereas moderating meetings, consulting and auditing are instruments of organizational development of projects and programmes.

Training of persons and teams in projects and programmes

Training of staff members can be performed either generally for participants in different organizations or specifically for the requirements of a project or a programme. The objective of project- or programme-specific training is the qualification of members of a project or a programme organization.

The objective of training services can be to develop competencies for the performance of content processes or of the management process in a project or a programme. Target groups for training in project and programme management can be the owner, the manager, team members of and contributors to a project or a programme.

The objective of common project- or programme-specific training is collective learning. Members of the project or programme organization acquire common methodological skills and learn a common language. Agreements on rules for the cooperation in the project or programme can be reached.

---

1) *Titscher, S.,* (Professionelle Beratung) 1997, p. 33 ff.

Management coaching of persons and of teams in projects and programmes

Management coaching of individuals in a project can be useful for the project manager, the project team members but also for members of the project owner team. Equally, in a programme the programme manager, members of the programme team and members des programme owner team can be clients of management coaching.

The objective of project-related management coaching is to support a person in meeting the demands of the project role to be performed. The competencies of the person are to be further developed to assure the project quality and to ensure compliance with the guidelines for project and programme management.

The person is the client of the management coaching and should therefore assign the coach. He/she will be supported in dealing with the complexity and the dynamics of the project, in designing the project management process, in implementing project management methods and in solving conflicts. The coach cooperates with the coached person in counseling meetings, in which, for example, feedback on project management documentations and on the designs of meetings is given. The coach also observes the coached person, for example in project meetings.

In project-oriented organizations coaching is required for project managers, because different kinds of projects are sometimes performed by new project managers with little experience. Since the project owner team's responsibility to lead the project managers is often not observed and the project manager lacks an "organizational home" (e. g. in a project management Expert Pool), a management coaching of the project managers by organization-internal or -external coaches is recommended.

In project-oriented organizations coaching situations are increasingly organized for the transfer of the training content into projects, too. In an Austrian production company there exists, for example, an organizational rule that project managers graduating from a project management training will be coached during implementation of the training content in the next suitable project.

Management coaching can be organized not only for individuals but also for project owner teams, project teams and sub-teams. The coaching of teams focuses on developing the respective team competence. Competencies are to be developed to enable joint designing of the "big project picture", to ensure commitment in the team, to solve conflicts and to utilize synergies in the team.

Moderating communication situations in projects and programmes

Communication situations in projects and programmes, which can be moderated by a consultant, are project owner team meetings, project or programme team meetings, workshops, presentations and project vernissages.

The moderation of communication situations in projects and programmes aims at optimizing the outcome of the respective communication situation. By introducing a consultant for moderation the project manager can delegate the control over the communication process and concentrate on content matters. The role of a moderator is team-external, however, he/she is part of the project or programme organization.

Moderation services include preparation, performance and follow-up of a project- or programme-related communication situation.

Content-related consulting and content-related auditing of projects and programmes

The objective of content-related consulting and auditing of projects and programmes is to contribute to solving a content problem – not a management problem – in the project or in the programme. Expert competence is provided by a team-external consultant or auditor.

In content consulting the consultant works relatively close with the project team or sub-team members. In any case he/she is awarded the status of an external expert with adequate detachment from the client system. He/she is therefore never a project team member.

The cooperation between the auditor and the project is not so close in content auditing, since the auditor does not develop a (common) solution, but gives feedback to existing results.

Management consulting and management auditing of projects and programmes

Management consulting and management auditing are central business processes of the project-oriented organization to assure management quality in projects and programmes.

Processes, roles and methods for management consulting are described in Chapter H2. Processes, roles and methods for management auditing are described in Chapter H3.

Temporary management of projects and programmes

Temporary management of projects and programmes can either refer to all management processes of the project or the programme or to individual processes, for example the process of resolving a project or programme discontinuity.

The temporary manager becomes a member of the project organization and is therefore part of the project. Temporary management is thus not a consulting but a management service.

H 2 Management consulting of projects and programmes

H 2.1 Management consulting of projects and programmes: Objectives and services

Management consulting of a project

The objective of the management consulting of a project is to (further) develop the management competence of the project. Consulting serves to implement the guidelines for project and programme management of the project-oriented organization and thereby assures management quality in the project. Not only the competencies of the project manager and/or the project team but that of the project as a temporary organization are developed. The project is the client of consulting.

Objects of consideration of management consulting of projects can be the project management process overall or the project management sub-processes. Only the project start process or the project controlling process, the management of a project discontinuity or the project close-down process can be the objects of consulting.

Management consulting of programmes

Management consulting of programmes differs from management consulting of projects with regard to the longer duration and higher complexity of programmes. Special services of management consulting of programmes are, for example, establishing a Programme Office, providing support in programme marketing and creating integrative programme standards (see Chapter G: Programme management). The management consulting of a programme should be performed in combination with the management consulting of projects of the programme to increase the efficiency of consulting.

The quality of management consulting can be assessed with regard to the management competence achieved in the project or programme. Improvements in the competencies of the members of the project or programme organization, in the efficiency of meetings, in the management documentation, in the image of the project or the programme and in the relationships to relevant environments can be observed.

Management consulting of projects and programmes must be differentiated from management consulting of a project-oriented organization, which aims at (further) developing the management competencies of the organization as a whole. Yet, frequently, management consulting of projects and programmes is performed in combination with the formal implementation of project and programme management in an organization (see Chapter J6: Development of project-oriented organizations).

CHAPTER

H

H 2.2 Management consulting process and application of methods

The process of management consulting of projects and programmes comprises the phases depicted in Figure H2.1. In each phase the cycle of information gathering, the development of hypotheses on the situation, the planning of interventions and the implementation of the planned interventions is run through.

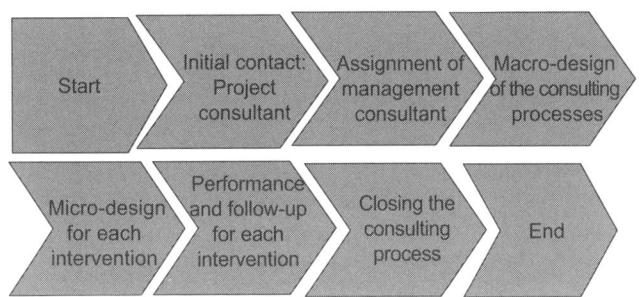

Fig. H2.1: Process of management consulting of projects and programmes

The phase of the initial contact between the project and the consultant is significant, because substantial relationships are established. The form of the request for consulting services (personal or per e-mail, by the project manager or a project team member), the extent of common history between the consultant and members of the project organization (cooperation experience, recommendation or picked from the "Yellow Pages") influence the expectations as regards the cooperation.

The information provided at initial contact enables the development of hypotheses regarding the consulting requirements and the client system project or programme. Potential hidden agendas may possibly be detected. It must be decided which project management sub-processes become the object of consulting. A rough consulting budget for the desired consulting services is to be determined (see Figure H2.2).

budgets for management consulting of projects		
• consulting of the project start process	5 – 7	consultant days
• consulting of the project controlling process	3 – 4	consultant days
• consulting of the management of a project discontinuity	8 – 10	consultant days
• consulting of the project close-down	3 – 5	consultant days

Fig. H2.2: Consulting budgets for different services of management consulting of projects

The owner of a management consulting of a project should be the project owner team, not the management board of a company or the manager of a division. They may initiate the consulting, but the project owner team itself should be positive about the benefit of consulting, desire it and agree upon it with the members of the project organization. Appropriate owners of the consulting assignments of various client systems are depicted in Figure H2.3.

client system	owner of the consulting assignment	agreement with/information from
programme	programme owner team	members the programme organization
project	project owner team	members the project organization
project team	project manager, representing the project team	project owner team
sub-team	sub-team manager, representing the sub-team	project owner team, project manager
project manager	project manager	project owner team

Fig. H2.3: Owners of consulting assignments

The objective of the macro-design of the consulting process is its planning. Depending on the scope of the consulting services the forms of communication (interviews, one-on-one meetings, observations, workshops and presentations), their objectives, schedules, durations and participants, and the types and forms of documentation must be roughly planned and agreed.

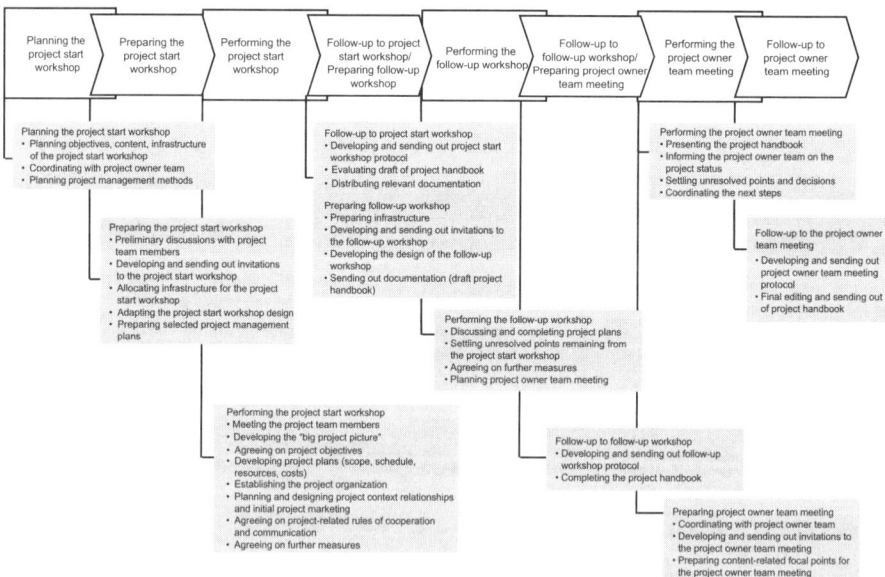

Fig. H2.4: Macro-design for project start consulting

In the micro-design for each individual intervention (interview, documentation analysis, workshop, etc.) the objectives, the schedule, the participants, the tools, etc. must be planned and prepared in detail. The planned interventions for further development of the project management competence of the project are to be implemented and follow-up work to be performed in each case.

The closing of the consulting process comprises filing the documentation, reflecting on and evaluating the consulting process, planning the next steps and the performing of a final meeting of the consultant with the consulting owner.

H 2.3   Roles in management consulting of projects and programmes

Management consulting of a project is not only performed by the consultant but also by a temporary consulting system, which comprises the consultant as well as members the project organization as project experts.

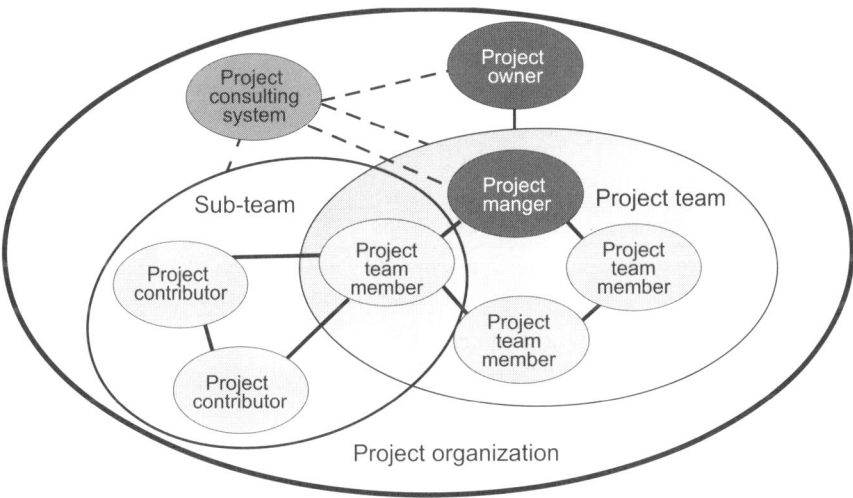

Fig. H2.5: Management consulting system as part of the project organization

The project representatives in the consulting system are experts with regard to the project, its content and problems, and they support the transfer of results of the consulting system into the project owner team and project team. The consultant, who is neither part of the project owner team nor of the project team, cooperates as project management expert with representatives of the project. The role of the management consultant is described in Figure H2.6.

Role: Management consultant of a project or a programme
objectives
• (further) develop the management competence of a project or a programme
position in the organization
• appointed by the project owner team
• part of the project organization
• not a member of the project owner team or project team
formal authority
• none
tasks
• gathering information, developing hypotheses on the project management competence of the project
• planning interventions for further developing the project management competence in the project
• implementing the interventions
• performing analyses, developing reports, providing project management expert opinion
• moderating project meetings, supporting the development of project management documentation
• management coaching of individual members of the project organization
environment relationships
• project owner team
• project team
• individual members of the project organization
• representatives of relevant project environments

Fig. H2.6: Role description: Management consultant of a project or a programme

To perform the consulting tasks, the consultant requires project and programme management competencies, consulting competencies, and knowledge about the project-oriented organizations that perform the project. The consulting competencies include competencies regarding the design of the consulting process as well as using methods of intervention and consulting standards. The consultant represents the values depicted in Figure H2.7.

The consultant is not an expert on the project content. He/she is not a leader in the project and therefore has no formal authority. The differences between the consultant and executives in projects are analyzed in Figure H2.8.

The management consultant of a project...
• .. does not have "better" but "different" perceptions compared to members of the project organization.
• ... does not keep secrets. Information he/she is provided with is adequately returned to the project organization.
• ... ensures the acceptability of interventions for the further development of the project management competence of the project.
• ... recognizes the expertise of the representatives of the project organization in the project management consulting system.
• ... does not balk at constructive conflicts.

Fig. H2.7: Values of the management consultant of a project

project management consultant	project owner or project manager
• has a co-responsibility • does not have formal authorities • performs consulting functions • stands outside the project team • provides different views	• has an overall responsibility • has formal authorities • performs project management functions • part of the project owner team or project team • has similar views as other team members

Fig. H2.8: Differences between management consultant and project leaders

Management consultants may come from within the project-performing organizations or be recruited externally. The internal consultant has the advantage of knowing his/her own organization and its agents and possibly costs less than the external consultant. The external consultant, on the other hand, has the advantage of having experience with other organizations and of being more detached from the client system.

The "home organization" the consultant represents strongly influences the potential consulting success. An internal consultant coming from a project management Expert Pool, for example, is bound to be accepted more easily than a consultant from the human resource department.

One can differentiate between "full-time" and "part-time" management consultants. Large companies sometimes employ full-time consultants. In most cases consulting tasks are perceived as a job enlargement of senior project managers or executives of the permanent organization. Since they are usually very busy, they are only available to a limited degree. A consulting pool combining (former) senior project managers and external consultants ensures the desired flexibility.

Frequently, people work simultaneously as management consultants and as management trainers. It is important to note that considerable differences exist between training and consulting (see Figure H2.9).

project management consultant	project management trainer
• contributes to concrete problem-solving • cooperates with members of the project organization • object of consideration: Project • works within the rules of the project	• general contribution to learning • mixed group of participants • object of consideration: Individuals, the group of participants • special training rules (working hours, venue)

Fig. H2.9: Differences between management consultant and management trainer

The degree of professionalism of the management consultant can be assessed through the compliance with formal processes in performing the consulting tasks. Professional consulting requires a clear definition of the objectives and the sequence of management consulting, the development of consulting documentation (minutes of meetings, workshop documentation, analyses, recommendations) and the calculation of consulting costs (number of days, day rate). In order to ensure the acceptance of the consulting company-internal consultants should not be assigned on an informal basis.

H 3 Management auditing of projects and programmes
(in cooperation with Martina Huemann)

H 3.1 Management auditing of projects and programmes: Objectives and services

Auditing is a quality assurance method which investigates the compliance with specified approaches and standards as well as their efficiency and usefulness. Auditing is a systematic business process which can be performed for projects and programmes.

Project auditing considers both the management quality and the content quality of a project. Management auditing of a project considers only the project management quality, i.e. the organizational competence of the project, the team competencies of the project team an the project owner team, and the individual management competencies of the members of the project organization.

The objective of management auditing is to further develop the project management competencies in the project. It is a learning opportunity for the audited project.[2]

Management audits are performed routinely or prompted by a specific reason. In order to give a project the opportunity to improve its project management competencies, the management auditing should be performed in a relatively early phase of the project process. Appropriate points in time for the management auditing are after the project start process or after some project controlling cycles have already been executed. It can be performed repeatedly during the performance of a project.

H 3.2 Management auditing process and application of methods

The management auditing process of projects and programmes comprises the following phases:

- situation analysis,
- planning the management auditing,
- preparing the management auditing,
- performing the analyses,
- developing the audit report,
- presenting the audit results,
- closing the management audit.

In Figure H3.1 the management auditing process is depicted in detail.

---

2) Management auditing of a project is more formal and binding than a project review. In a peer review, for example, a project manager gives feedback to another project manager. A "project health check" is also less formal than a project-auditing.

Tasks / **Responsibilities**	Auditing owner	Auditor(s)	Representatives of the audited project	Representatives of relevant project environments	Documents
Situation analysis					
• Description of situation and initial hypotheses		P			1)
Planning management auditing					
• Agreement on audit scope, methods	C	P	C		
• Development audit plan		P			2)
• Coordination, information audit plan	I	P	C	(I)	
Preparation management auditing					
• Provision of relevant project management documents for documentation analysis		P			
• Preparation interviews: Project organization		P	C		3)
• Preparation interviews: Relevant project environments		P	C	C	
• Preparation documentation analysis		P			
• Preparation self-assessments		P			
• Preparation observations (possible)		P			4)
• Preparation management auditing presentation		P			
• Documentation: General information on management auditing		P			5)
Performance of analyses					
• Performance of documentation analysis		P			6)
• Performance of interviews: Project organization	C	P	C		7)
• Information on intermediary results	I	P	I		
• Performance of interviews: Relevant environments		P			8)
• Performance of observations (possible)		P	C	(C)	9)
• Performance of self-assessment: team project management competence		P	C		10)
• Performance of self-assessment: individual project management competences		P	C		11)
Development of reports about audit results					
• Draft management audit report		P			12)
• Coordination, revision of management audit report	I	P	C		
• Completion of management audit report		P			
Presentation management auditing					
• Preparation management auditing presentation	I	P	C		
• Performance management auditing presentation	C	P	C	(C)	13)
Closing of management auditing		P	C		
• Distribution of management audit report		P		C	
• Feedback to management auditors	P	C			
• Approval of results	P	C			
• Management auditing follow-up agreement	P			C	14)

Legend:

P ... Performance
I ... Contribution
I ... Information

Documents:

1) situation analysis
2) management audit plan
3) interview plan
4) observation plan
5) questionnaire: General information on management auditin
6) questionnaire: Analysis of organizational project management competence
7) questionnaire: Project status and project ratios
8) interview protocol
9) observation protocol
10) self-assessment: Team project management competence
11) self-assessment: Individual project management competence
12) audit report
13) audit presentation
14) follow-up agreement

Fig. H3.1: Description of the management auditing process

The management auditing of a project starts with the assignment of the management auditing by the management auditing owner to the auditor(s). The owner of a management auditing of a project should be the project owner team. Yet the audit can be initiated by other roles in a project-oriented organization (e.g. manager of a profit center).

The approach or the guidelines, which form the basis for the management auditing of a project or a programme is to be agreed upon in the course of the assignment of the auditors. This basis may be a generic or a company-specific project and programme management approach. The forms and tools for the project management auditing depicted here are based on *ROLAND GAREIS Project and Programme Management*®.

The phases of the management auditing of a project are described below. The application of auditing methods is described in Case Study H1.

Situation analysis

It is the objective of a situation analysis by auditors to identify the reason for the auditing and the situation in the project. First assumptions on describing the situation and the project management competencies in the project are developed.

Planning the management auditing

Planning the management auditing results in the macro-design of the auditing process. The application of auditing methods (interviews, one-on-one meetings, observations, workshops and presentations) are specified. Finally, the auditing plan is arranged and agreed upon with the project manager as a representative of the project to be audited and the auditing owner.

Preparing for the management auditing

To prepare for the management auditing the auditors are provided project management documentations (e.g. the work breakdown structure, project environment analysis, project organization chart, start workshop protocol, project controlling reports) and other documents (correspondence) relevant for documentation analysis. The planned interviews, the observation and the self-assessment are prepared.

Performing the analyses

The analyses are performed cyclically. Based on the information gathered the hypotheses advanced in the situation analysis are extended or new ones are developed. If necessary, detailed analyses are planned. Based on the documentation analysis questions for interviews can be compiled.

A questionnaire to grasp the organizational project management competence can be used as a tool for performing the analyses. Figure H3.2 shows an extract of the questionnaire.

Project management audit questionnaire: **Analysis of the organizational project management competence** To be filled in by the project management auditor		
name of the audited project		
date of the project management audit		
name(s) of auditor(s)		
filled in by		
legend: • Existent: 0 = no document, 1 = information available, 2 = document available • Quality: 1 = not adequate, 2 = low quality, 3 = average quality, 4 = good quality, 5 = very good quality		
1 project start process		
1.1 project planning methods in the project start process	**existent**	**quality**
project objectives plan		
objects of consideration plan		
work breakdown structure		
work package specifications (for selected WP)		
project bar chart		
CPM schedule		
project resource plan		
project financing plan		
project cost plan		
project risk analysis		
project scenario analysis		
interpretation		
1.2. methods for designing the project context relationships in the project start process	**existent**	**quality**
pre- and post-project phase – analysis		
relationships to the corporate strategy – analysis		
project and other projects – analysis		
business case analysis		
project environment analysis		
interpretation		

Fig. H3.2: Extract: Questionnaire for management competence of a project

Developing the audit report

The audit report is developed by the auditors. Figure H3.3 shows the structure of an audit report. The report summarizes the results of the analyses and the recommendations for the further development of the project management competence of the audited project.

It is recommended to coordinate the report with the project manager of the audited project. This is not about any possible changes in the results but about ensuring the traceability of the audit results. The report serves as a basis for the measures for further development to be agreed between the project owner team and the project manager.

table of contents: Audit report	
1.	executive summary
2.	situation analysis, context and description of the auditing process
3.	description of the audited project
4.	analysis of the project management competencies in the project
	4.1 analysis project start: Organizational, team and individual project management competencies
	4.2 analysis project coordination: Organizational, team and individual project management competencies
	4.3 analysis project controlling: Organizational, team and individual project management competencies
5.	recommendations for further development of the project management competencies in the project
6.	general recommendations for further development of the competencies of the project-oriented organization
enclosures	

Fig. H3.3: Table of contents of the audit report

Presenting the audit results

Finally, the analysis results and the recommendations for further development of the project management competence of the audited project are presented by the auditors. The auditing owner, the project manager and members of the project organization should take part in the presentation. Representatives of relevant environments may be invited, too. The results can also be presented in the form of a workshop.

Closing the management audit

The formal closing between owner and auditors takes place in a closing meeting.

The auditors are not responsible for the implementation of the audit results. Implementation arrangements are made between the project owner team and the project manager or project team. The project owner team undertakes the controlling of the implementation of the agreed measures.

⋯⟩ **CASE STUDY H1: Management auditing of the contracting project "Implementation of an ERP system"**

company performing the project:
- international IT company as IT-contractor

project objective:
- performance of the customer contract for the implementation of an ERP (Enterprise Resource Planning)[a] system in a service company ("Customer")

strategic importance of the project for the IT supplier:
- very high, reference project

project budget of the IT supplier:
- € 700,000

project duration:
- 1,5 years

project status:
- project start performed

management auditing:
- initiator: PM Office of IT supplier
- auditing owner: Project owner
- management auditor: External project management consultant

Management auditing assignment for the project "Implementation of an ERP system"	
Project: Implementation of an ERP system	**Start date of auditing:** 23.02.2003
Reason for and timing of auditing: • Routine audit • After project start	**End date of auditing:** 05.03.2003
Objectives: • Analysis of the project management competencies in the project after the project start, • Analysis of the organizational, team and individual project management competencies, • Basis for an agreement between project owner and project team for further development the project management competencies.	**Non-objectives:** • Analyzing the content processes, • Interviews with all relevant environments.
Auditing methods: • Documentation analysis, • Interviews with members of the project organization and with representatives of selected relevant environments, • Self-assessment of the competencies of members of the project organization, • Observation of a project team meeting.	**Auditing budget:** • Auditor: 7 person days, • Members the project organization: 6 person days.
Initiator auditing: PM Office	**Project manager:** Mr. Z.
Auditing owner: Mr. M. = project owner	**Auditor:** Huemann
Management auditor	Owner
Version: 1.0	Date: 23.02.2003	Author: MH

Fig. H 3.4: Management auditing assignment for the project "Implementation of an ERP system")

Situation analysis: Iinitial Hypotheses

- The project is a strategically important project for the IT supplier, since it includes a new technological solution and is performed for an important customer. It is planned to be used as a reference project.
- The project start has been performed. Extensive project management documentation exists.
- Emphasis is placed on the technological solutions for the implementation of the ERP system. The necessary organizational changes in the customer's company are barely considered. There is no holistic consideration of the project objectives.
- There is a project manager of the project-performing IT supplier and a project manager of the customer. Optimization potential exists in the project organization.

Planning and preparing the management auditing:

- For information gathering purposes documentation analysis, interviews, an observation and self-assessments were used.
- The auditing plan is depicted in Figure H3.5.

CHAPTER

H

AUDITING PLAN

Working form	Content	Participants	Date	Place
Meeting: Start management auditing	▪ Assignment auditors	▪ Auditing owner ▪ Auditor	23.02 Duration: 1 hour	Room: 2.22
Meeting	▪ Clarifying auditing plan ▪ Clarifying self-assessments ▪ Handing over project management documentation	▪ Project manager ▪ Auditor	25.02 Duration: 1.5-2 hours	Room: 2.22
Self-assessment: Project management competence project manager	▪ Performing self-assessment project management competence project manager	▪ Project manager	27.02 Duration: 1 hour	
Self-assessment: Team project management competence	▪ Performing self-assessment team project management competence	▪ Project team	27.02 Duration: 3 hours	Room: 2.22
Group interview: Project manager and project team members	▪	▪ Project manager ▪ Project team member 1- 4 ▪ Auditor	01.03 Duration: 2 hours	Room: 1.12
One-on-one meeting: Project owner	▪	▪ Auditing owner ▪ Auditor	01.03 Duration: 1 hour	Room: 1.12
Group interview with customer representatives	▪	▪ "Project manager" of the customer ▪ Another representative of the customer ▪ Auditor	01.03 Duration: 1.5 hours	At customer's site
Observing the project team meeting	▪	▪ Project team ▪ Project management auditor	02.03 Duration: approx. 1.5 hours	Room: 1.22
Management auditing presentation	▪ Presenting auditing results	▪ Auditing owner ▪ Project manager ▪ Project team ▪ Interviewed customer representatives ▪ Auditor	03.03 Duration: 2 hours	Room: 1.10
Meeting: Auditing end	Approval of audit results	Auditor Auditing owner	05.03 Duration: 1 hour	Room: 2.22
Version: 1.0	Date: 23.02.2003	Author: MH		

Fig. H3.5: Management auditing plan for the project "Implementation of an ERP System"

Performing the analyses
Documentation analysis

- Extensive project management documentation was available, which was handed over by the project manager.
- As a tool for performing the analyses, the "organizational project management competence" questionnaire was used.
- The availability of project management documents was checked and the quality of the available project management documents was assessed.
- Figure H3.6 shows a sample analysis of availability and quality of project management documents after the project start process. Figure H3.7 shows the detailed analysis results of the work breakdown structure.

Interpretation the documentation analysis results

- Many project management plans were available, however, the quality of the documents, was average to low.
- The project was not perceived holistically. The work breakdown structure, for example, did not display the deliverables to be performed by the customer.

1.1 Project planning methods in the project start process	Document	Quality
Project objectives plan	2	2
Objects of consideration plan	1	-
Work breakdown structure	2	2
Work package specifications (for selected WP)	2	3
Project bar chart	2	3
CPM schedule	No demand	-
Project resource plan	0	-
Project financing plan	no demand	-
Project cost plan	2	4
Project risk analysis	2	3
Project scenario analysis	1	-
... etc.	...	...
Version: 1.0	Date: 26.02.2003	Author: MH

Legend:
Document: 0 = no document, 1 = information available, 2 = document available
Quality: 1 = not adequate, 2 = low quality, 3 = average quality, 4 = good quality, 5 = very good quality

Fig. H3.6: Analysis of the organizational project management competence of the project "Implementation of an ERP system"

PROJECT MANAGEMENT AUDITING: WORK BREAKDOWN STRUCTURE				
Project: Implementation ERP system				
Criterion	**Weight**	**Score**	**Weighted score**	**Interpretation**
Completeness	35	2	0.7	Not complete; deliverables to be performed by the customer are not depicted; only technical solution is displayed
Structuring	35	2	0.7	Structure is too rough to be used as a planning and controlling instrument; object-oriented, not process-oriented
Representation	20	2	0.4	OK; legend displayed, WBS code depicted; terms partly not task-oriented
Meeting formal criteria	10	4	0.4	Good, revision number indicated, date, author indicated
Total	**100**	**11**	**2.2**	**Low quality**

Quality: 1 = not adequate, 2 = low quality, 3 = average quality, 4 = good quality, 5 = very good quality

Fig. H3.7: Analysis of the quality of the WBS of the project "Implementation of an ERP system"

Self-assessment: Project management competencies

- Both a self-assessment of the project management competencies of the project manager and a self-assessment of the project management competencies of the project team were performed.
- The assessment of the project management competencies of the project team was based on questions regarding the development of the "big project picture" in the team, the establishment of commitment in the team, the utilization of synergies in the team, the solving of conflicts in the team, the learning in the team, and the joint designing of the project management process.
- Figure H3.8 shows a sample question.
- All project team members participated in the self-assessment of the project management team competence.

Interpretation of the self-assessment results

- Project management competence of the project manager: The result shows a great deal of knowledge and experience in the traditional project management methods, little knowledge and experience in designing project organizations.
- At the time of the management audit (shortly after the project start had been performed) the team project management competence of the project team of the IT supplier was existent to an average to low extent.

Development of team commitment	Knowledge	Experience
1 = none, 2 = low, 3 = average, 4 = high, 5 = very high		
Awareness of the importance of commitment	3	2
Visualizing results of meetings	2	2
Developing and checking TO DO-lists	4	3
Accepting joint responsibility for the project results	3	2
Interpretation on knowledge: The team is aware of the importance of commitment in the team, etc., some methods for developing commitment are known.		
Interpretation on experience: The team has consciously defined the objective of creating commitment in the team – the team accepts joint responsibility for the objectives. Results of meetings are rarely visualized, TO DO-lists are frequently used.		

Fig. H3.8: Self-assessment project management competence of the project team

Interviews

- Group interviews with the project team, a group interview with customer representatives and a meeting with the project owner were performed.
- Figure H3.9 shows the interview protocol of the group interview with the customer representatives.

Interpretation of the results of the group interview with customer representatives

- It became clear that the customer had already established a parallel "project organization". It was headed by the manager of the customer's IT department, Mr. W., who could only spend little time on the project due to overwork.
- Mr. W. was present at the project start workshop. However, he does not consider himself responsible for project management. In his view, this is solely the project-performing IT supplier's task.
- Mr. W.'s deputy, Ms. G., had developed a schedule for the tasks which in her view needed to be performed.
- There was little awareness of the necessity of organizational measures in connection with the implementation of the ERP system.

INTERVIEW PROTOCOL – group interview with customer representatives	
Project	Implementation ERP system
Date interview	01. 03. 2003; 4 p.m. to 6 p.m.
Place	Vienna, customer's office
Interview partner	Mr. W. (project manager of the customer and IT manager at the customer's), Ms. G. (Mr. W.'s deputy)
Interviewer	Huemann
Objectives of the interview	
• Analysis of the management of the project "Implementation of an ERP system" from the customer's point of view • Clarification of the customer's "project manager" role	
Process, observations and disruptions during the interview	
• Process OK, no disruptions • IT manager dominates	
Interview results	
Project start – involvement	• Participation of Mr. W. and Mr. F. at the start workshop to give input for the technical IT solution • No contribution to project management plans; this is Mr. Z.'s task
Role perception of the "project manager" of the customer in the project	• Coordination of the necessary IT tasks in his department • Controlling the deliverables to be performed
… etc.	…
Version: 1.0 Date: 01.03.2003 Author: MH	

Fig. H3.9: Extract from an interview protocol of a group interview

Developing the report and presentation

- A report was developed and a presentation of the audit results was performed.
- The IT supplier's project owner, project team and PM Office manager attended the presentation. On the customer's side the manager of the IT department and his deputy took part.
- The following results of the analyses were summarized:
  - The project shows a low degree of project management competence.
  - No holistic view of the project exists. The project is only perceived from the IT supplier's point of view. The deliverables of the customer are neither planned nor controlled.
  - Organizational and personnel-related changes necessary for the implementation of the ERP system in the customer's company are not considered.
  - Two project organizations exist simultaneously (the IT supplier's, and the customer's).
- The following recommendations for further developing the management competencies of the project were given:

- Optimizing the existing project management plans, holistic description of the project (including all customer deliverables),
- Establishing an integrated project organization, i. e. project owner and a representative of the IT company and a representative of the customer, only one project manager, project team members from both IT supplier and customer.

MANAGEMENT AUDITING FOLLOW-UP AGREEMENT		
Project: Implementation ERP system	**Project manager:** Mr. Z.	
End date auditing: 05.03.2003	**Project owner:** Mr. M.	
Auditor: Huemann		
Measures	**Responsible**	**Deadline**
Identifying a member for the project owner team of the customer	Customer representative	14.03.2003
Identifying project team members of the customer	Customer representative	14.03.2003
Performing joint workshops	Project manager	20.03.2003
… etc.	…	
Project manager	Project owner	
Version: 1.0	Date: 07.03.2003	Author: MH

Fig. H3.10: Extract of the management auditing follow-up agreement

(a) An ERP system covers all central accounting areas. It comprises the areas cost accounting, purchasing, disposition, stock-keeping and sales, as well as managing product information.

H 3.3   Roles in management auditing of projects and programmes

Management auditing of a project is performed within a temporary auditing system, in which the auditing owner, the auditors, the representatives of the project or the representatives of relevant environments of the project to be audited, cooperate.

Fig. H3.11: Management auditing system

In the auditing system rules for cooperation are to be agreed. To design a management auditing in a cooperative way, the auditor should periodically inform the project manager and the owner of the audited project about intermediary results. The communication policy is to be determined at the beginning of the auditing.

The operability and the quality of the auditing depend on the willingness to cooperate of the representatives of the project to be audited, the available time and the available resources.

Management auditing owner

A distinction can be made between the initiator of the management auditing and the owner of the management auditing. The initiator of the management auditing can be the project owner team, a representative of the PM Office, a representative of a profit center or an external customer. The owner of the management auditing should always be the project owner team, since the project is the "client" of the management auditing. The management auditing cannot be performed without the consent and the contribution of the project owner team. The role of the auditing owner is described in Figure H3.1.

Role: Management auditing owner
objectives
• ensuring the project management quality of the project to be audited • ensuring a learning opportunity for the project to be audited • ensuring the operability of the management auditing
position in the organization
• part of the auditing systems
tasks
• contributing to clarifying the situation • making contact between auditors and representatives of the project to be audited • agreeing with auditors on scope and application of methods • being available as interview partner (possibly) • participating in the audit presentation • giving feedback and approving audit results • follow-up agreement with representatives of the audited project
environments
• management auditors • representatives of relevant environments of the project
formal authority
• assigning of auditors • stopping the management audit • closing-down the management audit

Fig. H3.12: Role description: Management auditing owner

Representative of the project

The members of the project organization as representatives of the project in the management auditing system are experts with regard to the project. The role of the representative of the project is described in Figure H3.13.

Role: Representative of the project
objectives
• cooperating and providing information for the management audit • providing resources for the management audit
position in the organization
• part of the management auditing system
tasks
• contributing to the clarification of the situation • giving feedback to audit plan • providing project management documents for documentation analysis • being available as interview partner • performing a self-assessment of the individual project management competence • contributing to self-assessment of the project management competence of the project team • giving feedback to audit report • participating in the audit presentation • follow-up agreements with the project owner team
environments
• auditors • management auditing owner • project organization of the project to be audited • representatives of relevant project environments
formal authority
• none

Fig. H3.13: Role description: Representative of the project in management auditing

Management auditors

Management auditors are always project- or programme-external persons. If the management audit is performed by several auditors, they are called team of auditors. The team of auditors is headed by the lead auditor. The role of the management auditor is described in Figure H3.14.

Role: Management auditor	
objectives	
• analyzing and giving recommendations for further development of the project management competencies of the project to be audited	
position in the organization	
• part of the auditing system • assigned by management auditing owner	
tasks	
• clarifying the situation • clarifying the information policy with owner and project manager of the project to be audited • agreeing on auditing scope and application of methods • planning the communication structures of the management auditing • developing and coordinating the audit plan • preparing and performing the analyses • informing the project manager of the audited project • developing and coordinating the audit report • presenting the audit results	
environments	
• pool of management auditors • representatives of relevant environments of the audited project	
formal authority	
• requesting project management documents	

Fig. H3.14: Role description: Management auditor

To fulfill the tasks of management auditing the auditor requires project and programme management and auditing competencies. The auditing competencies include competencies for designing the audit process and for using audit methods. The management auditor is not an expert regarding the project content.

Auditors can come from the project-oriented organization performing the project or be recruited externally. In large project-oriented organizations there is a (virtual) pool of management auditors. In most cases management audit tasks are perceived as a job enlargement of senior project managers. For a good performance of the auditing tasks it is important for the auditor to have an appropriate degree of detachment from the client system.

H 4   Methods of management consulting and management auditing of projects and programmes

The methods described below are to be used in management consulting and management auditing of projects and programmes. A professional consultant and auditor of projects and programmes has competencies in the adequate use of these intervention methods. However, knowledge of the described methods is also beneficial to the project and programme manager. In project and programme management, too, the use of documentation analysis or of interviews is recommended, for example.

The methods used in management consulting and auditing serve to gather information and are at the same time interventions. An intervention is defined as a goal-oriented communication. A certain reaction from the communication partner is wanted. An intervention is an attempt to direct, to cause a significant difference.[3]

The intervention tree depicted in Figure H4.1 gives an overview of written, oral and analogous intervention methods.

A selection of intervention methods is described in the following.

H 4.1   Documentation analysis

Documentation analysis is a task which considers communications (letters, protocols, minutes, etc.) which have become manifest. "When people speak and write, they express their intentions, attitudes, their interpretations of situations, their knowledge and their tacit assumptions about their environment."[4] Through the content analysis of documents, such as papers, but also of photographs and movies, the characteristics apparent in language and form are identified and described[5]

In management consulting and management auditing documentation analysis is used to assess the project management quality of the project. Relevant project management documents for a documentation analysis are, for example, the work breakdown structure, the bar chart, the project environment analysis, the project organization chart and the project progress report, but also correspondence and project meeting minutes. Documentation analyses can be performed regarding the content and the form of the documents.

Documentation analysis is a process consisting of planning, preparation, performance and follow-up. Planning of the documentation analysis includes, for example, planning of the analysis criteria, the form and the scope of the documents to be analyzed.

---

3) *Willke, H.,* (Interventionstheorie) 1996, p. 12 f.
4) *Mayntz, R.,* (empirische Soziologie) 1974, p. 151.
5) Cp. *Lamnek, S.,* (Qualitative Sozialforschung 2) 1995, p. 172 f.

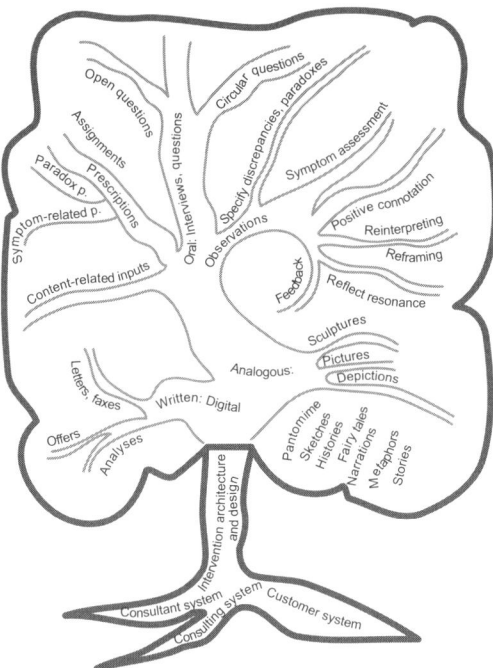

Fig. H4.1: Intervention tree[a]

[a] *Königswieser, R., Exner, A.,* (Systemische Intervention) 1998, p. 36.

Sometimes the challenge is to get the correct documents from the project manager. In many cases no distinction is made between content-related project documents and project management documents. The documents can be handed over in person, via mail or via e-mail. When the project manager hands the documents over in person, the auditor or consultant can get an overview of the existing documents together with the project manager. The return of the documents is to be agreed with the project manager.

The documentation analysis uses checklists as tools. Figure H4.2 shows a checklist describing the criteria for assessing the quality of a work breakdown structure.

Audit: Work breakdown structure				
project:			**company:**	
criterion	weight	score	weighted score	interpretation
completeness	30			• all phases to enable realization of project objectives, • project management as "phase", • objects of consideration used throughout.
structuring	30			• process orientation (phases), • adequate degree of detail: number of phases (6-8), • number of WP per phase (6-8)
representation	30			• Use of SW to develop a basis of data for scheduling, etc. and to enable professional project controlling. • Task orientation in WP relationships, consistency, adequate codification, objects of consideration visible.
meeting formal criteria	10			• author, version, date, page numbers
total:	100			
project management auditor:		Date:		Page:

Fig. H.4.2: Checklist for assessing the quality of a work breakdown structure

H 4.2   Interview

An Interview is an inquiry. Various types of interviews can be differentiated:[6]

- according to the degree of standardization (structured – partly structured – not structured),

- according to the degree of authority of the interview (soft – neutral – hard),

- according to the form of contact (direct – by phone – in writing),

- according to the number of persons interviewed (one-on-one interview – group interview – survey),

---

6) *Bortz, J., Döring, N.,* (Forschungsmethoden) 1995, p. 217.

- according to the number of interviewers (one interviewer – two interviewers – more than two interviewers),
- according to the function (investigating – mediating).

The qualitative interview can be described as an oral and personal form of inquiry which seeks to approximate an everyday communication situation. "This includes noticeable diffidence on the part of the interviewer during the discussion as well as the interviewer's possibility to be responsive to his/her respective interview partner."[7]

Characteristics of a qualitative interview are:[8]

- oral and personal,
- non-standardized,
- with open questions only,
- neutral to soft form of interview,
- with mediating but also investigating intention,

In management auditing and consulting of projects and programmes mainly qualitative interviews are used. In the course of the interview the interviewer seeks to get an idea of the project management competencies in the project. It is the respondent, not the interviewer, who is an expert in this. The gathering of information is foregrounded in an interview.

The interview partners are representatives of the project organization (e. g. project manager, project owner team, project team members) and representatives of relevant environments (e. g. representatives of suppliers, representatives of the customer). In management auditing and consulting one-on-one interviews as well as group interviews are used. In group interviews it is recommended to appoint two interviewers in order to comply with the complexity of the situation. The interviewers divide roles. One interviewer can, for example, ask questions and keep eye contact with the interview partners, while the other interviewer concentrates on the documentation.

An interview is a process consisting of planning, preparation, performance and follow-up. The planning of the interview includes, for example, planning the objectives, interview partners, sequence, place, documentation, and planning the questions. The performance of the interview consists of individual process steps, too: Introduction, sequence of questions, and end. In the introduction of the interview, for example the objectives, the context, the method of the interview and the form of documentation must be clarified, and the interview partner is to get an overview of the questions.

In an interview, different techniques are used, for example circular questions. Circular questions thematize the relationships between respondents and people who are not present, and the impact of these relationships is visualized.[9] Typical circular questions to ask a project manager, are: "How do you think the project owner team assesses the

---

7) Cp. *Kepper, G.,* (Qualitative Marktforschung) 1996, p. 35.
8) *Lamnek, H.* (Qualitative Sozialforschung 1), 1995, p. 38.
9) *Exner, A..,* (Unternehmensberatung), 1992, p. 210.

project management competencies in the project?" or: "How would you describe the relationship of the customer to the project?"

To support the interview, tools, such as an interview plan or an interview protocol, are used. An extract of an interview protocol is depicted in Case Study H1, Figure H3.9.

H 4.3 Observations of meeting

Observation is defined as "constituting a significant difference".[10] The observer is to determine differences to deduce significances from them.[11] This requires selection[12].

- Selective allocation: Definition of the content to be observed.
- Selective perception: Definition of what is to be attended to with regard to the selected content, when the observation starts and how long it lasts.
- Selective memory: The observation must be documented (observation protocol or audiovisual tools).

Management consulting and auditing of projects and programmes uses observations of project owner team meetings, project team meetings and project sub-team meetings for assessing the project management competencies in the project.

The observer is always part of the observation system. The observation is a process, consisting of planning, preparation, performance and follow-up to the observation. The planning of the observation includes, for example, planning the observation situation, number of observers, observation criteria, sequence, place and observation documentation.

The performance of an observation as such is a process, too, consisting of the following steps: Introduction, observation sequence and end. In the introduction the observer introduces himself/herself and explains the context, the method, the objectives, and the form of documentation.

The observer should be seated separately from the persons acting in the meeting. Observation is a method rarely used in the management context. Whether the observation can be used in a management consulting and auditing is thus highly dependent on the culture of the project or programme.

As tools for an observation checklists and observation protocols are used.

H 4.4 Reports

A report is a written form of intervention. In a report the results of analyses and recommendations are documented.

Like interviews and observations, reports require planning, preparation, performance and follow-up. The planning of the report includes, for example, planning the objectives,

10) *Willke, H.,* (Interventionstheorie), 1996, p. 12.
11) *Willke, H.,* (Interventionstheorie), 1996, p. 13 ff.
12) *Weick, K. E.,* (Systematic Observation Methods) 1968, p. 357 ff.

CHAPTER
H

the target group, the structure of the report, the number of pages and the process of developing it. The client system, for example, can be involved in the process of developing the report by providing the client with a draft of the report, with the request for feedback. That way, understanding and acceptance of the result are increased.

For reports in management consulting and auditing tables of contents and sample reports can be used as tools. Figure H3.3 shows a table of contents of a management auditing report.

H 4.5   Presentations and workshops

Presentations and workshops serve, on the one hand, to present (intermediary) results, and on the other hand, to gather further information.

In management consulting and auditing of projects and programmes presentations and workshops are used to design a picture of the project management competencies in the project together with representatives of the project organization and representatives of relevant environments. A presentation of the results enhances common understanding and acceptance of the results in the client system.

As opposed to workshops, presentations are shorter and allow less interaction. Presentations and workshops require planning, preparation, performance and follow-up. Planning the presentation includes, for example, planning the objectives, the sequence, the duration, the participants, the moderation, the tools and the documentation.

For workshops and presentations standard agendas can be used.

H 5 Institutionalization of management consulting and management auditing of projects and programmes in the project-oriented organization

Management consulting and management auditing of projects and programmes are gaining importance to assure the management quality in project-oriented organizations. Both processes are important integration instruments in project-oriented organizations.

Management consulting and auditing develop into new external, but also organization-internal services. Organizations develop internal consultants and auditors and establish "Project Management" Expert Pools, members of which are internal and possibly also external consultants and auditors. "Management consultant and auditor" is a new role in the project management career path.

The development of internal management consultants and auditors includes the following steps:

- assuring project management competence (training and certification in project management, gaining experience as a project manager),
- gaining social competence through group dynamics trainings, team work, etc.,
- gaining an understanding of the role of the consultant (consulting approach, consulting standards) through instruction to become a project management trainer and consultant and through networking with project management consultants,
- gaining experience as a project management trainer,
- cooperating in management consulting and auditing with senior consultants,
- management consulting and auditing supervised by a senior consultant,
- autonomous management consulting and auditing.

In addition to individual competencies, organizational standards for management consulting and auditing are to be established, too. Agreements on internal service charges are to be made, specifications of the services are to be undertaken, the processes and methods are to be described and tools are to be provided.

A large Austrian petroleum and gasoline company, for example, has established special guidelines for management auditing of projects and programmes (see Figure H4.3) to this end, and has trained auditors in management auditing of projects and programmes.

Table of contents:
Guidelines for management auditing of projects and programmes
1. Introduction
2. Definition: Management auditing of projects and programmes
3. Description of the process of management auditing
4. Description of the roles in management auditing
5. Methods for management auditing
6. Forms and checklists for management auditing

Fig. H4.3: Table of contents: Guidelines for management auditing of projects and programmes

References

Seaver, M. (Ed.), (Quality Management) Gower Handbook of Quality Management, Gower, Aldershot, 2003

Krczal, A., (Von Qualitätskontrolle zu Qualitätsverbesserung), Von der Qualitätskontrolle zur kontinuierlichen Qualitätsverbesserung, in: *Eckardstein, D., Kasper, H. Mayerhofer, W.* (Ed.), Management, Thoerien – Führung – Veränderung, Schäffer-Poeschel, Stuttgart, 1999

Deming, W. E., (Crisis) Out of the Crisis, Cambridge University Press, Cambridge, 1992

Ishikawa, K., Lu, D., (Quality Control) What is Total Quality Control? The Japanese Way, Prentice Hall, Englewood Cliffs, NJ, 1985

Bounds, G. M., Dobbins, G. H., Fowler, O. S., (Quality Perspective) Management – A total Quality Perspective, South-Western Publication, Cincinnati, Ohio, 1995

Huemann, M. (Improving Quality) Improving Quality in Projects and Programs, in: *Morris, P. W., Pinto, J.,* The Handbook of Managing Projects, Wiley & Sons, 2004

Titscher, S., (Professionelle Beratung) Professionelle Beratung – Was beide Seiten vorher wissen sollten…, Ueberreuter, Wien, 1997

Königswieser, R., Exner, A,. (Systemische Intervention) Systemische Interventionen, Architekturen und Designs für Berater und Veränderungsmanager, Klett-Cotta, Stuttgart, 1998

Mayntz, R.; Holm, K.; Hübner, P., (empirische Soziologie) Einführung in die Methoden der empirischen Soziologie, Westdeutscher Verlag, Opladen, 1974

Lamnek, S., (Qualitative Sozialforschung 1) Qualitative Sozialforschung. Band 1. Methoden und Techniken, Psychologie Verlags-Union, Weinheim, 1995

Lamnek, S., (Qualitative Sozialforschung 2) Qualitative Sozialforschung. Band 2. Methoden und Techniken, Psychologie Verlags-Union, Weinheim, 1996

Bortz, J., Döring, N. (Forschungsmethoden) Forschungsmethoden und Evaluation, Springer-Verlag, Berlin, 1995

Kepper, G.: (Qualitative Marktforschung) Qualitative Marktforschung: Methoden, Einsatzmöglichkeiten und Beurteilungskriterien, 2. überarb. Auflage, Dt. Univ.-Verl., Wiesbaden, 1996

Exner, A., Königswieser, R., Titscher, S., (Unternehmensberatung) Unternehmensberatung – systemisch, in: Das systemisch – evolutionäre Management, Orac Verlag, Wien, 1992

Weick, K. E., (Systematic Observation Methods) Systematic Observation Methods, in: *Gardner, L.,* The Handbook of Social Psychology, Volume Two: Research Methods, Addison-Wesley, Reading (Massachusetts), 1968

Wilke, H. (Interventionstheorie) Systemtheorie II: Interventionstheorie, Lucius & Lucius, Stuttgart, 1996

CHAPTER

H

I

Project portfolio management

The clustering of projects and programmes into project portfolios, networks of projects and chains of projects can create synergies for the management of project-oriented organizations.

The assigning of a project or a programme, the coordination of a project portfolio and the networking of projects are business processes which are part of the project portfolio management of the project-oriented organization.

Important methods for complying with these business processes are the investment proposal and the project proposal, the investment portfolio score card, the business case analysis, the project assignment, the project portfolio database, the project portfolio score card and the network of projects graph.

I Project portfolio management

Contents

I 1 Project clusters and business processes for project portfolio management

I 1.1 Project clusters: Project portfolio, networks of projects and chains of projects

Project-oriented organizations simultaneously perform a multitude of various projects and programmes. They are therefore highly differentiated organizations. To comply with integration tasks, projects (and programmes) can be clustered into project portfolios, networks of projects and chains of projects. Synergies can be created by means of clustering several projects and considering relationships between the projects. Furthermore, this clustering can promote the realization of the strategies and objectives of the project-oriented organization. Considering relationships between the projects in the relevant cluster of projects develops optimization opportunities.

A project portfolio is the set of all projects of a project-oriented organization. A project portfolio takes into consideration all the current (and planned) projects and programmes at a given point in time. If a project-oriented organization holds many projects in its portfolio (e.g. more than 100 projects), it is sensible to define several project portfolios for different types of projects (e.g. contracting projects, product development projects, etc.). A project portfolio presents a point in time-related analysis.

A network of projects is a set of closely-coupled projects held within the project portfolio. Various criteria can be used for coupling projects into networks of projects, such as the use of a joint technology for projects, the performance of projects in the same geographical region or the performance of projects for a common customer.

A chain of projects is a set of sequential projects for the performance of several related business processes. It represents a specific form of a network of projects. A chain of projects is analyzed over a period of time. Chains of a project and a programme are also possible (e.g. a conception project followed by a realization programme).

In comparison to projects and programmes, project portfolios, networks of projects and chains of projects are not organizations but rather are clusters of organizations. As such, they are objects of management consideration.

Relationships between projects		
Set of successive projects	Set of all projects of a project-oriented organization	Set of closely-coupled projects
Observation of a period	Point in time observation	Point in time observation
↓	↓	↓
Chain of projects	Project portfolio	Network of projects
Clusters		

Fig. I1.1: Clusters of projects

I 1.2   Objectives of project portfolio management

The general objective of project portfolio management is the optimization of project portfolio results. From the project-oriented organization's point of view, it is not the optimization of the results of individual projects or programmes which is striven for, but rather the optimization of the results of the project portfolio. This objective can conflict with the optimization of the objectives of individual projects.

Different objectives are pursued in the various business processes of project portfolio management. The central objectives of assigning a project or a programme are the selection of a favorable investment and the adequate organization form for initializing the investment. The objectives of the project portfolio coordination are the coordination of the objectives of the projects of the portfolio, the coordination of the internal and external resources used in these projects, the organization of learning from and between the projects, and the determination of the projects' priorities. The objective of the networking of projects is the creation of synergies between the projects of a network.

I 1.3   Business processes and methods for project portfolio management

The assigning of a project or a programme, the coordination of a project portfolio and the networking of projects are business processes for the project portfolio management. The managing of a chain of projects is to be seen as a specific form of networking of projects.

The assigning of a project or a programme is to be seen as a business process for the project portfolio management since a decision to start a new project or programme ought not to be taken as an isolated decision but rather in the context of the thus newly-created project portfolio.

The methods for the different business processes of project portfolio management are depicted in Figure I1.1 and are described in the following chapters.

methods for assigning a project or a programme	
• investment proposal	must
• investment portfolio score card	can
• business case analysis or cost-benefit analysis	must
• project definition	must
• project proposal	must
• project assignment	must
methods of project portfolio coordination	
• project portfolio database	must
• project portfolio score card	must
• other project portfolio reports	can
• project proposals	must
• project progress reports	must
methods for networking of projects	
• networking workshop	can
• network of projects graphics	can
• project portfolio reports	can
• project progress reports	can

Fig. I1.2: Methods of project portfolio management

The tasks involved in integrating projects and programmes in project portfolios, networks of projects and chains of projects are fulfilled by specific, permanent organization units of the project-oriented organization, i.e. the Project Portfolio Group, the PM Office and Expert Pools (see Chapter J).

CHAPTER

I

I 2 Assigning a project or a programme

I 2.1 Business process: Assigning a project or a programme[1]

Objectives of assigning a project or a programme

Assigning a project or a programme is a business process of the project portfolio management.

It is the objective of the assigning process to make decisions regarding the realization of an investment and regarding the organization form of its initialization. The consequences of a proposed investment for the investment portfolio of the project-oriented organization are to be considered as part of the analysis of a proposed investment. Those investments which make an optimal contribution to the realization of the strategic objectives of the project-oriented organization are to be selected.

Possible organizational alternatives for the initialization of an investment are the project organization, the programme organization and the permanent organization.

By defining the assigning of a project or a programme as a business process, its tasks and the necessary decisions are formalized. The clear differentiation between an investment decision and an organization decision contributes to assuring the quality of decision-making in order to reduce risks of bad investments and inadequate organization forms for the initialization of investments.

The assigning process considers those investments whose initialization probably requires a project or a programme. For small investments, for example investments in a piece of equipment or furniture, no business case analysis needs to be performed. For such investments decision-making is less complex. In these cases, the investment decisions are within the authority of managers of profit centers or Expert Pools, and do not require the involvement of an investment decision committee or a Project Portfolio Group.

Tasks in assigning a project or a programme

The business process of assigning a project or a programme starts with the formulation of a problem or a reason for an investment, and ends with the assignment of a project. The following are the phases of the assignment process:

- developing the investment idea,
- developing an investment proposal and a project proposal,
- investment decision-making,

---

1) The terminology used for describing the assignment process is adequate for internal projects. For (external) contracting projects, this business process of assigning a project or a programme corresponds to "developing an offer". The decision of whether or not to perform a customer contract, and the selection of the adequate organization form comply with the decision situations in the assignment process.
Both the development of an offer and the assignment of an internal project can be so complex that they themselves are carried out in a project form. They are then an offer development project or a conception project.

- organization decision-making, and
- assigning the project or programme.

Methods for assigning a project or a programme are:

- investment proposal,
- investment portfolio score card,
- business case analysis or cost-benefit analysis,
- project definition,
- project proposal, and
- project assignment.

The business process "Assignment of a project or a programme" is depicted in detail in Figure I2.1.

Objectives of assigning a project or a programme

- Decision regarding the realization of an investment, considering the strategic objectives of the project-oriented organization,
- description of the investment (of the business case): Problem formulation, actual analysis, alternative solutions, investment analysis,
- decision regarding the organization form for the initialization of the investment,
- project definition (including a first draft of a project management documentation) for the project for initializing the investment,
- clarification of the availability of the required project resources,
- nomination of the project owner team,
- assigning of the project manager and the project team to perform the project.

Time boundaries of assigning a project or a programme

Start:	investment idea defined
End:	project or programme assigned
Duration:	from establishment of proposal team: 2-8 weeks

CHAPTER

I

Tasks and responsibilities in assigning a project or a programme

Tasks	Promoters	Proposal team	PM Office	Investment decision committee	Project Portfolio Group	Expert Pool manager	Project/programme owner team	Project/programme manager	Results/documents
Developing the investment idea									
• Documenting the idea	P								
• Concretizing the idea	P		C						1)
• Quick Check	C			(P)	P	C			2)
Developing the investment proposal and the project proposal									
• Forming the proposal team	C	C	C	(P)	P				3)
• Gathering information	I	P	C			C			
• Draft of proposals	I	P	C						
• Review of proposals	C	C	C	(P)	P	C			
• Revision of proposals		P	C						4)
Investment decision									
• Presenting the investment proposal	C	P	C						
• Investment decision-making	I	I		(P)	P	I			
Project decision									
• Presenting the project proposal for the project decision		P	C						
• Project decision	I	I	C	(I)	P	I			
• Nominating the project owner team					P				
Project assignment									
• Nominating the project manager						I	P	I	
• Formulating the project assignment						C	C	P	5)
• Assigning the project team						I	P	C	

Legend:
P ... Performance
I ... Contribution
I ... Information

Results/documents:
1) investment proposal for quick check
2) decision protocol quick check
3) proposal made to proposal team
4) investment proposal and project proposal
5) assignment of project or programme

Fig. I2.1: Assignment of a project or a programme: Responsibility matrix

The decision regarding the realization of an investment and the decision regarding the adequate organization form for its initialization can be taken by the Project Portfolio Group. Many project-oriented organizations employ an investment decision committee for investment decision-making purposes, and use the Project Portfolio Group only for

CHAPTER

I

project portfolio coordination. In the decision-making process the managers of Expert Pools are to be involved in clarifying the availability of the necessary project resources.

Results of assigning a project or a programme

The results of the assignment process are the investment proposal and the project proposal, the business case analysis, the decision protocol and any project or programme assignment. The decision regarding the performance of the investment and the decision regarding the organization of its initialization are documented in the decision protocols.

In order to assess the quality of the assignment process, in addition to these results, the planning of the business process, i.e. its duration, costs and organization, must be considered. Experience has shown that the assigning process frequently takes a long time and that the responsibilities and decisional authorities are not clear. There are great potentials for optimization in this field.

The business process of assigning a project forms the basis for the start of a project. The project objectives defined in the course of the assignment should, therefore, be as complete and operational as possible. The first draft of the project management documentation developed during the assignment process forms the basis for project management.

CHAPTER

I

Investment decision versus organization decision

A fundamental objective of the assigning process described above is the distinction between investment decision and organization decision. It is only once the decision for an investment has been taken that the precondition for the performance of a successful project is created!

Both experience and theory have shown that, frequently, no clear distinction is made between the investment and the project initializing the investment as well as between the investment and organization decision. In the project portfolio selection framework depicted in Figure I2.2, which is basically interesting from a structural point of view, sometimes the project term is used instead of the investment term (e.g. "project proposal" instead of "investment proposal" and "individual analysis of projects" instead of "individual analysis of investments"). The portfolio selection must also refer to the investment portfolio.

449

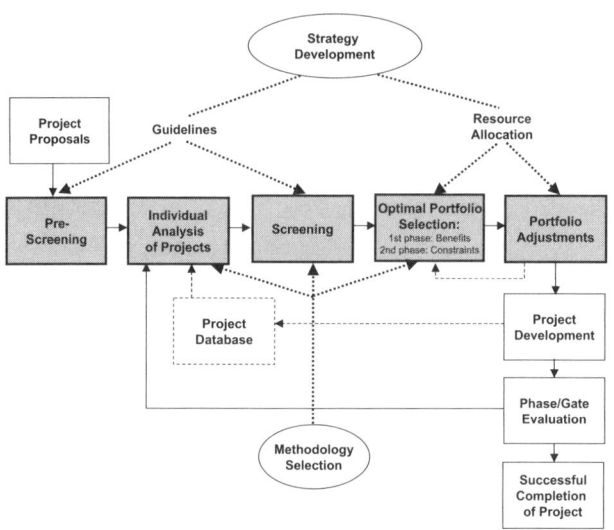

Fig. I2.2: Project portfolio selection framework by Archer and Ghasemzadeh[2)]

I 2.2 Investment proposal and investment portfolio score card

Investment proposal

The investment proposal serves the purpose of describing an investment to be applied for. The investment proposal summarizes the problem formulation and the reason for the investment, the investment objectives, a description of the investment object, the required first payments for the investment, the contributions of the investment for realizing the financial objectives, the contributions to realizing the other objectives of the project-oriented organization and the organization form for the initialization of the investment (see Figure I2.3).

---

2) *Aarto, K., Martinsuo, M., Aalto, T.,* (Strategic Management), 2001, p. 29.

INVESTMENT PROPOSAL
Name of the investment:
Type of investment: ❑ services ❑ organization ❑ marketing ❑ personnel ❑ infrastructure ❑ environment relationships
Problem formulation, reasons for the investment:
Investment objectives:
Description of the investment objects:
First payments for the investment:
Contribution for realizing financial objectives:
• net present value:
• ROI:
• benefit-cost ratio:
• amortization period:
• risks:
Contribution for realizing customers' objectives:
Contribution for realizing environment-related objectives:
Contribution for realizing innovation objectives:
Contribution for realizing business process and resource objectives:
Organization form for initializing the investment: ❑ small project ❑ project ❑ programme ❑ permanent organi- zation
Appendix: • Business case analysis or cost-benefit analysis, project proposal
Promoter Proposal team
Version: Date: Name: Page 1 of 1

Fig. I2.3: Investment proposal form

The investment proposal forms the basis for the decision by the investment committee or the Project Portfolio Group regarding the performance of an investment.

The description of the problem formulation or of the reasons for an investment constructs a common point of view regarding the actual situation before an investment is implemented. The description of the investment objectives serves to depict the situation striven for by the investment. The investment object (e.g. a building or a product) is roughly specified in the description of the investment. The contributions towards the realization of the objectives of the project-oriented organization can be differentiated according to financial objectives, customer relationship objectives, business process and resource objectives and innovation objectives.

A crucial criterion for decision-making regarding the performance of an investment is its financial success. The business case analysis or the cost-benefit analysis (see Chapter F3) are thus to be enclosed with the investment proposal. The investment proposal is to be supplemented by a project or programme proposal for the initialization of the investment (see Chapter I2.4). The investment proposal is therefore to be differentiated from the project proposal.

CHAPTER

I

Investment portfolio score card

The decision to perform an investment can either be taken by an isolated analysis of the investment or by an integrated analysis of the investment portfolio. In the case of an isolated analysis of the investment, the contribution of the investment toward the realization of the strategic objectives of the project-oriented organization is assessed. If considering several investments the results of the individual investments are compared with each other and the "best" investments are selected. As a tool for this, a scoring table such as that from the ABN AMRO bank (see Figure I2.4) can be used.

Investment	Financial Perspective	Customer Perspective	Internal Processes	Innovation	Score	Approved/ Denied
A					90	Approved
B					86	Approved
C					81	Approved
D					70	Denied
E					65	Denied

Fig. I2.4: Table for scoring investments (example from ABN AMRO bank)

In the case of an analysis of the investment portfolio, the investments and their relationships are analyzed. Complementary and competing relationships between investments are analyzed and integrated into the overall assessment of the portfolio. Alternative investment portfolios can be compared with the aid of several investment portfolio score cards. Considering the strategic objectives of the project-oriented organization, the selection of investments is made as a decision for the optimal investment portfolio.

An investment portfolio score card is a method for analyzing the investment portfolio. The total of all planned investments and of investments currently initialized can be defined as the investment portfolio of an organization.

In accordance with the balance score card model (see Excursus F2), various factors can be used to analyze the investment portfolio. Basically, the realization of financial objectives, objectives concerning relevant environments, innovation objectives and business process and resource objectives are considered. These factors must be operationalized to enable the analysis of the compliance with individual objectives. This is done in the investment portfolio score card depicted in Figure I2.5.

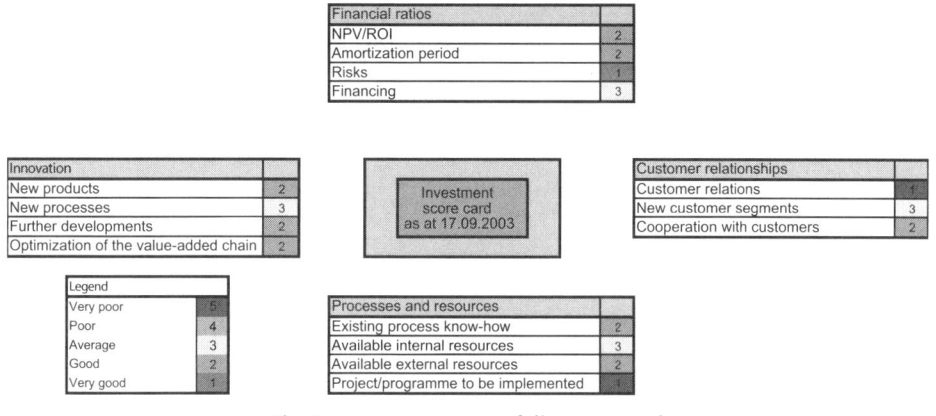

Fig. I2.5: Investment portfolio score card

I 2.3   Business case analysis

Definitions and examples

The term "business case" is a relatively new term for an investment. The "business case analysis" is thus an analysis of the investment.

The business case analysis encompasses the description of the costs and benefits of an investment, a calculation of these costs and benefits, possibly a financial investment analysis and financial simulations as well as the presentation of the financial ratios (net present value, benefit-cost difference, ROI, amortization period, risks) of the investment.

A form for the documentation of the results of the business case analysis or the cost-benefit analysis is depicted in Figure I2.6.

BUSINESS CASE ANALYSIS/ COST-BENEFIT ANALYSIS		
Name of the investment:		
Type of investment: ❑ services ❑ marketing ❑ infrastructure	❑ organization ❑ personnel ❑ environment relationships	
Period of the analysis: start date: end date:	**Required rate of return: x %**	
Description and evaluation of the costs:		
period	description of the costs	calculation of the costs
Description and evaluation of the benefits:		
period	description of the costs	calculation of the costs
Financial investment analysis (optional): Table of cash flows		
Contribution for realizing financial objectives:		
• net present value:		
• ROI:		
• benefit-cost ratio:		
• amortization period:		
• risks:		
Results of the financial simulation (optional):		
Promoter	Proposal team	
Version: Date:	Name: Page 1 of 1	

Fig. I2.6: Form: Business case analysis

Objectives of the business case analysis

The business case analysis forms the basis for the decision-making regarding the implementation of an investment. The business case analysis represents an important communication instrument in this process.

Figure I2.7 depicts the phase of developing the business case analysis in the product innovation process of an international technology group. Following a rough qualitative evaluation, a detailed quantitative evaluation of the investment in a new product is performed in the form of a business case analysis.

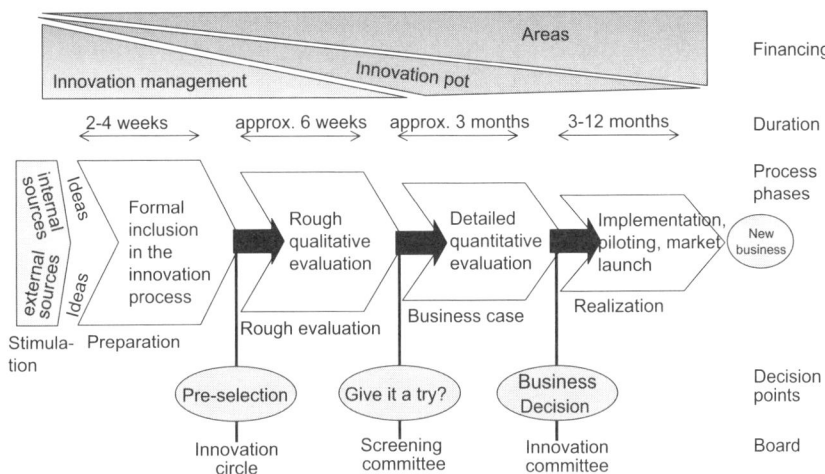

Fig. I2.7: Business case analysis in the product innovation process

Since the original assumptions can change during the performance of an investment, the business case analysis is also an instrument for controlling the investment.

Developing a business case analysis

The description of the problem formulation and the reasons for an investment, the investment objectives and the investment object form the basis for developing a business case analysis. These descriptions are made during the developing of an investment proposal. The analysis of the contributions toward the realization of environment-related objectives, innovation objectives and business process and resource objectives should also be performed as part of the development of the investment proposal.

For performing the business case analysis, the analysis period to be considered for the costs and benefits or cash flows both in and out, and the required rate of return, are to be determined. The costs and benefits per period for this period then have to be described and (monetarily) evaluated. A monetary evaluation enables the performance of a financial investment analysis.

⋯⟶ **PROJECT PORTFOLIO MANAGEMENT**

....

It has to be decided whether the business case analysis is to be performed based on a cash flow analysis or a cost-benefit analysis. In the case of a cash flow analysis, only cash in- and outflows caused by the investment are considered. In the case of a cost-benefit analysis, non-cash flow consequences of an investment are also taken into consideration. The cost-benefit analysis therefore enables a more holistic approach to the analysis of an investment.

1. Definition of the analysis period and the required rate of return

The life cycle of an investment presents the analysis period of the business case analysis (see Figure I2.8). In order to define the period of an investment, the investment decision is to be determined as the start event and the completed disinvestment is to be determined as the end event[3]. All costs and benefits until the point in time of the decision being taken are to be considered as "sunk costs" with regard to the investment decision and, hence, do not need to be considered as part of the business case.

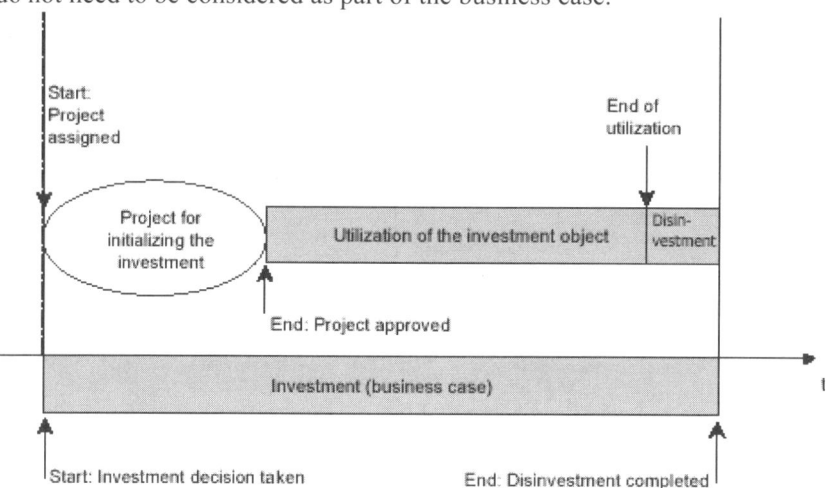

Fig. I2.8: Relationship between the project and the investment life cycle

Should a method of dynamic investment analysis be applied, a required rate of return must be determined. These possibilities are described in Excursus I1: Dynamic investment analysis.

2. Description of the costs and benefits of an investment

The description of the costs and benefits caused by the investment forms the basis for the analysis of an investment. They are to be verbally described per period. Basically, a distinction can be made between direct and indirect costs and benefits and between tangible and intangible costs and benefits (see Figure I2.9).

---

3) It may also be necessary to analyze chains of several investments in order to compare several investments.

direct costs and benefits	"indirect" costs and benefits
"Direct" costs and benefits are those caused by the pursuit of the (actual) investment objectives.	"Indirect" costs and benefits are those resulting as side effects of an investment. They have the following joint characteristics: They occur unintentionally and cannot be prevented (without costs).
tangible costs and benefits	**intangible costs and benefits**
Tangible costs and benefits are costs and benefits which have a market price. In the case of an incomplete market, no market prices can be used. The use of opportunity costs becomes necessary.	Intangible costs and benefits: Costs and benefits which do not have a market price.

Fig. I2.9: Basic distinction between types of costs and benefits

type of cost/benefit	costs	benefits
direct, tangible	construction costs	construction company's profits
direct, intangible	–	reduced path time
indirect, tangible	loss of agricultural naturalness	–
indirect, intangible	noise pollution	the location becomes more interesting for the residents

Fig. I2.10: Examples of different types of costs and benefits

Depending on the type of investment, different types of costs and benefits can be considered (e.g. personnel costs and consulting costs or benefits caused by reducing risks, benefits caused by faster procedures). An example of a cost-benefit analysis for an investment in the further development of the project costs calculation system at an engineering construction company is depicted in Figure I2.11. It distinguishes between the costs and the benefits of the project for the initialization of the investment and between the follow-up costs and follow-up benefits of the application of the new project costs calculation system.

benefits	interpretation	benefits per year
• project benefits • personnel develop- ment	• a qualification leap of 5% (moderation technique, PM, teamwork, company view) Salary costs for 12 people at EUR 29,000 = EUR 348,000	2002: 17,400 2003: 17,400
• follow-up benefits • risk reduction by project controlling and by realistic ob- jectives • marketing of "project management" as a service	• 0,5 % of the turnover from contract han- dling projects • CM = EUR 36,000 per annum	2002: 750,000 2003: 1.500,000 2004: 1.500,000 2005: 1.500,000 2006: 36,000 2007: 36,000 2008: 36,000

costs	interpretation	costs per year
• project costs • personnel costs • consulting costs • EDP costs • personnel costs for training • training costs	• for planning and implementation 1,400 person-days at € 180 = € 252,000 • for planning and implementation 40 person-days at € 910 = € 36,400 • additional hardware, increased re- quirements, adaptations € 36,000 • directly affected: 50 people at 5 days, indirectly affected: 200 people at ½ day = 350 person-days at € 910 = 318,500 • 15 person-days at € 910 = € 13,650	2002: 108,000 2003: 144,000 2002: 18,200 2003: 18,200 2003: 36,000 2003: 318,500 2003: 13,650
• follow-up costs • personnel require- ments due to decen- tralization • costs caused by com- prehensive planning and controlling (project analysis and structuring) • EDP follow-up costs	• average 1 ¾ people at € 29,000 = € 50,750 • during the start phase ½ per affected department (PM, assembly, calcula- tion, sales, etc.) 5 people at = 5 Mann á € 29,000 = € 145,000 • € 72,700 p. a.	2002: 50,750 2003: 50,750 2004: 50,750 2005: 50,750 2003: 145,000 2004: 72,500 2005: 0 2003: 72,700 2004: 72,700 2005: 72,700

Fig. I2.11: Benefits and costs of further development of the project costs calculation system in an engineering construction company

3. Calculation of the costs and benefits of an investment

If an investment analysis is based on cash flows, the profitability of an investment is determined by its difference of cash in- and cash outflows.

For analysis purposes, the cash flows during a certain period (month or year) are gen-
erally added at the end of each period. The resulting cash flows can be depicted in cash flow tables (see Figure I2.12) or in cash flow diagrams.

Monetary analyses are also performed in the cost-benefit analysis to determine the costs and benefits. If, for the individual types of costs and benefits, market prices of a complete market are available, they are included in the calculation. If no complete market exists (e.g. subsidized prices, import controls, unemployment, etc.), then opportunity costs have to be used. Opportunity costs are costs which are assumed for goods or services since either no market price exists for the good to be priced or, due to an incomplete market, the existing market price appears inappropriate for the analysis.

Types of costs and benefits which appear either difficult or impossible to subject to monetary evaluation can be considered by means of verbal descriptions, of specifying social indicators or of sensitivity analyses.

4. Performance of the financial investment analysis

In order to determine the profitability of an investment, an overall result for the analysis period is to be calculated from the net cash flows (or cost-benefit differences) per period. This can be done either with or without considering the current market value of the money.

Without considering the current market value of the money, the periodic net cash flows are simply added up in order to evaluate the profitability of an investment. If the current market value of the money is considered, the net cash flows from various periods cannot be directly compared. They are to undergo a discount process by using dynamic investment calculations (see Excursus I1: Dynamic investment analysis). An example of using the net present value method for the evaluation of a product investment of a telecommunications company is depicted in Figure I2.12.

Cash flow – in €k	Startup	Year 1	Year 2	Year 3	Year 4	Year 5	Total
Investment cash outflow	-6.000	0	0	0	0	0	-6.000
Change in current assets		-200	-200	-100	100	400	0
Income/cost savings							
Detail		2.000	5.000	4.000	4.000	3.500	18.500
Detail		1.200	2.400	3.200	3.000	2.800	12.600
Income/cost savings		3.200	7.400	7.200	7.000	6.300	31.100
Costs							
Detail		-500	-1.000	-1.000	-1.000	-1.000	-4.500
Detail		-2.000	-2.000	-2.500	-2.500	-1.000	-10.000
Detail		-300	-800	-700	-500	-500	-2.800
Costs	0	-2.800	-3.800	-4.200	-4.000	-2.500	-17.300
Cash flow	-6.000	200	3.400	2.900	3.100	4.200	7.800
Discounted cash flow	-6.000	185	2.915	2.302	2.279	2.858	4.539
Discounted cash flow	-6.000	-5.815	-2.900	-598	1.681	4.539	Net present value

Interpretation:

Regulations of the telecommunications company for developing business case analyses

- application of the net present value method
- negative sign: Cash outflow
- analysis period: Period in which the investment is practically usable. The standard is 5 years; re-investments are to be considered.
- current assets: Consider build-up of inventories and accounts receivable (capital tie-up).
- required rate of return: 8%, composite interest rate includes rate of return on equity
- discounted cash flow accumulated: Defines the amortization period

Fig. I2.12: Example of a business case analysis of a telecommunications company

···⫶ EXCURSUS I1: Dynamische Investitionsrechnung

In accordance with the investment term, the profitability of an investment is determined by its cash flows.

Normally, at the beginning of an investment, the cash outflows are predominant whereas, at a later date, the cash inflows are predominant. The cash flows relevant to the evaluation of an investment are its future cash flows. Historical cash flows, i.e. cash flows of the past, are not to be considered in the investment analysis.

The estimation of cash flows can be based on the assumption of certain or uncertain expectations (deterministic or stochastic cash flows).

Due to the different temporal accumulation of the individual payments both in and out, two or several cash flows are not directly comparable. Due to the principle of equivalence, however, it is possible to calculate, for a specific point in time, a value which is equivalent to a cash flow. The value of a cash flow which is calculated for the starting point of the first period of an investment is called the present value. The choice between two or several investments then becomes a choice between the present values of the investments.

The internal interest rate is that required interest rate which sets the capital value of a payment flow to zero.

Method	Calculation	Decision-making regulation
Net present value method:	$$T = \sum_{t=0}^{T} Z_t (1 + i)^{-t}$$ i = required interest rate T = life cycle K = net present value Z = cash flow	An investment is beneficial if its net present value is zero or positive. In the case of two or several alternative investment plans, the one with the largest value is the most beneficial.

Method of internal interest rate	$K = \sum_{t=1}^{T} Z_t(1+r)^{-t}$ R = internal interest rate T = life cycle K = net present value Z = cash flow	An investment is beneficial if the internal interest rate is the same as or greater than the minimum rate of return required by the company. In the case of two or several alternative investment plans, the one with the highest internal interest rate is the most beneficial.
Dynamic amortization period	$Z_0 = \sum_{t=1}^{n} Z_t(1+i)^{-t}$ Z_0 = Initial payment Z_t = cash flow n = amortization period	In the case of two or several alternative investment plans, the one with the shortest amortization period is the most beneficial.

5. Performance of a sensitivity analysis and of simulations

Due to their long-term character, investment decisions are more subject to uncertainty than other company decisions. The consideration of the risks associated with an investment can be performed in two different ways. On the one hand, in a sensitivity analysis, parameters on which the profitability of an investment is dependent, can be determined. In the case of an investment in a production facility these are, for example, the volumes of products which can be produced annually, the market price of the products and the costs of the facility's operating personnel.

In the case of the simplest form of sensitivity analysis, a parameter is analyzed with regard to determining the extent to which deviations within a certain fluctuation range affect the results of the investment analysis. A sensitivity analysis can be performed with reference to several parameters.

As a result of the sensitivity analysis, critical points are determined to establish within which limits the parameters can vary without an investment considered as optimal not becoming non-optimal.

In the simulation analysis, the uncertain parameters are defined as stochastic variables for which probability distributions are established, and which are considered in the investment calculation as stochastic variables (see also Monte Carlo simulation in Chapter F1).

> ### Tips for the business case analysis
>
> - The business case analysis already has a high practical relevance for product and infrastructure investments. It increases in importance for organization investments. It has a low relevance for contracting projects. However, the company which performs the contract can create an analysis from a customer's point of view as a marketing instrument.
> - Investments are to be described as holistically as possible, i.e. not only the direct investment objectives are to be considered, but also the indirect objectives of marketing, organization and personnel management.
> - Wherever possible, monetary evaluations are to be performed for the costs and benefits. These evaluations present important foundations for communication in the investment decision process.
> - In evaluating the costs and benefits, frequently both the knowledge of internal and external experts as regards empirical values and knowledge databases can be used.
> - The business case analysis should also undergo a controlling process in order to control the investment during its life cycle.
> - "Don't take a sledgehammer to crack a nut": Depending on the period, volume, strategic importance, etc. of an investment, the degree of detail of the business case analysis can be varied.

I 2.4   Project proposal

A project proposal presents the basis for the decision to initialize an investment by a project (or a programme).

The project proposal is developed by defining the boundaries and the context of a project. The results of this project definition are documented in the project proposal form.

The project definition determines what a project includes. Clarity regarding the project boundaries only arises upon explicit exclusion of those things which do not belong to the project.[4] What does not belong to a project, i.e. excluded tasks and decisions, the social project environment, the pre- and post-project phase, is the project context.

The project definition is performed by roughly planning the project objectives, the project phases, the project start and end dates, the project costs and project income as well as central project roles. In order to enable a holistic project point of view, a project must be considered from these various perspectives. The compatibility of the defined project boundaries must be guaranteed by coordinating the results achieved.

The project context is defined by a rough representation of the relationships between the project and the company strategies, a rough description of the activities of the pre- and the post-project phase, a list of relevant project environments and a list of those projects in the project portfolio which have relationships with the project.

Project portfolio reports can be used to analyze the relationships between the proposed project and other projects of the project-oriented organization (see Chapter I3.4).

---

4) Points excluded during the project definition phase can be "included" in the project at a later date.

Project boundaries cannot be defined as right or wrong, but rather as the result of agreements within the project team and between the project owner team and the project team.

However, the tips presented below should be considered in order to define practical project boundaries.

Tips for defining the project boundaries
• Project boundaries are to be defined such that a holistic result arises which also enables the analysis of a business case.
• The relationship between the project owner team and the project team is to remain the same for the project.
• The definition of the project boundaries is to enable the agreement on operational project objectives.
• The definition of project boundaries enables actions and decisions specific to the project. Only that which is definable can also be planned!
• Boundaries between various projects which relate to the same object and/or to the same customer (e.g. developing an offer, performing a contract, providing technical assistance) are to be set.

In order to develop crucial information for the project definition, it is necessary to develop initial, rough project plans (work breakdown structure, bar chart, cost plan, project environment analysis, etc.). These project plans form the basis for a detailed project planning during the project start process.

For developing a project proposal a project proposal form should be used. The project proposal form is to be designed according to the project assignment form (to be used later). An example of a project proposal form is depicted in Figure I2.13. Appendices to the project proposal form should represent the first drafts of the project management documentation.

The basic structures of the proposed project become apparent from the project proposal. This information enables the Project Portfolio Group to take a decision regarding the assignment of the project.

PROJECT PROPOSAL		
Name of the investment:		
Project start date:	**Project end date:**	
Project objectives:	**Non-objectives of the project:**	
Project phases:	**Project costs:**	**Project income:**
Project owner team:	**Project manager:**	
Project team members:		
Decisions and documents from the pre-project phase:		
Expectations regarding the post-project phase:		
Relationship to the business case:		
Relationships to other projects:		
Relevant project environments:		
Appendix: • Project objectives plan • Work breakdown structure • Project milestone plan • Project cost plan • Project environment analysis		
_____ **Promoter**		_____ **Proposal team**
Version:	Date:	Name:	Page 1 of 1

Fig. I2.13: Project proposal form

I 2.5   Project assignment

The basis for the assignment of the project by the project owner team to the project manager and the project team is the decision taken by the Project Portfolio Group to initialize an investment in a project form. A project assignment is to be made in writing and be signed by the project manager and the project owner team.

Developing the project assignment during the assignment process is performed by the designated project manager, possibly also in cooperation with future project team members, in coordination with the project owner team.

The documentation developed for the project proposal, i.e. the project definition as well as the drafts of project management documents (work breakdown structure, milestone plan, project cost plan, project objectives plan and project environment analysis), forms the basis for developing the project proposal.

A project assignment serves the purpose of summarizing the agreements on objectives between the project owner team and the project team. The project assignment should contain the information apparent in the project assignment form (see Figure I2.14).

CHAPTER

I

PROJECT ASSIGNMENT		
Project name:		
Project start date:	**Project end date:**	
Project objectives:	**Non-objectives of the project:**	
Project phases:	**Project costs:**	**Project income:**
Project owner team:	**Project manager:**	
Project team members:		
Decisions and documents from the pre-project phase:		
Expectations regarding the post-project phase:		
Relationship to the business case:		
Relationships to other projects:		
Relevant project environments:		
Appendix: • Project objectives plan • Work breakdown structure • Project milestone plan • Project cost plan • Project environment analysis		
_____ **Promoter**		_____ **Proposal team**
Version:	Date:	Name:	Page 1 of 1

Fig. I2.14: Project assignment form

I 3   Project portfolio coordination

I 3.1   Business process: Project portfolio coordination

Structure of project portfolios

The set of all projects and programmes which are simultaneously managed in a project-oriented organization can be perceived as the project portfolio. Either all the projects of an organization or subsets of the projects can be clustered in a project portfolio. Subsets of projects differentiated by project types can be used for the definition of project portfolios, such as the portfolio of offer development projects, of contracting projects, of product development projects, etc.

The number of projects in project portfolios, their objectives and variables and the phases at which the individual projects are, continually change. The start of new projects and the close-down of projects result in high dynamics of project portfolios. The structures of project portfolios, i.e. the contained types of projects, the types of relationships between projects, the environment-related relationships of the individual projects, the durations, costs and risks of the projects, etc., remain relatively constant, however. Since project-oriented organizations continually perform projects, project portfolios are not limited in time, but rather extend over the lifetime of the project-oriented organization.

The project portfolio is a central integration instrument of project-oriented organizations. The organizational differentiation resulting from the performance of projects and programmes is complemented by an integrative point of view.

Objectives of project portfolio coordination

Objects of consideration in the project portfolio coordination process are the projects and programmes and their relationships to each other at a certain point in time.

Objectives of the coordination of the project portfolio are to coordinate projects with regard to the strategic objectives of the project-oriented organization and to optimize the project portfolio results. The objectives of project portfolio coordination can conflict with the objectives of individual projects.

Project portfolio coordination pursues internal- and external-oriented objectives. Various projects performed for the same customer are to be coordinated with regard to long-term customer strategies and fundamental behavior patterns towards the customer. Purchases and services from various projects by the same supplier can be optimized with regard to the purchasing conditions. The internal and external resources used in the projects are to be coordinated, projects' priorities regarding access to scarce resources are to be established.

Controlling of the structure of the project portfolio, control of the project portfolio risk and the organization of the learning process of and between projects, are further objectives. The possible stopping or interrupting of projects for strategic reasons is also an objective of project portfolio coordination.

CHAPTER

I

Tasks in project portfolio coordination

The business process of project portfolio coordination is described in detail in Figure I3.1. The business process encompasses the phases of preparation, performance and follow-up of the periodical coordination meeting by the Project Portfolio Group. Depending on the size and the dynamics of the project portfolio to be coordinated by a Project Portfolio Group, it may be necessary to hold a coordination meeting each week or every two weeks. This meeting is to last 2-4 hours.

In order to prepare the coordination meeting, the project portfolio database is to be updated, project portfolio reports are to be developed, new investment and project proposals are to be collected and checked for completeness. Project progress reports of selected projects are to be compiled. The documents resulting from the project portfolio coordination are project portfolio reports, investment and project proposals, project progress reports and the minutes of the coordination meeting held by the Project Portfolio Group.

Objectives of project portfolio coordination
- Optimization of the project portfolio results,
- Controlling of the structure of the project portfolio (number of projects per type of project, durations and budgets of the projects, etc.),
- Control of the project portfolio risk,
- Coordination of the project objectives with the strategic objectives of the project-oriented organization,
- Coordination of the objectives of individual projects,
- Coordination of the resources used both internally and externally in the projects, and determination of project priorities,
- Organization of learning of and between projects,
- Any potential stop or interruption of projects (for strategic reasons).

Time boundaries of project portfolio coordination
- Start: Assignment to organize a coordination meeting of the Project Portfolio Group
- End: Employees of the project-oriented organization are informed about essential decisions taken by the Project Portfolio Group
- Duration: 3 – 4 days.

Tasks and responsibilities in project portfolio coordination

Tasks / **Responsibilities**	Investment decision committee	Project Portfolio Group	PM Office manager	Nominated project owner	Employees, project manager	Proposal teams	Expert Pool manager	Results/documents
Preparing project portfolio coordination meeting								
• Updating project portfolio database	P							
• Developing project portfolio reports			P					1)
• Compiling current investment and project proposals			P			C	C	2)
• Compiling selected project progress reports			P		C			3)
• Invitation of participants		I	P	I				4)
Project portfolio coordination meeting								
• Distributing information material		I	P	I				
• Decisions: Project portfolio structure		P						
• Coordinating objectives of projects	I	P						
• Decisions: Project portfolio risk		P						
• Decisions: Environment strategies		P						
• Organizing learning		P						
• Determining project priorities		P						
• Developing the minutes	I		P					5)
Follow-up to the project portfolio coordination meeting								
• Adapting the project portfolio database			P					6)
• Informing employees about important decisions		C	P				I	I

Legend:
P ... Performance
C ... Contribution
I ... Information

Results/documents:
1) Project portfolio reports
2) Current investment and project proposals
3) Selected project progress reports
4) Invitation to the coordination meeting of the Project Portfolio Group
5) Minutes of the coordination meeting of the Project Portfolio Group
6) Adapted project portfolio database

Fig. I3.1: Project portfolio coordination responsibility matrix

CHAPTER

I

Results of the project portfolio coordination

With regard to the design of the project portfolio of a project-oriented organization, various quality criteria can be applied, such as the ideal number of projects per type of project, the minimum/maximum project portfolio budget, the maximum use of scarce resources in the project portfolio, the maximum number of cooperation projects with one supplier, the maximum number of projects with the same person as project manager, etc.

Results of the project portfolio coordination are, on the one hand, optimized structures of the project portfolio regarding these quality criteria, and an optimized project portfolio risk.

On the other hand, the overall optimization of the relationships to the project and programme environments, the optimization of the project and programme results with regard to progress, costs and income, etc., the assurance of the realization of the investment strategies of the project-oriented organization and the optimization of the implementation of organizational and personnel strategies in the project portfolio coordination, present the results of the project portfolio coordination.

Measures for implementing organization and personnel strategies are, for example, the assignment of multi-role performers to several projects, the application of guidelines for project and programme management, the use of project management consulting and project management auditing for quality assurance purposes in projects and programmes, and the targeted development of project management personnel.

The current status of a project portfolio considering these different criteria can be visualized in a project portfolio score card (see Figure I3.2).

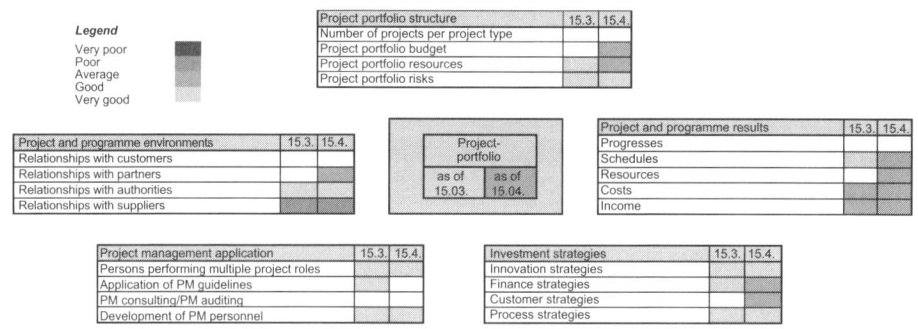

Fig. I3.2: Projekt portfolio score card

Organization of the project portfolio coordination

The necessary decisions regarding the optimization of the project portfolio are to be taken by the Project Portfolio Group. The preparation of the decision-making process is to be performed by the PM Office by maintaining the project portfolio database, analyzing the project portfolio and developing project portfolio reports.

I 3.2   Project portfolio database

The project portfolio database forms the basis for project portfolio management. A project portfolio database is a database of information accumulated from projects and programmes. It is not a project information system which contains detailed information about all projects. The project portfolio database relies on the data of the individual projects. In order to compare and accumulate data from projects, standardized minimum requirements are to be established for the projects' documentation. The project portfolio database should include the following information:

- Information regarding the project organization, such as project owner team, project manager, selected project team members,
- Information regarding relevant project environments, such as customers, suppliers and partners,
- Information regarding products and markets, such as type of product, technology and region,
- Information regarding the type of project and relationships of the considered project with other projects, for example type of project, affiliation to a programme, and
- Information regarding project ratios, such as project start date, project end date, net present value of the investment initialized by the project, project profit, project risk, project progress and the level of criticalness of the project.

The project portfolio database can encompass data concerning current, planned, interrupted, stopped and completed projects. As an example, Figure I3.3 depicts the project portfolio database of a regional health insurance company.

Short term	Project names	Start date	Planned end date	Size	Type	Project manager	Owner
GROSSKL	Conversion to capitalization and lower case printing in the EDP area	02.12.02	30.06.03	G	ER	OP, Franz	Huber
BÜROEXT	Office automation in the external branches, including OFIS take-over	01.01.02	31.12.03	G	OR	OP, Schubert	Huber
DEPROG	Redesigning of the entire data collection programs	01.03.02	31.03.03	K	ER	EDP RZ, Heinz	Huber
WANLAN	EDP network to and within external branches	15.04.02	05.12.03	G	OR	OP, Bauer	Hofer
EXZAHL	EX payments	01.03.02	30.05.03	K	ER	EDP-E, Reich	Huber
KOSTAUFL	Redesigning the generating and entry of reversals	03.03.03	15.04.03	K	ER	EDP-E, Steiner	Huber
BAP	EDP solutions for the Balneo and physiotherapy fields	01.01.02	31.07.03	K	ER	G, Groß	Müller

Legend
CC = construction conception project
CR = construction realization project
EC = EDP conception project
ER = EDP realization project
OC = organization conception project
OR = organization realization project

Fig. I3.3: Project portfolio database of a regional health insurance company

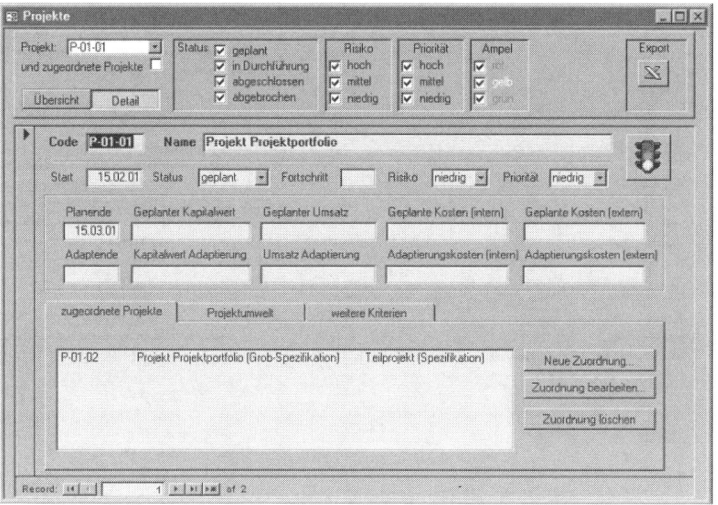

Fig. I3.4: Input mask of the RGC projektportfolio manager

I 3.3 Project portfolio analyses and project portfolio reports

The decisions to optimize the project portfolio of a project-oriented organization can be supported by means of project portfolio analyses and project portfolio reports. In addition to the project portfolio score card depicted in Figure I3.2, detailed analyses can be carried out and reports can be developed for the individual criteria of the score card.

Typical project portfolio reports are, for example, a project portfolio bar chart, a project portfolio budget, a project portfolio personnel timetable or a profit-risk matrix. Figure I3.5 depicts an example of a project portfolio bar chart of the regional health insurance company which was developed based on the project portfolio database depicted in Figure I3.3.

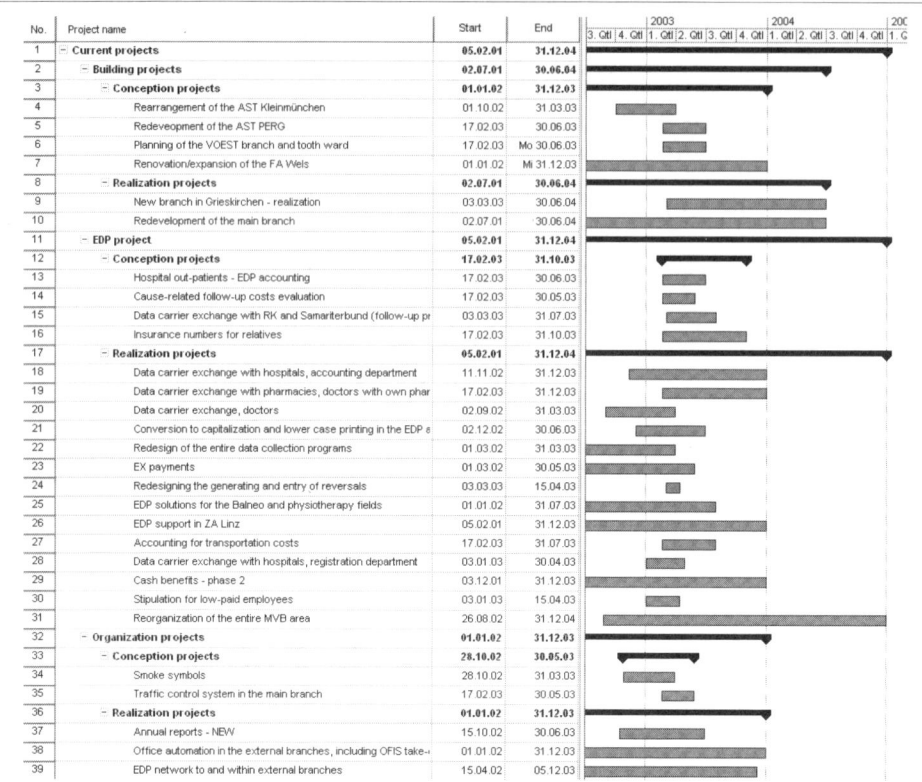

No.	Project name	Start	End
1	– Current projects	05.02.01	31.12.04
2	– Building projects	02.07.01	30.06.04
3	– Conception projects	01.01.02	31.12.03
4	Rearrangement of the AST Kleinmünchen	01.10.02	31.03.03
5	Redeveopment of the AST PERG	17.02.03	30.06.03
6	Planning of the VOEST branch and tooth ward	17.02.03	Mo 30.06.03
7	Renovation/expansion of the FA Wels	01.01.02	Mi 31.12.03
8	– Realization projects	02.07.01	30.06.04
9	New branch in Grieskirchen - realization	03.03.03	30.06.04
10	Redevelopment of the main branch	02.07.01	30.06.04
11	– EDP project	05.02.01	31.12.04
12	– Conception projects	17.02.03	31.10.03
13	Hospital out-patients - EDP accounting	17.02.03	30.06.03
14	Cause-related follow-up costs evaluation	17.02.03	30.05.03
15	Data carrier exchange with RK and Samariterbund (follow-up pr	03.03.03	31.07.03
16	Insurance numbers for relatives	17.02.03	31.10.03
17	– Realization projects	05.02.01	31.12.04
18	Data carrier exchange with hospitals, accounting department	11.11.02	31.12.03
19	Data carrier exchange with pharmacies, doctors with own phar	17.02.03	31.12.03
20	Data carrier exchange, doctors	02.09.02	31.03.03
21	Conversion to capitalization and lower case printing in the EDP a	02.12.02	30.06.03
22	Redesign of the entire data collection programs	01.03.02	31.03.03
23	EX payments	01.03.02	30.05.03
24	Redesigning the generating and entry of reversals	03.03.03	15.04.03
25	EDP solutions for the Balneo and physiotherapy fields	01.01.02	31.07.03
26	EDP support in ZA Linz	05.02.01	31.12.03
27	Accounting for transportation costs	17.02.03	31.07.03
28	Data carrier exchange with hospitals, registration department	03.01.03	30.04.03
29	Cash benefits - phase 2	03.12.01	31.12.03
30	Stipulation for low-paid employees	03.01.03	15.04.03
31	Reorganization of the entire MVB area	26.08.02	31.12.04
32	– Organization projects	01.01.02	31.12.03
33	– Conception projects	28.10.02	30.05.03
34	Smoke symbols	28.10.02	31.03.03
35	Traffic control system in the main branch	17.02.03	30.05.03
36	– Realization projects	01.01.02	31.12.03
37	Annual reports - NEW	15.10.02	30.06.03
38	Office automation in the external branches, including OFIS take-i	01.01.02	31.12.03
39	EDP network to and within external branches	15.04.02	05.12.03

Interpretation:
- In total, the health insurance company holds a project portfolio of 29 projects on the controlling day (6 of which are construction projects, 18 are EDP projects and 5 are organization projects).
- Some conception projects take longer than 5 months.
- Several projects are due for close-down in the period between May and August 2003. Then resources are once again available for new projects.

Fig. I3.5: Project portfolio bar chart of the regional health insurance company

Analysis of the project portfolio

The analysis of a project portfolio serves the following purposes:

- To create an overview of all projects which form one portfolio,
- To recognize familiarities and differences between these projects, and
- To recognize synergetic and/or competing relationships between the projects.

Analyses of project portfolios are to be performed periodically depending on the dynamics of the project portfolio. Furthermore, analyses of a project portfolio are to be performed ahead of the start, close-down or stopping of a project in order to determine the associated overall consequences for the portfolio. The analysis of the consequences of the stopping or close-down of a project enables transfer of know-how, reorganization of personnel, allocation of released resources for other projects, etc.

In the form of project portfolio reports, the results of the analyses can be processed differently for various target groups. This processing is performed either by the PM Office or by the respective user himself/herself. In order to communicate the relatively complex relationships in project portfolios, it is recommended that various visualization options be used (e.g. lists, portfolio matrices, bar charts, tables, etc.).

Project portfolio score card

The project portfolio score card presents an essential project portfolio report. It is an integrative instrument that compiles all the important criteria regarding the management of the project portfolio (see Figure 13.2).

By using traffic light colors for evaluating the status of the project portfolio on a control date, the attention of the Project Portfolio Group can be focused on critical developments. By depicting several statuses of the project portfolio over time the success of management measures can be evaluated.

The evaluations of the individual criteria require an appropriate interpretation and an expansion in the form of detailed project portfolio reports, for example on the structure of the project portfolio, on the project portfolio resources or the project portfolio budget.

Project portfolio list

In the project portfolio list, projects can be grouped according to various criteria. Relevant distinctions are, for example, various project contents (e.g. offer, contracting and product development projects) and different phases in the investment life cycle (e.g. conception or realization projects).

Project portfolio bar chart

It is the objective of the schedule analysis of a project portfolio to present the durations and statuses of the projects with regard to schedule as well as essential dependencies between projects. The schedule analysis can be depicted in a project portfolio bar chart, arranged according to types of projects (see Figure 13.5).

Project portfolio resource plan

A resource analysis of the project portfolio is necessary if several projects use the same (scarce) resource. In this case, it is no longer possible to carry out an independent resource planning of the projects. The coordination of plans and the determination of priorities are necessary.

In the case of external projects, resource shortages which arise are generally those related to the personnel performing the job (e.g. designers, programmers, etc.) and/or necessary infrastructure (e.g. equipment, space). These can frequently be solved by outsourcing the performance of services to third parties or by the short-term obtaining of additional resources from the market.

Since, in the case of internal projects, the involvement of the employees of the project-oriented organization is usually required for problem-solving and ensuring the acceptance of the results, there are only limited opportunities for the short-term obtaining of the necessary resources from external sources. The project employees and the management capacities therefore present bottlenecks of internal projects.

Traditional methods of "multi-project planning" take as a basis a CPM network-supported optimization of the use of resources. Planning of the resource requirements is performed at the activity level of the CPM schedule. Since the network planning technique is only used for a few projects by project-oriented organizations or is frequently only used for parts of projects, CPM network-supported methods can only partially comply with the requirements of the resource analysis. Global methods of resource planning and scheduling are therefore to be used. Available resource capacities can be divided as a percentage over individual projects, for example. Estimations regarding the resource requirements per period for the individual projects form the basis for this planning process.

Project portfolio costs and project portfolio income

In the analysis of the costs and income of project portfolios, a distinction is to be made between external contracting projects and internal projects.

The costs and income of contracting projects can be quantified using business ratios such as turnover, costs and contribution. These ratios can be differentiated according to products and technologies or according to markets and regions, for example.

The costs and benefits of internal projects are to be analyzed based on the business case analyses or cost-benefit analyses.

Portfolio matrix models

The projects of the project portfolio can be visualized in portfolio matrix models. Possible combinations of criteria for the depiction of matrices are, for example, contribution and risk, net present value and risk, costs and benefits, or duration and risk.

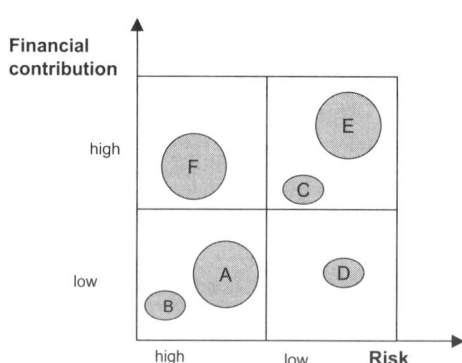

Fig. I3.6: Contribution risk matrix of the project portfolio of contracting projects

Project portfolio progress report

For repetitive projects, such as product development projects or contracting projects, project milestones can be standardized. The projects' progress on a control day can be determined and depicted according to the type of project.

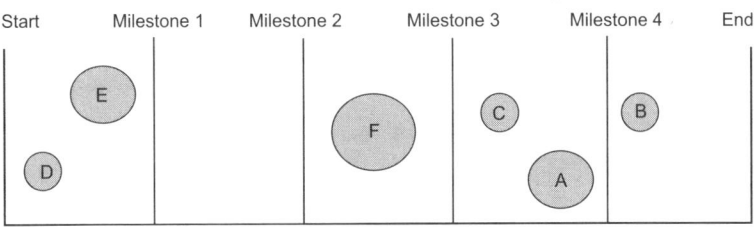

Fig. 13.7: Project portfolio progress report for contracting projects

Interpretation of project portfolio reports

The meaningfulness of project portfolio reports depends on the quality of the data in the project portfolio database. The provision of appropriate data by the individual projects places high demands on the maturity of a project-oriented organization.

The designing of relevant project portfolio reports and their interpretation require specific competencies. The use of symbols, for example the "traffic light colors" of green, yellow and red for depicting the projects' degree of criticalness, increase the reports' informational value.

The appropriate interpretation, for example of the project portfolio progress report for contracting projects, enables the use of this report as an early-warning system with regard to the acquisition of customer contracts. An international IT company has developed the acquisition potential report depicted in Figure 13.8 for the current portfolio of contracting projects. On the one hand, this report depicts the progress of the current contracting projects on the x-axis. On the other hand – symbolized by the size of the project circles – the acquisition potential for follow-up contracts is depicted. In this figure, the y-axis does not have any informational value. This project portfolio report also serves the purpose of realizing strategic sales objectives.

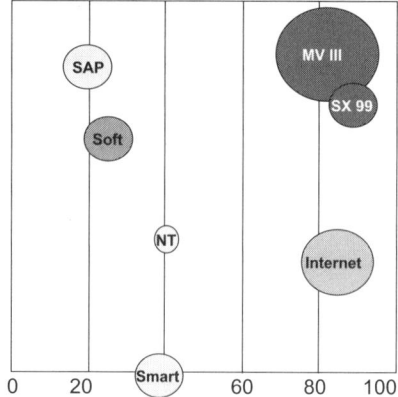

Fig. 13.8: Acquisition potential report

The processing of project portfolio ratios, such as the budget for product developments or the average duration of product developments, enables strategic control of the project portfolio.

Comparison of the number of projects – managed by one person with the role of project manager or project owner – combined with the relevant maximum values, enables quality assurance within the project-oriented organization.

Analysis of the dynamics of the project portfolio

The current status of a project portfolio is to be seen in its temporal context. Changes to the project portfolio over the course of time can be analyzed. For two or several control dates, the following applies:

- The number of newly-started projects and the number of closed down/stopped projects can be determined,
- The progresses in the project portfolio can be compared,
- The changes in the contribution risk matrix can be observed,
- etc.

Project portfolios are dynamic. Their developments can be documented and anticipated (see Figures 13.9 and 13.10).

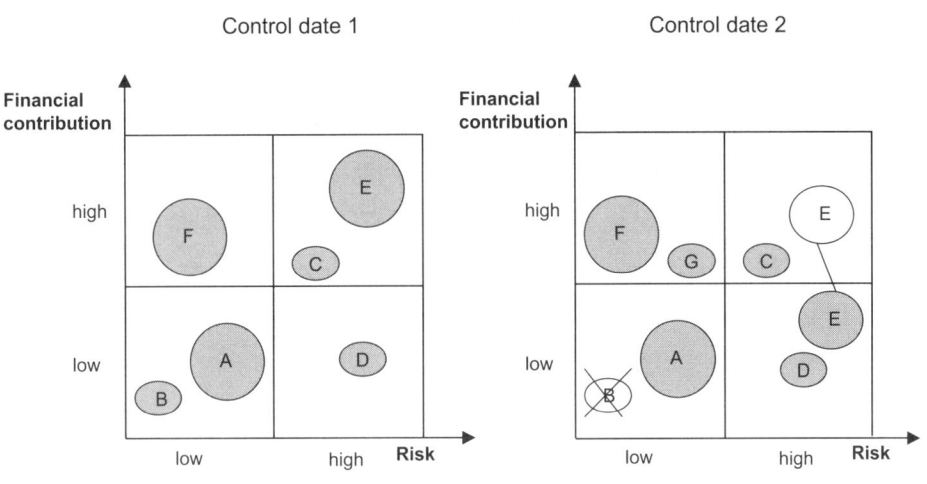

Fig. I3.9: Contribution risk matrices of contracting projects on various control dates

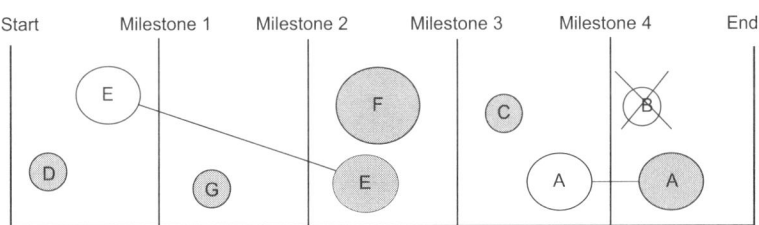

Fig. I3.10: Progress of contracting projects on various control dates

⋯⋗ **CASE STUDY I1: Project portfolio coordination of an engineering construction company**

Project-oriented organization
- Austrian engineering construction company. Products: Hydraulic machinery and systems. Services: Hydraulic developments, sales, construction, commissioning, training and consulting, maintenance.

Types of projects:
- Offer projects, contracting projects, product development projects and organization projects.

Analyzed project portfolio:
- Portfolio of contracting projects.

Database of the portfolio of contracting projects:
- The database is depicted and interpreted in Figure I3.11.

CODE	PROJECT	PROJECT MANAGER	PROJECT STEERING COMMITTEE	PROJECT OWNER	PROJECT CONTENTS	PROJECT START	PROJECT CLOSE-DOWN	PROJECT STATUS	%	PROJECT COSTS (ATS m)	CUSTOMER
AP-91/1	PAK MUN	Has Haslinger	Fu, Mo, Son, Sto, Str	Str.	T,R,H,StWB	Mar 91	Nov 94	Warranty period	10	xxx	EGAT
AP-91/2	ANDELSBUCH	Sen Sensenb.	Enz	Enz	T,A	Apr 91	Mar 95	Commissioning	90	xxx	VKW
AP-92/1	SINGKARAK	Dem Demel	Fu, Mo, Son, Sto, Str	Str.	T,R,A,H	Jul 92	Mai 98	Manufacturing	0	xxx	PLN
AP-93/1	TEDZANI III	Sen Sensenb.	Enz	Enz	T,A	Mar 93	Jän 96	Assembly	50	xxx	ESCOM
AP-93/2	SATSUNAIGAWA	Bo Bohm	Enz	Enz	T,R	Jun 93	Jul 97	Konstruktion	90	xxx	EPDC
AP-93/3	BELLEVILLE	Sei Seisenb	Str, Ang, Son	Str.	T	Jul 93	Jun 97	Manufacturing	25	xxx	AMP/Ohio
AP-93/4	BRUEGG	Sem Semper	Enz	Enz	T,R	Nov 93	Nov 95	Assembly	50	xxx	PKW
AP-94/1	DORNACHBRUGG	Gu Gupf	Enz	Enz	T	Nov 93	Nov 95	Manufacturing	50	xxx	EL.Birseck
AP-94/2	AMSTEG	Gu Gupf	Enz	Enz	T	Nov 94	Mai 98	Design	10	xxx	SBB
AP-94/3	FREUDENAU	Has Haslinger	Str	Str.	T,R,A,H	Nov 94	Mai 98	Design	60	xxx	ARGE Freudenau
AP-94/4	CIRATA II	Los Losbichler	Fu, Mo, Son, Sto, Str	Str.	T	Dez 93	Feb 98	Design	75	xxx	PLN
AP-94/5	YBBS	Has Haslinger	Str, Sto	Str.	REHAB T,R	Mar 93	Feb 96	Manufacturing	95	xxx	ARGE Freudenau
AP-94/6	KARNAFUL	Sem Semper	Enz	Enz	T,R	Jun 94	Jul 96	Design	60	xxx	BPDB
AP-94/7	JUBILÄUMSWERK	Gu Gupf	Enz	Enz	T	Dez 94	Apr 97	Design	30	xxx	EVN
AP-94/8	SCHWELLÖD	Gu Gupf	Enz	Enz	T,R,H,StWB	Dez 94	Apr 97	Design	30	xxx	EVN
AP-94/9	KLOSTERLE	Sen Sensenb	Enz	Enz	REHAB T,H,R	Nov 94	Mai 97	Design	5	xxx	VKW
AP-95/1	CHENDEROH	Sen Sensenb	Str, No, DC	Str.	T,R	Mar 95	Dez 99	Design	0	xxx	TNB
AP-95/2	DIEMLACH	Gu Gupf	Enz	Enz	T,R,A,E,L	Jän 95	Apr 96	Design	30	xxx	Stw. Kapfenberg
AP-95/3	STARKENBACH	Pres Preslmayr	Enz	Enz	REHAB T,H,R	Feb 95	Apr 96	Design	5	xxx	Handl
AP-95/4	BHUMIBOL 3+4	Sem Semper	Enz	Enz	REHAB T	Mar 95	Nov 97	Preparatory phase	70	xxx	EGAT

Interpretation of the database:

The project portfolio database of the contracting projects includes information regarding project environments and project ratios such as project costs, contract value and progress. Information regarding project costs and contract value are obviously not made available to all users of the database.

Fig. I3.11: Project portfolio database of the contracting projects of a engineering construction company

Project portfolio reports:
- The project portfolio bar chart and the project portfolio progress report are depicted and interpreted in Figures I3.12 and I3.13.

CODE	PROJECT	START	END	1995	1996	1997	1998	1999
		1.3.91	31.12.99					
AP-91/1	PAK MUN	1.3.91	17.11.94					
AP-91/2	ANDELSBUCH	1.4.91	15.3.95					
AP-92/1	SINGKARAK	20.7.92	1.5.98					
AP-93/1	TEDZANI III	1.3.93	1.1.96					
AP-93/2	SATSUNAIGAWA	1.6.93	1.7.97					
AP-93/3	BELLEVILLE	15.7.93	1.6.97					
AP-93/4	BRUEGG	1.11.93	1.11.95					
AP-94/1	DORNACHBRUGG	24.11.93	1.11.95					
AP-94/2	AMSTEG	7.11.94	1.5.98					
AP-94/3	FREUDENAU	1.11.94	1.5.98					
AP-94/4	CIRATA II	24.12.93	1.2.98					
AP-94/5	YBBS	1.3.93	1.2.96					
AP-94/6	KARNAFUL	30.6.94	1.7.96					
AP-94/7	JUBILÄUMSWERK	1.12.94	1.4.97					
AP-94/8	SCHWELLÖD	1.12.94	1.4.97					
AP-94/9	KLÖSTERLE	23.11.94	1.5.97					
AP-95/1	CHENDEROH	15.3.95	31.12.99					
AP-95/2	DIEMLACH	2.1.95	1.4.96					
AP-95/3	STARKENBACH	1.2.95	1.4.96					
AP-95/4	BHUMIBOL 3+4	1.3.95	1.11.97					

Interpretation of the project portfolio bar chart:
- 20 contracting projects are being performed on the control day.
- The projects last between 1.5 and 8 years. The projects started recently are shorter.
- Several projects will be closed down at the end of 1995. There may already be the need for new contracts. The acquisition situation is to be analyzed.

Fig. I3.12 Project portfolio bar chart

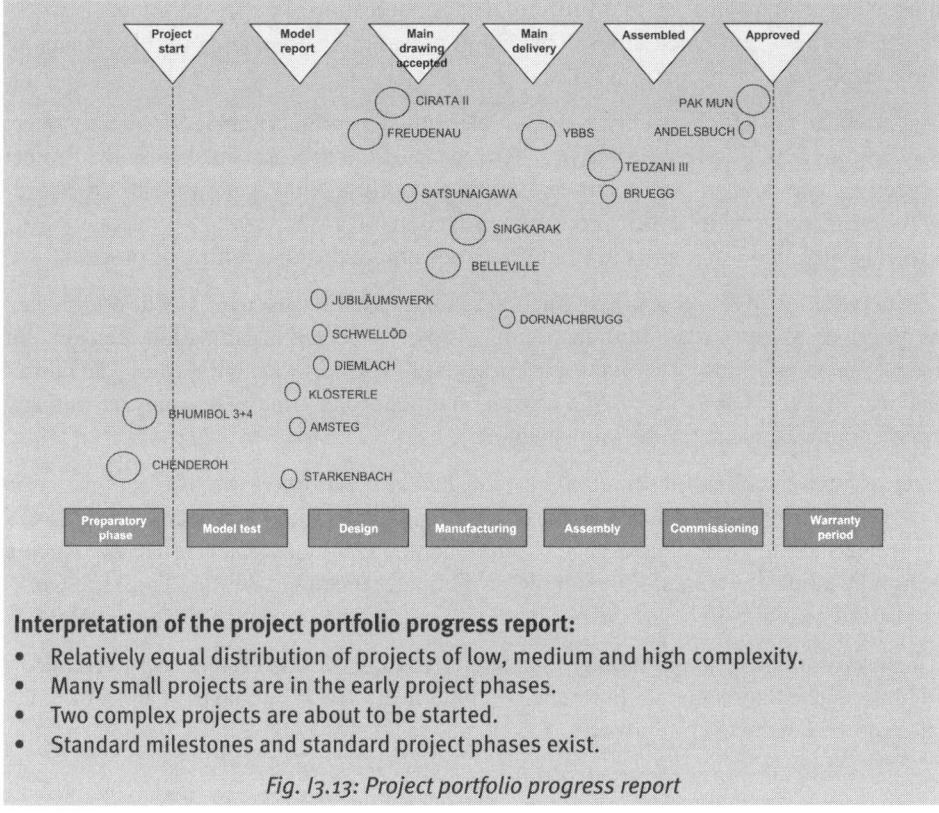

Interpretation of the project portfolio progress report:
- Relatively equal distribution of projects of low, medium and high complexity.
- Many small projects are in the early project phases.
- Two complex projects are about to be started.
- Standard milestones and standard project phases exist.

Fig. I3.13: Project portfolio progress report

I 4 Networking of projects

I 4.1 Business process: Networking of projects

Objectives of the networking of projects

Networks of projects are social networks of closely-coupled projects (see Excursus I2: Social networks). The coupling of projects into networks of projects can result from the cooperation of the individual projects with the same partners or suppliers, the performance of services for the same customer, the performance of project services in the same region, the use of the same technology, etc.

Networks of projects do not have any clear boundaries, i.e. projects from customers, partners or suppliers of the project-oriented organization can also be considered in so far as their consideration can contribute towards the realization of the joint objectives.

The network partners of networks of projects are organizations, i.e. projects and programmes, which communicate with each other. The joint intention of the network partners is the creation of synergies in the network of projects and the organization of lear-

483

ning. Networking comes about by means of communication of network partners in workshops, meetings and discussions. This communication can be supported by means of joint databases, chatrooms and links.

The projects' need for networking can be determined by project managers, by the Project Portfolio Group or by the PM Office. By means of the necessary focus on the project objectives and deliverables, the potentials of networking with other projects are not always clear to the members of project organizations. It is therefore also the task of the Project Portfolio Group and the PM Office to promote the networking of projects.

The networking of projects can be performed for a specific reason only or be established as a periodical form of communication in a project-oriented organization. Reasons for the ad hoc networking of projects are, for example, project discontinuities. The consequences of a crisis or a chance of a project on other projects are to be analyzed and any necessary measures are to be implemented.

Even if the networking of projects is established as a periodic form of communication in a project-oriented organization, the criteria for the coupling of projects into networks of projects are to be defined ad hoc. Communication can take place either via existing communication structures, for example profit center meetings, or in networking workshops and meetings to be arranged.

The networking of projects requires a cooperative organization culture. An active information policy, the chance for horizontal communication and mutual trust are central values for the networking of projects.

···⟩ EXCURSUS 12: Social networks

The word "net" stands for a set of nodes and connections between these nodes. Each node is connected to another node within the network – either directly or indirectly. All links in the network are movable. That which is seen as part of a network does not stand alone as an entity, but rather can influence the other parts of the network via the various connections, and vice versa: It can also be influenced by the other parts.

The "social network" applies the network model to a network of social systems.

"A social network describes a quantity of social units with the social relationships existing between these units. ... The social units can be persons, positions or roles, groups, organizations or even whole societies. Accordingly, the social relationships can vary, for example at the personal level, such as sympathy, communication or role obligations and, at group level, overlapping memberships, capital interconnections or even trade relationships between countries."[a]

With social networks, the boundary with the environment is indistinct and blurred. Where interaction ends, where and how it arises, changes and fades out, is not recognizable by network participants in its complexity. Individual interaction has ad hoc ends, yet they are only the end of the communication chain. It simply stops when a further connection no longer appears practical or possible, yet, if circumstances change, it can expand or retract at any time.

"The boundaries of networks are often blurred and their activity often seems to turn on and off with no discernible regularity. ... Instead of being held together within a boundary, a network coheres from shared values, goals and objectives. A network is recognized by its clusters of interaction and channels of communication, rather than by a fixed boundary that includes and excludes."[b]

This uncertainty further applies to the internal order (responsibilities, levels, roles) of a social network. The blurred boundaries and subdivisions make it difficult for common sense to recognize social networks as coherent, functioning social entities. This is the reason why objects without any fixed boundaries, like social networks, cause orientation problems. [c]

"Unlike a hierarchy, whose internal parts and external boundaries can be crisply mapped on a flow chart, a network has few inner divisions and has indistinct borderlines. A network makes a virtue out of its characteristic fuzziness, frustrating outside observers determined to figure out where a network begins and ends."[d]

The cohesion of a social network is brought about by the joint intentions and values of its members. A social network must represent certain basic intentions. These intentions form the basis of a social network. They are the putty which holds the network together.

"If a network could be drawn on paper, its lines of coherence would consist of the ideas that the participants agree upon, manifested in commitments to similar ideals."[e]

These collective basic intentions describe the network. If these intentions, the purpose of the network, are represented by a name or a symbol, this makes it easier to both identify the network and for participants to identify with the network.

"In contrast to bureaucracies, whose existence hinges on members who perform highly-specialized tasks and who are totally dependent on one another, networks are composed of self-reliant and autonomous participants."[f]

It is a special advantage of social networks that they not only permit, but also promote ... these individual interests of their members.[g]

For the various intentions and the autonomy of the network participants do not, as you might expect, endanger the survival of a network. Actually, quite the opposite is true: They make it more resistant.

"For governance, 'network' implies a non-hierarchical system of equal, self-sustaining members. Unlike a bureaucracy, a network is dependent on no one of its parts. No organ performs a specialized task necessary for the function of the whole. A net has no center. It is made up of links between parts. … It is precisely this attribute of self-sustaining parts that gives the network form its remarkable resiliency and its adaptability to stress. Segmentation explains why, for example, underground political movements are so difficult to suppress. Squashing one node does little to impair the effectiveness of the net as a whole."[h]

The relationships in a social network are based on three characteristics which distinguish social networks from social systems: Autonomous participants, flexible, decentralized organization and the participants' willingness to take part in the network.

[a] *Endruweit, G.* (ed.), (Soziologie) 1989, p. 465
[b] *Lipnack, J., Stamps, J.,* (Networking) 1982, p. 229 f
[c] Cp. *Lipnack, J., Stamps, J.,* (Networking) 1982, Chapter 1.1.3
[d] *Lipnack, J., Stamps, J.,* (Networking) 1982, p. 8
[e] *Lipnack, J., Stamps, J.,* (Networking) 1982, p. 9
[f] *Lipnack, J., Stamps, J.,* (Networking) 1982, p. 7
[g] Cp. *Lipnack, J., Stamps, J.,* (Networking) 1982, p. 8
[h] *Lipnack, J., Stamps, J.,* (Networking) 1982, p. 223 and 225

CHAPTER

I

Tasks in the networking of projects

The networking of projects commences with the definition of the requirement for networking. Several project managers, the Project Portfolio Group or the PM Office can define the requirement of networking. These can also initialize the networking of projects.

As a preparatory measure for the networking of projects, the networking requirement is to be concretized, members of the individual project organizations are to be invited and relevant information regarding the projects to be networked is to be supplied.

It is recommended that a networking workshop be carried out for the purpose of networking of projects. As a form of communication, the workshop enables direct interaction between the members of various project organizations. Based on an appropriate exchange of information, the relationships between the projects which make up the network of projects are to be analyzed. A network of projects graph can be used to visualize these relationships. This supports the joint design of the network of projects.

Based on the analysis of conflicting and complementary relationships between the projects, strategies and measures for the management of the network of projects can be determined. The arrangement of the agreed measures is the responsibility of the project managers of the projects to be networked.

Objectives of the networking of projects

- Communication between closely-coupled projects of the project portfolio of a project-oriented organization,
- Analysis of conflicting and complementary relationships between the projects of the network of projects,
- Use of synergies in the network of projects,
- Organization of learning in the network of projects.

Time boundaries of the networking of projects

- Start: Requirement for the networking of projects is defined
- End: Measures for the use of synergies between the projects are arranged
- Duration: 1-2 weeks

Tasks and responsibilities in the networking of projects

Tasks \ Responsibilities	Project Portfolio Group	PM Office	Various representatives of project owner team	Various project managers	Project team members	Representatives of relevant project environments	Expert Pool managers	Results/documents
Initializing and planning the networking								
• Initializing the networking	P	P	P	P	C			
• Planning the networking	P			C				
• Concretizing the need for networking	P		C	C	C			
Preparing the networking								
• Invitation of participants to the networking workshop	P		I	I	I	I	I	1)
• Gathering information on projects to be considered	P			C	C	C	C	
Networking workshop								
• Information exchange between projects		C	C	P	C	C	C	2)
• Developing the network of projects graph		C	C	P	C	C	C	
• Planning measures		C	C	P	C	C	C	
• Coordinating the measures	I	C	C	P		I	C	3)
Follow-up to the networking								
• Initializing the measures		P	I	P	I	I		

Legend:
P ... Performance
C ... Contribution
I ... Information

Results/documents:
1) Invitation to the networking workshop
2) Network of projects graph
3) List of measures

Fig. I4.1: Description of the business process "Networking of projects"

CHAPTER
I

Results of the networking of projects

Results of the networking of projects are, on the one hand, the analysis of the relationships between the projects of the network of projects (possibly visualized in a network of projects graph) and, on the other hand, a list of measures for utilizing the synergies in the network of projects. The following basic measures are possible:

- Redefinition of project objectives based on conflicts of objectives between projects,
- Rearrangement of personnel based on resource conflicts between projects, potential changes to project priorities,
- Redesigning of project-environment relationships based on a more holistic point of view by considering several projects,
- Balancing-out of risks between projects by changing contractual relationships with customers, partners and suppliers,
- Transfer of know-how between the projects of the network of projects, and
- Establishment of communication structures for the periodical coordination between two or more projects.

In an extreme case, the information gleaned from the analysis of the relationships in a network of projects can also result in stopping or interrupting projects.

I 4.2   Methods for networking of projects

Networking workshop and networking meetings

It is recommended that a networking workshop be carried out for the purpose of networking of projects. The working forms applied in the workshops enable direct interaction of the workshop participants.

Participants of a networking workshop are, on the one hand, members of the networking project organizations and, on the other hand, representatives of relevant project environments.

A networking workshop lasts between half a day and a full day. Only one workshop is usually required for the networking of projects. If needed, networking workshops can be supplemented by periodical networking meetings.

Network of projects graph

The relationships between the projects of a network of projects can be visualized in the form of a network of projects graph. This diagram depicts, on the one hand, the networking projects and, on the other hand, the relationships between these projects. Circles of varying sizes and colors can be used to depict the projects. The symbols used are to be described in the form of a legend.

Several projects with common features can be grouped together by using various frames. Lines between the projects (NOT targeted arrows) can be used to depict the relationships. Objects to be considered in networking are, above all, the relationships between the projects.

The relationships between individual projects can be described in the form of qualitative statements (see Figure I4.2), for example.

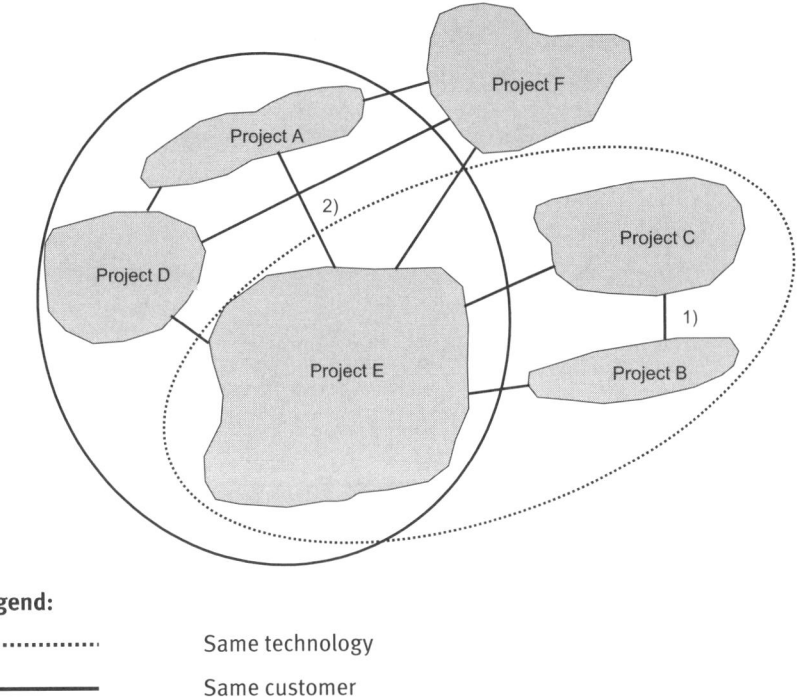

Legend:

....   Same technology

──────────   Same customer

Interpretation:

1) There is the danger of bottlenecks in capacity occurring at the supplier used in projects B and C.
2) Milestone 3 must have been reached for project A in order to be able to close down project E.

Fig. I4.2: Symbolism and interpretations in a network of projects graph

The use of the network of projects graph and an impact matrix for the analysis of the relationships between projects in a network of projects are described in the following case study.

···❯ **CASE STUDY I2: Networking of the bulb-type turbine projects of an engineering construction company[(a)]**

Project-oriented organization:
- See Case Study I1: Project portfolio coordination of an engineering construction company.

Networking of bulb-type turbine projects:
- Objectives of networking: Developing market strategies, main focuses of development, and sourcing strategies.
- The choice of projects from the project portfolio is made according to the "same product" criterion.
- The network of projects graph depicted in I4.3 presents all projects associated with the "bulb-type turbine" product and their relationships with each other.

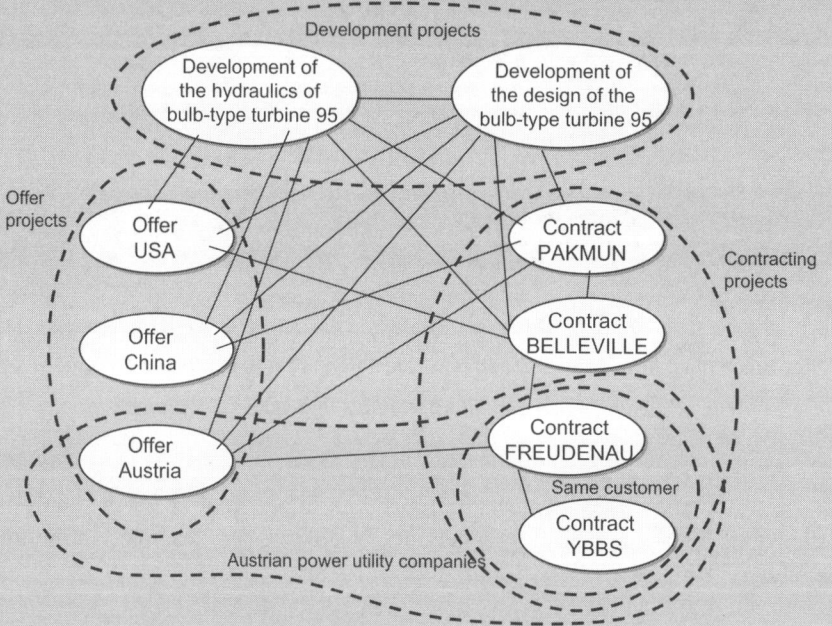

Fig. I4.3: Network of bulb-type turbine projects

- The network of projects graph is the basis for analyzing the relationships between the projects in an impact matrix (see Figure I4.4).
- Evaluations can determine the degree of "passiveness", i.e. the influence on a project by the others, and the degree of "activeness", i.e. the influence exerted by a project.

- From this, critical projects are determined, i.e. those projects which, on the one hand, have many active effects yet also absorb many effects from the network of projects.
- Strategies for optimizing the network of projects can be derived from this.
- The impact matrix depicted in Figure I4.4 documents both the importance of the mutual effects and the types of relationships between the projects.

IMPACT from	Further development of the hydraulics of bulb-type turbine 95	Further development of the design of bulb-type turbine 95	Offer USA	Offer China	Offer Austria	Contract PAK MUN	Contract BELLEVILLE	Contracts FREUDENAU YBBS	"active"	"critical"
Further development of the hydraulics of bulb-type turbine 95	Feedback 2	Limit to design 3	Secured offer data 0	Secured offer data 2	3	0	0	0	10	140
Further development of the design of bulb-type turbine 95	Alternative hydr. design 2	Alternative concepts 0		3	1	0	0	0	6	84
Offer USA	Competition specification 3	Manufacturing costs 1	Price specification 0	0	0	3	0	0	7	0
Offer CHINA	Market-specific concepts 0	Synergies 3	0	2	0	0	0	0	5	60
Offer AUSTRIA	High level 3	High quality 1	Power utility company standards 0	0	3	0	0	0	7	77
Contract PAK MUN	Plant measurements 2	Operational experience 2	Same model as 0	Operational experience 2	Sourcing strategy 1	0	2	0	9	0
Contract BELLEVILLE	Competitors' feedback 2	Largest diameter 2	Sourcing strategy 0	Same diameter 3	0	0	0	1	8	48
Contracts FREUDENAU YBBS	Greatest capacity + diameter 0	Austria's largest BTT 2	Same diameter 0	Pilot task 0	3	0	1	3	9	36
"PASSIVE"	14	14	0	12	11	0	6	4		

Importance: 0 ... none, 1 ... little, 2 ... average, 3 ... high impact

Fig. I4.4: Impact matrix of the bulb-type turbine projects

Experiences made by the engineering construction company with regard to the networking of projects:

- The networking of projects puts existing, formal structures into a new context, and informal perceptions are formalized. By means of the holistic network point of view, synergies are used and an added value is developed.
- The thus newly-created EDP software applications enable the evaluation of the project portfolio database according to various criteria. Since all this data is available, each project manager and each team member has access to it. The subsidiaries are also included in the exchange of information.
- The newly-available diversity of data also includes the danger of there being too much information. That is why, during the design phase of the database, it was important to manage with few yet similar data common to all types of projects.
- By means of the networked depictions of projects and the thus achieved integrative point of view, synergies (sourcing, design concepts, marketing, etc.) and new potentials for saving costs can be detected.

[a] Cp. *Enzenhofer, D., Semper, W.,* (Controlling von Projekte-Netzwerken)

I 4.3   Management of chains of projects

The projects within a chain of projects are closely linked by means of their affiliation to the same investment process. Typical chains of projects are the chains of a conception and a realization project, of an offer and a contracting project, and of a pilot and a follow-up project.

The objective of the management of chains of projects is to ensure the continuity of the management of two or several successive projects or programmes. The management of chains of projects does not present its own business process, but rather a special case of networking between (successive) projects. The management of relationships between the projects of a chain of projects is to be performed in the project management processes of the individual projects.

For the management of chains of projects personnel policy and organizational measures must be fulfilled. An essential measure with regard to personnel is the inclusion of members of the project organization of the current project into the project organization of a follow-up project, for example overlapping amongst the members of project owner teams and project teams. The Project Portfolio Group takes the decision regarding the composition of the members of the project owner team as part of the project portfolio coordination.

The objective of integrating two successive projects into a chain of projects can be realized in the project close-down process of the current project and in the project start process of the follow-up project. The planning of the structures of the follow-up project and their documentation in a draft of a project manual present a central content of the project performed beforehand. Potential members of the project organization of the follow-up project are to be included in the project close-down process in order to participate in the structuring of the follow-up project and to get to know representatives of relevant project environments. In order to ensure the transfer of know-how into the follow-up project, members of the project organization and representatives of relevant project environments of the project performed beforehand, are to be invited during the project start process.

No specific methods are required for the management of chains of projects. The (organizational) measures for integrating the two projects are depicted in the project management documents, in particular in the documents of the project organization.

I 5   Investment controlling

I 5.1   Business process: Investment controlling

Objectives of investment controlling

An investment initialized by a project or programme can be subjected to controlling after the close-down of the project or programme. Objects of the investment controlling can be both the business process of initializing the investment (i.e. the project or programme) and its results.

The objectives of investment controlling are the determination of the economic success of a current investment (profitability, amortization period, etc.), the evaluation of its technical and organizational success, the planning and arrangement of any necessary controlling measures, and the development of learning points which could be useful for future projects and investments.

The business processes of assigning a project or a programme and the investment controlling present the context for the project management process. The investment controlling can be performed one or several times, depending on the lifetime of an investment. In the case of a product investment with a product lifespan of, for example, 3 years, the first controlling can be appropriate six months after close-down of the product development and launch project.

Investment controlling is particularly recommendable in the case of strategically-important investments whose results are subject to relatively high dynamics and whose controlling measures can contribute to the optimization of the investment results.

Investment controlling is to be distinguished from project evaluation and investment evaluation. In the case of an ex-post project evaluation, only the project which has initialized an investment, and not the investment, is considered. The objective of a project evaluation is to learn for similar projects since there are no further controlling measures possible regarding a project which has already been closed down. However, it is uncustomary to perform an exclusive analysis of the closed down project without also analyzing the business case.

At the end of an investment process, i.e. at the end of the utilization of an investment object, a final evaluation of an investment can be performed. In contrast to investment controlling, the ex-post investment evaluation offers no further possible controlling measures for the investment in question. However, the objective of the investment evaluation is as a learning process for similar investments.

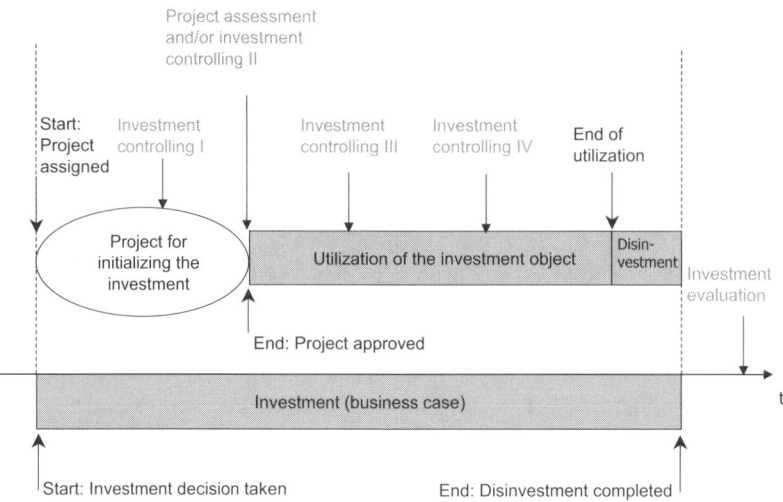

Fig. I5.1: Investment controlling, project evaluation and investment evaluation over time

Tasks in investment controlling

The business process of investment controlling encompasses the phases of planning the investment controlling, analyzing the investment and the project, planning and arrangement of controlling measures, and developing the investment controlling documentation.

During the planning of the investment controlling, the subject of the controlling is determined (investment and project), as well as which methods and which forms of communication are to be used. The objective of the planning of controlling measures is to optimize the investment in question. One possible measure is the premature stopping of an investment. This is necessary, should negative results be expected from the adapted business case analysis. The results of the investment controlling are summarized in documentation and communicated to the target group during a workshop or a presentation.

The business process "Investment controlling" is described in detail in Figure I5.2.

Objectives of investment controlling

- Establishment of the success of an investment initialized by a project,
- Planning and arrangement of any necessary measures for optimizing the investment,
- Organization of learning for future projects and investments.

Time boundaries of investment controlling

- Start: Assignment by the investment process owner for controlling the investment
- End: (Any necessary) controlling measures are arranged and investment controlling report is filed
- Duration: 2-5 weeks

Tasks and responsibilities in investment controlling

Tasks \ Responsibilities	Investment process owner	Investment controlling team	Employees	Project Portfolio Group	PM Office	Documents
Planning the investment controlling						
• Planning the application of methods	I	P				
• Planning the communication forms	I	P				
Analysis						
• Check: Project documentations	C	P		C		
• Check: Investment documentations	C	P				
• Gathering actual data		P	C			
• Planned versus actual performance analysis	I	P	C			
Planning (potential) controlling measures						
• Initial definition of controlling measures		P				
• Review of controlling measures	C	P	C			
• Adopting controlling measures	P	C	I	I	C	1)
• Information transfer to other projects		C		C	P	
Documentation and presentation						
• Defining learning points of the current investment		P				
• Defining learning points of the performed project		P				
• Adapting the business case analysis	C	P				
• Presenting the results	I	P	I	I	I	
• Developing an investment controlling report	I	P				
• Distributing and filing investment controlling documents	C	P	C	C	C	2)

Legend:
P ... Performance
C ... Contribution
I ... Information

Results/documents:
1) List of controlling measures
2) Investment controlling documents

Fig. I5.2: Description of the business process "Investment controlling"

Results of the investment controlling

Results of the investment controlling are an action plan for optimizing the investment (if required), an adapted business case analysis (if required), documentation of learning points about the current investment, documentation about the closed down project, and the investment controlling report.

I 5.2   Methods of investment controlling[5)]

Methods of investment controlling are:

- Documentation analyses, interviews, surveys and inspections,
- Workshops, presentations and investment controlling reports.

The subject matters of the documentation analysis are mainly the business case analysis and the investment's results reports, plus the project management documents.

The business case analysis is the document which accompanies the investment process: Starting with the development of the business case for the investment proposal, the current adaptation within the framework of the investment controlling during project performance and after project close-down right through to the application of the investment evaluation, the business case analysis is a central instrument.

The methods of investment controlling are the same as for management consulting and management auditing of projects and programmes. A description of these methods has already been given in Chapter H4: Methods of management consulting and management auditing of projects and programmes. An example of the structure of an investment controlling report is depicted in Figure I5.3.

Table of contents: investment controlling report	
1.	Objectives of the investment controlling
2.	Basic conditions and methods for the investment controlling
3.	Description of the investment
	3.2 Actual status
	3.2 Planned versus actual comparison
4.	Description of the project initializing the investment
5.	Controlling measures
6.	Learning points for performing investments
7.	Learning points for project management
8.	Summary

Fig. I5.3: Table of contents of an investment controlling report

---

5) The same methods can be used for the investment evaluation as for the investment controlling.

16 Organization and personnel management for project portfolio management

Organization for project portfolio management

As can be seen from the descriptions of the business processes of project portfolio management, specific roles within the project-oriented organization are responsible for performing the tasks which make up these business processes.

The Project Portfolio Group and the PM Office assume the central roles of project portfolio management. Upon the assignment of a project or a programme, the Project Portfolio Group takes decisions regarding the suitable organization form for initializing an investment, and nominates the project or the programme owner team. In the business process of project portfolio coordination, the Project Portfolio Group pursues the objective of optimizing the project portfolio results. The Project Portfolio Group further recommends a possible investment controlling. The PM Office supports the Project Portfolio Group in performing these tasks and promotes networking between projects.

CHAPTER

I

Further roles for performing tasks within the business process of project portfolio management are assumed by the investment and project proposal team and the investment controlling team.

The objectives, the organizational position and the tasks to be performed by these specific roles are described in Chapter J: Organizational design of the project-oriented organization.

Personnel management for project portfolio management

Personnel management is of central importance for promoting networking and communication within project-oriented organizations.

On the one hand, personnel disposition tasks, such as the allocation of several roles within various projects to individuals, are to be performed. On the other hand, special management tasks, such as communicating information regarding the company strategies and relationships between projects, and the organization of learning in the project portfolio, are to be performed.

By decentralizing integration tasks in project-oriented organizations, the qualifications required of persons performing a role increase. Holistic ways of thinking, conflict management potential and multiple qualifications are required.

The requirement of "holistic ways of thinking" is particularly important in project-oriented organizations, since projects and programmes, project portfolios and networks of projects all present entities yet are at the same time parts of the overall organization. The need for a mutual way of thinking at different levels of abstraction, for an intellectual interplay between the part and the entirety, is derived from this.

Managers of project-oriented organizations require a high conflict management potential. The organizational distinction between projects and programmes creates diverse structural conflicts which are to be managed by performing integrative tasks.

The complexity of projects and programmes and the simultaneous assumption of several roles require that the employees are qualified to perform multi-faceted tasks. The way a project manager, for example, sees himself/herself, can not only be limited to the performance of project management tasks, but frequently includes the performance of sales and market research tasks for the organization, possibly arising from the project.

References

Endruweit, G. (Ed.), (Soziologie) Wörterbuch der Soziologie, Dt. Taschenbuch-Verlag, München, 1989

Enzenhofer, D., Semper, W., (Controlling von Projekte-Netzwerken) Controlling von Projekte-Netzwerken in projektorientierten Unternehmen, in: Projekt Journal

Lipnack, J., Stamps, J., (Networking) Networking – The first report and directory, Doebleday, New York, 1982

Aarto, K., Martinsuo, M., Aalto, T., (Strategic Management) Strategic Management through Projects, First Edition, PMA Finland, Helsinki, 2001

J

Organizational design of the project-oriented organization

Project-oriented organizations apply Management by Projects as an organizational strategy. They have specific organizational structures and business processes, specific IT, telecommunications and spatial infrastructures as well as specific cultures for the management of projects, programmes and project portfolios. New, permanent organizational units of project-oriented organizations are Expert Pools, Project Portfolio Groups and the PM Office.

The "flowing equilibrium" of organizations is ensured by means of continuous and discontinuous developments. The cycles in which discontinuities occur become shorter due to the increasing environmental dynamics. Organizations require competencies in order to deal with changes of their identity. In order to manage discontinuities, project-oriented organizations can use their project and programme management competencies.

The establishment of a project-oriented organization and its further development can be perceived as discontinuous developments.

J Organizational design of the project-oriented organization

Contents

J 1 Organizational structure of the project-oriented organization

J 1.1 Organizational strategy of the project-oriented organization

By applying Management by Projects as an organizational strategy, the project-oriented organization pursues the following objectives:

- Creating organizational flexibility by using temporary organizations in addition to the permanent organization,
- Delegating management responsibility in projects and programmes,
- Setting of objectives by defining project and programme objectives, and
- Ensuring organizational learning by using the monitoring potentials of projects and programmes.

The use of projects and programmes for the organizational design promotes the differentiation between organizations. With the high degree of differentiation of project-oriented organizations can be coped by means of integrative organization measures, such as guidelines for project and programme management and a project portfolio database.

The integration functions are to be performed by means of specific, permanent organizational units, i.e. by one or several Project Portfolio Group(s), by Expert Pools and the PM Office. It is their task to guarantee that the relatively autonomous projects and programmes comply with the general objectives and rules of the project-oriented organization.

The organizational structure of project-oriented organizations can be visualized in organization charts. Examples used to visualize project-oriented organizations are depicted in the following. The objectives, the position in the organization, the tasks, the environment relationships and the formal authorities of Expert Pools, the Project Portfolio Group and the PM Office are described in the following.

J 1.2 Organization charts of project-oriented organizations

The combination of permanent and temporary organizations within the project-oriented organization can be visualized in an organization chart. Since projects and programmes have a high strategic importance for the project-oriented organization, they should also be depicted symbolically in the organization chart. It is not necessary to present each individual project, but rather groups of projects (e.g. arranged according to types of project) and programmes.

When it comes to designing the organization chart of a project-oriented organization, there are design options regarding the use of symbols for distinguishing between permanent and temporary organizations, regarding the terminology used for the organizational units and regarding the organizational incorporation of the Expert Pools, the Project Portfolio Group and the PM Office. Boxes and ellipses can be used to depict permanent and temporary organizations, for example (see Figure J1.1). The committee for coordinating the project portfolio can be referred to as the Project Portfolio Group or as the Project Portfolio Steering Committee. In practice, the PM Office is also referred to as "PM Service" or "PM Support".

The organizational incorporation of the Expert Pools, the Project Portfolio Group and the PM Office are dependent mainly on the size of the project-oriented organization. Basically, there are no right or wrong solutions.

Different possible designs used for organization charts are depicted in Figures J1.1 and J1.2. .

CHAPTER

J

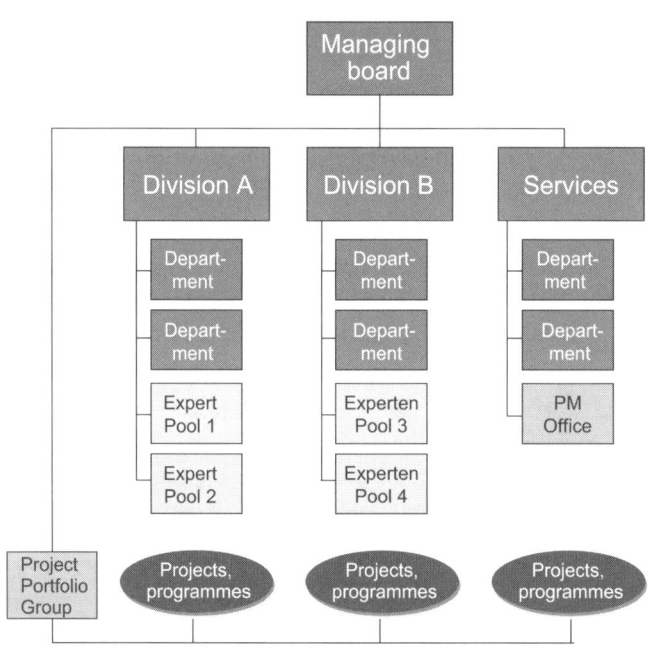

Fig. J1.1: Organization chart of the project-oriented organization (according to ROLAND GAREIS Management of the Project-oriented Company®)

Managing board

Division A	Division B	Services
Department	Department	Department
Department	Department	Department
Expert Pools	Expert Pools	PM Office

Regions

Project Portfolio Group

Projects and programmes

Fig. J1.2: Alternative design for an organization chart of a project-oriented organization

As an example of a project-oriented organization, the organization chart of an Ericsson regional organization is depicted in Figure J1.3. .

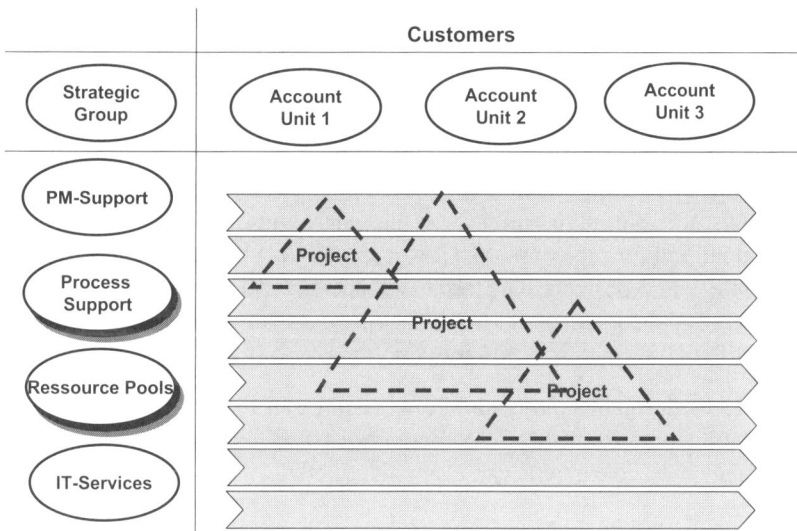

Customers

Strategic Group

Account Unit 1

Account Unit 2

Account Unit 3

PM-Support

Process Support

Ressource Pools

IT-Services

Project

Project

Project

Interpretations:
•„Accounts" are „Project Owners" and responsible for the business cases
•All ressources are organized in pools
•Project managers are assigned from the PM-Support Pool

Fig. J1.3: Organization chart of an Ericsson regional organization

J 1.3 Expert Pools

Experts from various Expert Pools of the project-oriented organization perform the business processes in projects and programmes.[1]

---

1) Instead of "Expert Pool", in practice other labels, such as "Resource Pool" and "Center of Excellence" are also used..

It is the objective of Expert Pools to provide sufficient, appropriately qualified experts for projects and programmes. It is a further objective to ensure the prerequisites for efficient implementation of the business processes in the projects by means of professional process management.

The tasks to be fulfilled in the Expert Pool are personnel management, process management and knowledge management tasks. Project-related tasks and their qualitative control are not performed in the Expert Pool but rather in the projects. In accordance with the model of the "empowered" project organization (see Chapter C4.1), the project team members and the project contributors are personally responsible for the manner and method of the performance – within the defined guidelines and standards – and for the quality of the performance.

Depending on the business operations of a project-oriented organization several types of Expert Pools can be differentiated. In an engineering construction company, for example, there are various technical Expert Pools (mechanical engineering, electrical engineering, etc.), an Expert Pool "Procurement", an Expert Pool "Installation", etc. Also in an IT company various Expert Pools can be differentiated, e.g. designers, programmers, testers, etc.

However, it is not only a sensible measure to organize Expert Pools in contracting companies. Persons qualified to work in projects should be allocated to Expert Pools for internal projects, too. Not all employees in the marketing department of a bank, for example, are qualified for project work. However, people who can also represent their department's interests in projects, who also have specialized knowledge and have sufficient project management competencies to work as project team members should be assigned to an Expert Pool "Marketing", possibly only in a virtual form.

A project-oriented organization should always have an Expert Pool "Project Management". In larger organizations, an Expert Pool of project management trainers and project management consultants can also be defined.

Various different roles are to be performed in an Expert Pool. A distinction is to be made between the pool manager and the pool members. The members of an Expert Pool can represent various qualifications. A distinction between junior experts, experts and senior experts enables the establishment of a career path as an expert in the project-oriented organization (see Chapter K). One person can be the manager of several Expert Pools at the same time. In the international Leica concern, for example, one manager is responsible for two Expert Pools, "Sensor technology" and "Optics".

The difference between an Expert Pool and a "traditional" department lies in the "empowerment" of the experts and in the perception of the management roles. The "traditional" department manager considers himself/herself above all as an expert responsible for his/her employees' performance with regard to content. The Expert Pool manager considers himself/herself above all as the manager of the pool with personnel and organization responsibilities, not as a content expert. This way the Expert Pool manager sees himself/herself, does not exclude him/her undertaking expert tasks in projects, however. However, he/she is not responsible for the work performed by the other experts which make up the pool.

The Expert Pool manager should be assessed according to the performance of his/her management tasks in the Expert Pool, and not according to the project and programme results.

The role of the Expert Pool manager is described in Figure J1.4.

Role: Expert Pool manager
Objectives
<ul><li>Lead the Expert Pool members</li><li>Assess the availability of resources for performing a project or a programme (in the assignment process)</li><li>Provide qualified employees for the projects and programmes</li><li>Provide guidelines and standards for performing the business processes of the Expert Pool</li><li>Knowledge management in the Expert Pool</li></ul>
Position in the organization
<ul><li>Reports to the manager of the profit center the Expert Pool belongs to</li><li>Manages various qualification groups in the Expert Pool: Junior experts, experts and senior experts</li><li>Contributes to the process of assigning a project or a programme, to the start process and to the close-down process of a project or a programme</li></ul>
Tasks
<ul><li>Recruit and develop the Expert Pool personnel</li><li>Dispose the Expert Pool personnel (into a project or a programme)</li><li>Develop and ensure know-how (methods, tools)</li><li>Maintain a knowledge database with information regarding the business processes of the Expert Pool</li><li>Obtain and provide the infrastructure necessary for performing the business processes of the Expert Pool</li><li>Develop and provide guidelines and standards for performing the business processes of the Expert Pool</li><li>Ensure compliance with the guidelines and standards for performing the business processes of the Expert Pool (e.g. by means of exchange of experience, supervision, audits, etc.)</li><li>Define ethical standards and ensure professional ethics</li></ul>
Non-tasks
<ul><li>Perform project-related work packages</li><li>Project-related quality control of the performance of the work packages</li></ul>

CHAPTER

J

Environment relationships
• Members of the Expert Pool
• Other Expert Pools, PM Office
• Management, managers of the profit centers
• Projects and programmes
• Customers, partners and suppliers

Formal authority
• Decisions regarding the organization, infrastructure and budget of the PM Office
• Decisions regarding the recruitment and the development of personnel for the Expert Pool
• Decisions regarding the disposition of Expert Pool members into a project or a programme
• Approval of guidelines and standards for performing business processes
• Initialization of supervision or audits regarding the business processes of the Expert Pool

Fig. J1.4: Description of the role: Expert Pool manager

J 1.4 Project Portfolio Group

The Project Portfolio Group presents a specific, permanent communication structure of medium-sized and large project-oriented organizations. Only organizations with at least 200 to 300 employees which perform at least 15 to 20 projects simultaneously, achieve a complexity, which makes a separate communication structure for the project portfolio management a sensible measure. It is only then that the management of a project portfolio should be given appropriate attention by means of establishing a Project Portfolio Group. In smaller project-oriented organizations, these tasks can be performed by the management in the course of management meetings.

In the case of larger project-oriented organizations, the Project Portfolio Group should be responsible for the assignment of a project or a programme and for project portfolio coordination. The Project Portfolio Group is authorized to take decisions and is thus not to be considered a staff unit.

Although integrative structures, such as the management circle of a company, are frequently not depicted in the organization charts of organizations, its strategic significance makes it advisable to depict the Project Portfolio Group in the organization chart.

Members of the Project Portfolio Group should be managers of the project-oriented organization with holistic responsibility and strategic orientation. Typical members of the Project Portfolio Group are, for example, the heads of marketing, finances, IT and organization. In medium-sized project-oriented organizations, members of the management board may also belong to the Project Portfolio Group. One member is to be nominated as the spokesperson for the Project Portfolio Group.

A Project Portfolio Group of an Austrian telecommunication company, for example, comprises five managers from those divisions particularly active in projects (finances, network set-up, marketing, IT and call center) and the PM Office manager. The spokesperson for the Project Portfolio Group is the manager of the finance division. This Project Portfolio Group meets once a week for 2-4 hours for project portfolio management.

Depending on the number and the complexity of the projects to be coordinated, several Project Portfolio Groups may be required in a project-oriented organization. At most, project portfolios with about 50 to 80 projects can be managed by one Project Portfolio Group. Project Portfolio Groups can be differentiated according to company divisions (or profit centers) and types of project.

An Austrian engineering construction company, for example, distinguishes between three Project Portfolio Groups: One for offer projects and contracting projects, one for product development projects and one for personnel, organizational and marketing projects. In order to ensure that the decisions taken by these three Project Portfolio Groups are coordinated, individual managers are members in several Project Portfolio Groups.

The PM Office is to support the Project Portfolio Group in its preparation, performance and follow-up of coordination meetings. In particular, the PM Office can develop the project portfolio reports. PM Offices therefore not only perform services regarding the project and programme management, but also regarding the project portfolio management.

A description of the role of the Project Portfolio Group is depicted in Figure J1.5.

CHAPTER

J

Role: Project Portfolio Group
Objectives
• Optimize the project portfolio results and the project portfolio risk • Design the project portfolio structures • implement the personnel and organization strategies in the project portfolio • Contribute to the optimization of the results of networks of projects • Contribute to the optimization of the management of chains of projects
Position in the organization
• Reports to the management of the project-oriented organization • Members are managers of the project-oriented organization (e.g. managers of profit centers, managers of IT, organization, marketing and finance divisions) and the PM Office manager • Meetings: Approx. twice per month, 3-4 hours each
Tasks
Tasks in assigning a project or a programme
• Possibly: Decide on investments to be selected (if no independent investment decision committee exists) • Coordinate the project objectives with the strategic objectives of the project-oriented organization • Decide on the organization form for initializing an investment (programme, project, small project, permanent organization) • Nominate the project owner team

Tasks in project portfolio coordination and networking of projects
• Coordinate both internal and external resources used in the project, determine project priorities
• Determine strategies for designing the relationships to project environments
• Brief analysis of selected (critical) projects
• Decide on the stopping or interruption of projects (for strategic reasons)
• Organize the learning of and between projects, the use of synergies
• Nominate the project owner team in chains of projects
Environment relationships
• Management, profit centers, PM Office, etc.
• Project proposal teams, project owner teams and project manager
• Employees of the project-oriented organization
• Customers, partners and suppliers
Formal authority
• Decisions regarding the designing of project portfolios
• Possibly: Decision regarding investment proposals
• Decision regarding project proposals
• Nomination of the project owner team
• Stopping and interruption of projects
• Initialization of the networking of projects

Fig. J1.5: Description of the role: Project Portfolio Group

J 1.5   PM Office

Organizations develop their extent of project orientation in phases. At the beginning, a few projects are performed as "exceptions". Then the number of projects and programmes performed simultaneously continuously increases. The practices in the projects and in project management are largely determined by the contributing employees. A project and programme management culture which is determined by individuals exists. There is no formal project portfolio management during this phase.

With the increasing importance of projects and programmes in organizations, there arises a need for standardizing the practices in projects and programmes and to ensure the quality of the project and programme management. There also arises the need for a project portfolio point of view and for project portfolio reports. A PM Office is frequently established in the course of the formal introduction of project and programme management. The PM Office represents the institutionalized competencies in project, programme and project portfolio management.

The objective of a PM Office is to ensure a professional project, programme and project portfolio management in the project-oriented organization. In this context, appropriate individual, collective and organizational competencies are to be developed by the PM Office.

By providing tools for project and programme management and by providing management support, the PM Office contributes to achieving the project and programme objectives. By means of regular maintenance of the project portfolio database, the development of project portfolio reports and their adequate communication, a contribution is also made toward optimizing the project portfolio.

Non-objectives of the PM Office are the performance of project owner team roles and project manager roles in projects.

The PM Office sees itself as a service provider, not as a controller. The services of the PM Office are depicted in Figure J1.6. The catalog presents the maximum amount of possible services by the PM Office. Various tools can be provided in virtual form.

CHAPTER

J

Services of the PM Office
Services for project and programme management
Provision of project management instruments • Providing instruments for project and programme management (guidelines, forms, etc.)
Management support and ensuring management quality • Management support for projects and programmes·Organizing management consulting, management auditing and possibly evaluating of projects and programmes
Individual and collective project management learning • Organizing training and further training in project and programme management and coaching of project and programme managers • Organizing the exchange of experience between project and programme managers
Organizational project management learning • ·Maintaining a knowledge database for project management ·Benchmarking the competencies for project and programme management
Project management marketing • Marketing the project and programme management in the organization ·Developing and maintaining a PM Office homepage
Services for project portfolio management
Provision of instruments for project portfolio management • Providing instruments for project portfolio management (guidelines, forms, etc.)
Assignment of a project or a programme • Review of the quality of investment and project proposals
Project portfolio coordination • Maintaining the project portfolio database • Developing project portfolio reports • Participating in and recording the coordination meetings of the Project Portfolio Group
Networking of projects • Initializing the networking and contributing to the networking of projects · • Ensuring continuity in the management of chains of projects

Fig. J1.6: Catalog of possible services of the PM Office

The "management support for projects and programmes" can contain the following services:

- Providing support in the use of project management software for the development of project plans,
- Moderating and recording project meetings,
- Developing project progress and project close-down reports,
- Developing a project homepage, etc.

Differentiated according to types of project, the project management knowledge database should contain the actual project management documentation of closed-down projects, project progress and project close-down reports and "lessons learned" from individual projects for project management.

The PM Office can use various instruments for performing services. Figure J1.7 depicts an overview of possible instruments for project and programme management and for project portfolio management. The individual instruments are described in Chapters J2 and K.

CHAPTER

J

Instruments for project and programme management

- Guidelines and forms for project and programme management
- Standard project plans
- Guidelines and forms for management consulting and management auditing of a project or a programme·Career path in project and programme management
- Standard training programmes in project management
- IT infrastructure (project management software, project portals) and moderation tools
- PM Office homepage
- Project management reference cards

Instruments for project portfolio management

- Guidelines and forms for project portfolio management
- Standards for investment and project proposals·
- IT infrastructure for project portfolio management (software for project portfolio management)
- Standard project portfolio reports

Fig. J1.7: Possible instruments of the PM Office

Examples of PM Office homepages are depicted in Figure J1.8. .

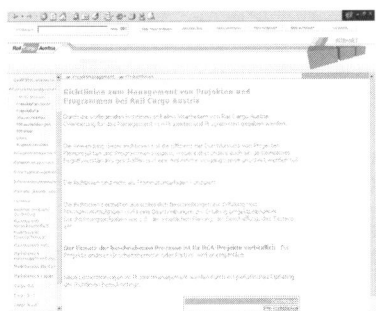 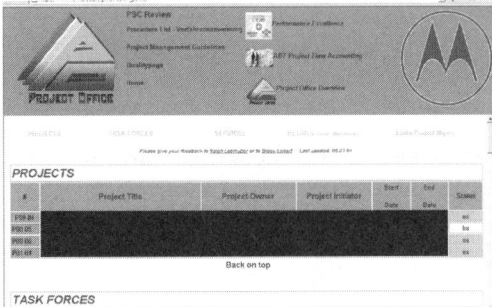

Fig. J1.8: Homepages of the PM Offices of Motorola Austria and Rail Cargo Austria

A distinction is to be made between the permanent PM Office and temporary Project Offices and Programme Offices. As a new, integrated organizational unit, the PM Office performs services for the project-oriented organization as a whole while the Project Office/Programme Office only performs services for one project/programme (see Chapter G). The differences between the PM Office and Project Office/Programme Office are summarized in Figure J1.9.

PM Office	Project Office, Programme Office
• Permanent services for project-oriented organizations • For all projects and for the project portfolio • Integrated in the permanent organization	• Temporary, various life-cycle durations • Services for a project/programme • Integrated in the project/programme organization

Fig. J1.9: Differences between PM Office and Project Office/Programme Office

Depending on the size of the project-oriented organization, the scope of the project portfolio and the services on offer, the PM Office can be made up of only one person or of several persons. A medium-sized production and trading firm with around 350 employees and a project portfolio averaging 25 projects, for example, employs two part-time employees for the PM Office.

The basic organizational structure of a PM Office is depicted in Figure J1.10. The role of the PM Office manager is depicted in Figure J1.11. Essential to the organization chart of the PM Office is the differentiation in positions which perform services for project and programme management and positions which perform services for project portfolio management. The Expert Pools "Project management" and "Project management trainers/Project management consultants" may belong to the PM Office but may also belong to other organizational units of the project-oriented organization.

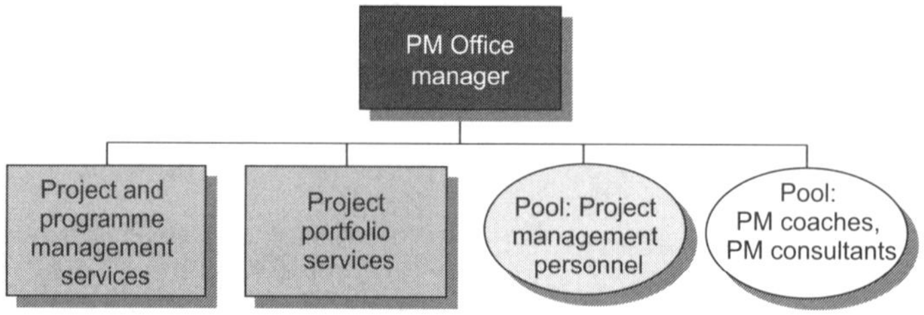

Fig. J1.10: Standard organization chart of a PM Office

Role: PM Office Manager
Objectives
• Ensure the professional project, programme and project portfolio management • Owner of the business processes of project and programme management and of project portfolio management • Lead the PM Office employees
Position in the organization
• Reports to the profit center manager or the service division the PM Office belongs to • PM Office employees report to the PM Office manager • Is a member of the Project Portfolio Group
Tasks
Tasks in project and programme management, quality assurance
• Work as an expert in project and programme management (management support, consulting, auditing)
Tasks in assigning a project or a programme
• Review the quality of investment and project proposals
Tasks in project portfolio coordination
• Develop project portfolio reports • Participate in coordination meetings of the Project Portfolio Group
Tasks in the networking of projects
• Initialize the networking of projects • Ensure continuity in the management of chains of projects
Tasks in organizational design
• Further develop the instruments for project, programme and project portfolio management
Tasks in personnel management
• Coach project and programme managers and project and programme owners

Tasks in managing the PM Office
• Organize the PM Office • Manage the PM Office personnel • Manage the PM Office budget • Possibly lead the Expert Pools "Project management" and "Project management training and project management consulting"
Environment relationships
• PM Office employees • Project Portfolio Group • Management, managers of profit centers, of Expert Pools • Project proposal teams, project owner teams and project managers • Employees of the project-oriented organization • Customers, partners and suppliers
Formal authority
• Approval of guidelines and standards for project and programme management and project portfolio management • Initialization of supervision or audits for project and programme management • Initialization of the networking of projects • Decisions regarding the organization, personnel, infrastructure and budget in the PM Office

Fig. J1.11: Description of the role: PM Office Manager

The structure of a PM Office of an international bank is depicted in Figure J1.12. This PM Office of the ABN AMRO bank had around 20 employees in 2003. Furthermore, pools of various project management experts (Support Pool, Project Manager Pool) belong to the PM Office. Services for the PPG (Project Portfolio Group), for example, are performed in the center of expertise. The PM Office is integrated in the European Division of the ABN AMRO bank. .

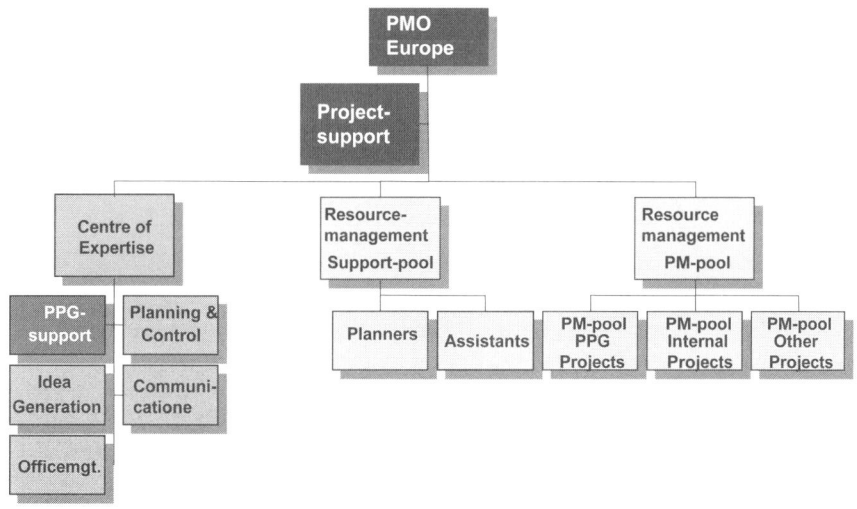

Fig. J1.12: Organization chart of the PM Office of the ABN AMRO bank

There are various options regarding the organizational incorporation of the PM Office into the project-oriented organization. The possible incorporation options into a business segment, into a service center and as a staff unit next to the management board, are depicted in Figure J1.13. .

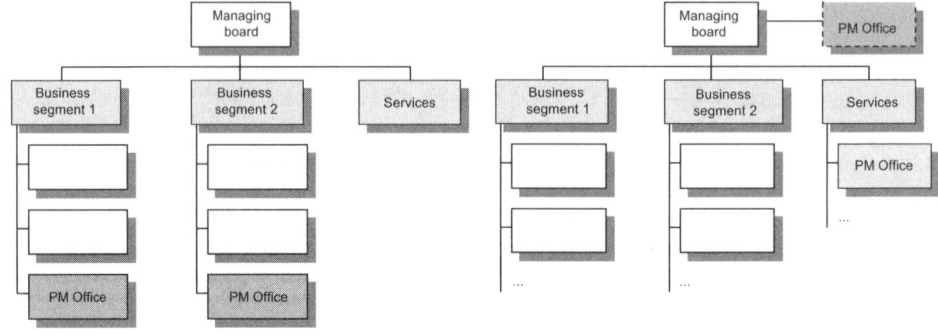

Fig J1.13: Options of the organizational incorporation of the PM Office

Should there be several PM Offices alongside each other in several business segments, they have to be coordinated formally or informally in order to ensure a standardized course of action in the projects in which several business segments cooperate. In large international concerns, e.g. at Ericsson, there is sometimes a PM Office at concern level whose operational implementation is supported by regional PM Offices in various different countries. The Ericsson PM Office at concern level with around 70 employees declared the year 2000 as the year of project management. PROPS, as the Ericsson project management approach, was provided to all subsidiaries throughout the world as general project management guidelines, training programmes were offered for project managers, Ericsson customer project managers and Ericsson senior project managers, and Ericsson project networking was organized. It can be seen from Figure J1.14 that members in 40 countries were provided with a web magazine, a website, conferences and mini-networks.

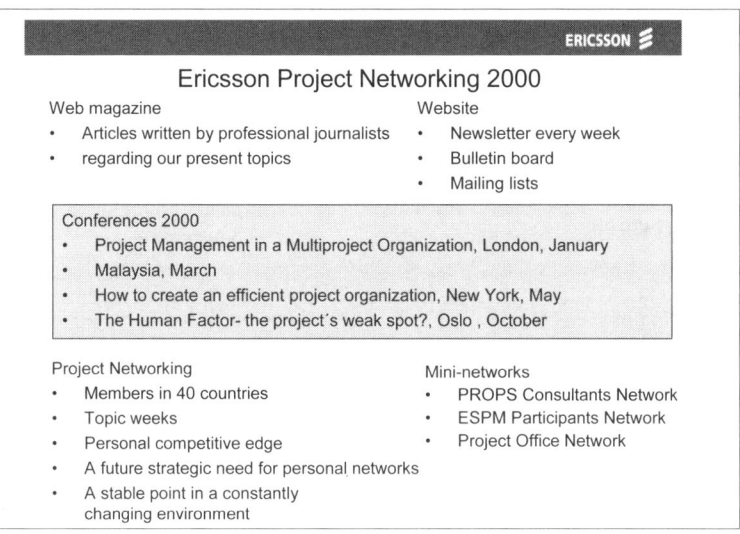

Fig. J1.14: Ericsson project networking 2000

J 2 Business processes of the project-oriented organization

J 2.1 Specific business processes of the project-oriented organization

The project-oriented organization is characterized by the following specific business processes:

- Project management,

- Programme management,

- Assurance of management quality in a project or a programme,

- Assignment of a project or a programme,

- Project portfolio coordination,

- Networking of projects,

- Personnel management in the project-oriented organization, and

- Organizational design of the project-oriented organization.

These business processes can be depicted in a "maturity model" of the project-oriented organization (see Figure J2.1).

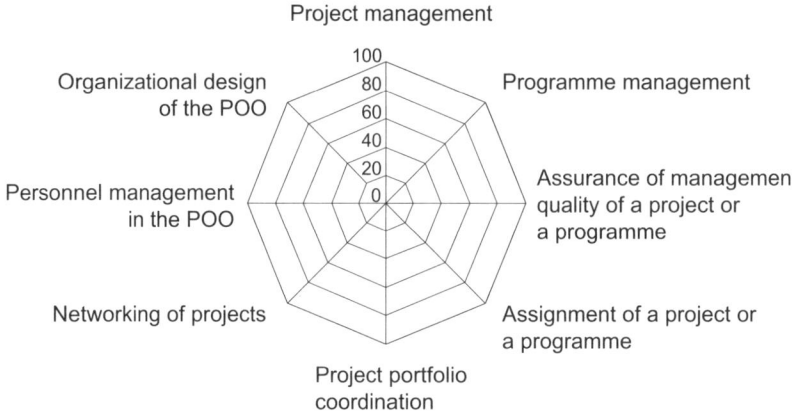

Fig. J2.1: Maturity model of the project-oriented organization

The specific business processes of the project-oriented organization are described in detail in Chapters E to K and do not require any further discussion at this point. Guidelines and forms of the project-oriented organization as well as standard project plans are described inChapter J2.3. The "maturity model" of the project-oriented organization is described in Chapter J6.

J 2.2   General business processes of the project-oriented organization

In accordance with an international trend, many project-oriented organizations are characterized by a process orientation. This guarantees the continuity of the process orientation in project and programme management and in the performance of other business processes.

Due to the importance of the business processes for the structuring of organizations, many companies also present their organization charts in a process-oriented manner. Thus, for example, in the organization chart by Telia, the Swedish telecommunications company, the most important business processes are visualized (see Figure J2.2). Projects and programmes may be necessary for performing business processes. As an example, the relationship between the phases of the business process "Product development" and projects is depicted in Figure J2.3.

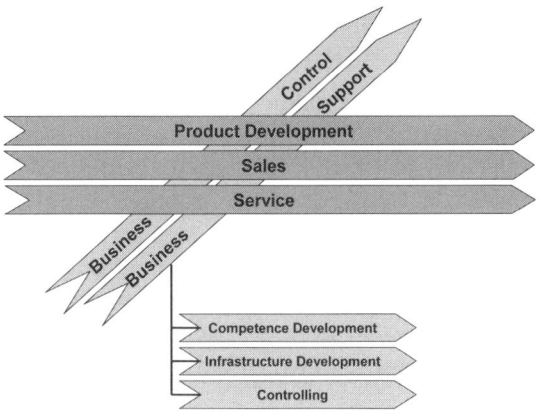

Fig. J2.2: Process-oriented organization chart of Telia

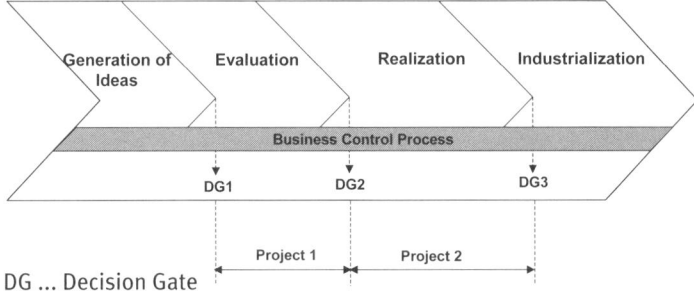

Fig. J2.3: Projects in the business process "Product development" of Telia

The general business processes of a project-oriented organization depend on the specific business operations of a project-oriented organization. The business processes of Telia presented above are, to some extent, specific to a telecommunications company and are fundamentally distinct from, for example, those of a construction or engineering construction company.

J 2.3 Guidelines, forms and standards of the project-oriented organization

Guidelines for project and programme management

In order to standardize the course of action and to assure the quality in project and programme management, project-oriented organizations apply guidelines to project and programme management. These guidelines are to provide employees with orientation for the management of a project or a programme and to assure the efficiency of the performance of a project or a programme. Guidelines for project and progamme management are an essential component of the organizational competence for the management of a project or a programme.

The guidelines for project and programme management are to define when business processes are to be organized as a project or as a programme, which methods are to be used for their management and which roles are to be performed in a project or a programme. The guidelines contain exclusively descriptions of management tasks and do not contain any descriptions of the business processes for performing a project, such as purchasing, testing, etc. Sample documentations of projects and standard project plans can be provided in order to support the use of guidelines for project and programme management.

An extract of a table of contents of guidelines for project and programme management of a public administration organization is depicted in Figure J2.4.

Table of contents	
1.	Introduction
1.1.	Objectives, content of the guidelines for project and programme management
1.2.	Updating of the guidelines
2	Definitions
2.1	Definition: Programme, project, small project
2.2.	Definition: Types of programmes and projects
2.3.	Definition: Project discontinuity
3	Project management
3.1.	Project management process
3.2.	Project management, sub-processes: Project start, project controlling, project coordination, resolution of a project discontinuity, project close-down
3.3.	Use of methods for project management
3.4.	Organization of projects
3.5.	Tools for project management
4	Programme management
4.1	Programme management process
4.2.	Programme management, sub-processes: Programme start, programme controlling, programme coordination, programme monitoring, resolution of a programme discontinuity, programme close-down
4.3.	Use of methods for programme management
4.4.	Organization of programmes
4.5.	Tools for programme management

5.	Appendix
5.1.	Forms for project management
5.2.	Forms for programme management
5.3.	Glossary
5.4.	Links to standard project plans and sample documentations

Fig. J2.4: Extract of the table of contents of the guidelines for project and programme management of a public administration organization

Guidelines for project and programme management are to be kept as brief as possible (approximately 30-40 pages, excluding appendix). They should not be designed as training documentation. Special training tools are necessary in order to communicate information regarding project and programme management .

The use of the business processes and methods described in the guidelines for project and programme management should be binding. The guidelines should form part of any possible quality management documentation certified according to ISO of the project-oriented organization.

Guidelines for project and programme management can apply to only one, several or all types of projects of a project-oriented organization. In addition to organization-specific guidelines, there are also guidelines used for several project-oriented organizations (e.g. the guidelines for project and programme management of the AUA concern), and generic guidelines which can be used for all types of projects and all types of companies. Generic guidelines, such as ROLAND GAREIS Project and Programme Management®, only need to be slightly adapted to be used according to the specific conditions of a company.

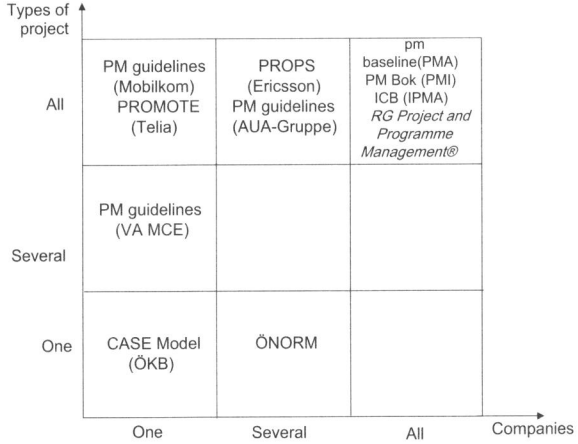

Fig. J2.5: Differentiation of guidelines according to possible uses (number of types of projects and number of companies)

Forms for project and programme management

As a rule, the guidelines for project and programme management also contain a set of forms which support the application of project and programme management methods. These forms can be provided to projects or programmes in electronic form by the PM Office.

In order to depict the application of project management methods in the project "Realization of an e-application", the forms of ROLAND GAREIS Project and Programm Management® were used in Chapter F.

Guidelines for management consulting and management auditing of a project or a programme

Since management consulting and auditing is becoming increasingly important for assuring the quality of projects and programmes in project-oriented organizations, guidelines and methods for consulting and auditing are to be provided.

The table of contents of guidelines for the management auditing of a project or a programme of an Austrian concern is depicted in Figure H5.1 in Chapter H5 "Institutionalization of management consulting and management auditing of projects and programmes in the project-oriented organization".

Guidelines for project portfolio management

The objective of the guidelines for project portfolio management is to create organizational competencies. The business processes of project portfolio management, the roles and methods used for project portfolio management are described in the guidelines to project portfolio management.

Business processes of project portfolio management are the assignment of a project or of a programme, project portfolio coordination and the networking of projects. Roles of project portfolio management are proposal teams, the Project Portfolio Group, the PM Office and Expert Pools. Essential methods used for project portfolio management are the investment portfolio score card, the business case analysis, the project definition, the project portfolio database and the project portfolio reports.

An example of the table of contents of guidelines for project portfolio management is depicted in Figure J2.6.

Table of contents	
1	Introduction
1.1	Objectives, content of the guidelines for project portfolio management
1.2	Updating of the guidelines
2	Definitions
2.1	Definition: Project and types of projects
2.2	Definition: Chain of projects, project portfolio, network of projects

3	Project portfolio processes
3.1	Process: Assignment of a project or a programme
3.2	Process: Project portfolio coordination
3.3	Process: Networking of projects
4	Roles of project portfolio management
4.1	Proposal team
4.2	Expert Pools
4.3	Project Portfolio Group
4.4	PM Office
5	Methods and tools for project portfolio management
5.1	Use of methods for project portfolio management
5.2	Tools for project portfolio management
6	Appendix
6.1	Forms for project portfolio management
6.2	Glossary

Fig. J2.6: Table of contents of the guidelines for project portfolio management

CHAPTER

J

J 2.4   Standard project plans

Standard project plans can be developed for performing repetitive projects, such as offer projects, contracting projects, product development projects, event organization projects, etc. Standard work breakdown structures, standard milestone plans, standard work package specifications, standard responsibility matrices and standard organization charts, for example, can be used as standards for performing such projects.

The use of standard plans reduces the amount of planning at the start of a new project, and makes it possible to revert to former experiences. The international IT concern Oracle, for example, provides the projects with standard plans for contracting projects, differentiated according to the implementation of various information technologies.

The use of standard project plans involves a risk due to its "linear" application, without appropriately considering the specifics of a new project. Attention must therefore be paid to the adequate adaptation of the standard project plans in order to account for the specifics of a project.

J 3   Infrastructure of the project-oriented organization

The infrastructure of a project-oriented organization can be divided into IT, telecommunications and spatial infrastructure. These three dimensions of the infrastructure are, in the following, treated differently, yet are also to be considered in their relationships to each other. The implementation of cabling in order to implement an IT network, for example, is dependent on the spatial infrastructure.

To implement temporary organizations, project-oriented organizations need infrastructures to be as flexible as possible.

J 3.1 IT infrastructure of the project-oriented organization

Essential elements of the IT infrastructure of the project-oriented organization are the system platform, the IT network and the application software.

System platform

The "system platform" of a project-oriented organization comprises the basic hardware solution and the operating systems.

The demands made by a project-oriented organization on a system platform comply with the needs of conventional organizations. However, a specific demand is made in that the system platform needs to be flexible in order to be able to comply with the dynamics of the project-oriented organization. Problems may occur, for example, should IT incompatibilities arise when cooperating with other project-oriented organizations.

In order to avoid this risk, the design of the system platform should consider the use of open standards and the use of flexible, easily-scalable components. Thus, the system platform can be relatively easily adapted to new demands.

Web-based applications based on widespread standards and which can use web browsers as universal user interfaces simplify cooperation between companies.

IT network

An efficient IT network is an essential prerequisite for efficient communication of the project-oriented organization. The network architectures for data communication can be depicted in a layer model (see Figure J3.1). An opening takes place from the inside to the outside which requires the use of various technologies.

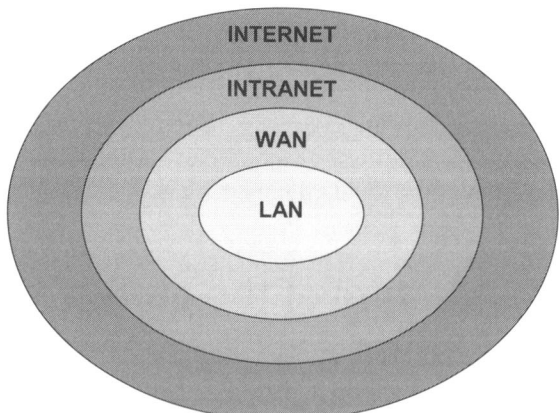

Fig. J3.1: Layer model of the network architectures [2]

---

[2] Grimm, R., Kozok. B., Lafos, F., (Kommunikationsinfrastruktur) 2001, p. 199

During the implementation of the IT network, a sufficiently-dimensioned, flexible and easily-extendible network architecture should be aimed at, e.g. by using flexible cabling and a wireless LAN. This offers members of a project organization an efficient network connection irrespective of their location.

By providing employees with notebooks, the flexibility of the IT network can be used. Project contributors can therefore work both at their workstations and at the customer's site.

Application software

The use of software for project management and project portfolio management as well as groupware and project portals is specific to project-oriented organizations.

Project management software supports the project planning and the project controlling. The standard software supports the planning and the controlling of project scope, project schedule, project resources and project costs. The tasks of designing the project organization and the project environment relations cannot essentially be performed using the project management software. Standard products available on the market are, for example, MS Project, Primavera Project Planner, Super Project, Project Scheduler, Open Plan, SAP PS.

Additions to the project management software are offered, on the one hand, in the form of graphics software, e.g. Graneda or WBS Chart Pro and, on the other hand, in the form of software for supporting special project management tasks. A list of software products which, for example, support project risk management, can be found in Figure F1.78, Chapter F1.10 "Risk management".

Software for project portfolio management supports the project-oriented organization in the maintenance of a project portfolio database and in the development of project portfolio reports. Products available on the market which also cover project portfolio management are, for example, Primavera, Planview, Artemis, OPX2, Planta PPMS, Niku Portfolio Manager.

Groupware supports communication and cooperation within projects. A relatively new aspect is the use of project portals for promoting Internet-based cooperation and communication within projects. A distinction is to be made between general "collaboration" portals, such as community zero, same-page, teamspace, and specific project portals, such as WelcomeHom, Primavera Teamplay, Pacific Edge Project Office, MS Project Central.

These specific applications of project-oriented organizations must also be compatible with the organization's other applications in order to enable standardized flows of data without media discontinuities. Therefore, interfaces between the project management software and, for example, frequently-used ERP software (e.g. SAP or Oracle i-Business Suite) are to be defined.

Peripheral devices

In addition to groupware, a web browser and video communication software should also be installed on the notebooks of members of virtual teams. The notebooks therefore also

CHAPTER

J

require appropriate peripheral devices, such as cameras, loudspeakers and microphones, in order to enable multimedia communication.

It is possible to equip either each notebook or just individual machines located in special multimedia rooms with peripheral devices.

J 3.2　Telecommunications infrastructure of the project-oriented organization

Definition: Telecommunications

The term "telecommunications" encompasses the transmission of speech, data and images by means of telephone, fax, e-mail, the Internet and video conferences.

Telecommunications as a competitive factor

By using an appropriate telecommunications infrastructure, project-oriented organizations can ensure both their organizational flexibility and their customer orientation. The use of a modern telecommunications infrastructure presents a competitive factor in the form of:

- Savings in time and costs:
  Members of project organizations employed at different locations cooperate increasingly virtually. With regard to the communication technology, this cooperation can be supported by means of corporate networks which can run quickly and cheaply via speech, data, video or multimedia communications.

- Efficient flow of information:
  Members of project organizations spend a major part of their working hours obtaining and forwarding information. Efficient networking of the computers and the telephone system ensures optimal organization of the communication.

- Fast feedback:
  Fast feedback to recommended solutions and achieved intermediary results is important to ensure the appropriate progress made in projects. The availability of the addressees can be supported by means of communication solutions, such as "call back upon engaged tone", cordless and mobile communication, voice mail, e-mail, video and multimedia communication.

J 3.3　Spatial infrastructure of the project-oriented organization

Architecture as a cultural element of the project-oriented organization

The architecture is a central element of the organizational culture. The architectural design of the offices influences the levels of satisfaction, creativity, effectiveness and productivity of the employees.

Flexible interior design options promote the project orientation of organizations. It should be possible to adapt the offices to the requirements of the current development phase of a project. Efficient communication in a project or a programme is only enabled with the appropriate architecture and interior design.

Interior design

If possible, the members of project teams should work together in open-plan offices even if the project is of a relatively short duration (e.g. 3-4 months). Communication between project teams working together in an open-plan office is more efficient than between project teams whose members work in several different office booths.

Both formal and informal communication is to be organized in a project or a programme. Appropriately sized and equipped meeting rooms are to be provided for project meetings and project workshops. However, the potential informal communication provided by jointly-useable cafeterias, break rooms, smoking rooms, etc. also contributes to the success of a project or a programme.

Furniture and desk sharing

Communication within projects is further promoted by means of the furniture, e.g. transparent walls, light tables and chairs, low-level cupboards which enable eye contact between the employees. Mobile, height-adjustable tables enable the organization of brief stand-up project meetings.

When working on a project or a programme, many employees spend a large part of their working hours not at any one fixed workplace. It is possible to provide such employees with workplaces based on the desk-sharing principle. These workplaces can be occupied flexibly. Representatives of relevant project environments could also use these workplaces.

For project workshops, rooms are to be provided which can be flexibly adapted to various numbers of participants and which have the appropriate medial equipment, such as video beamers, flipcharts, overhead projectors, pin boards, etc. The use of flexible partition elements enables the room to be adapted to different numbers of participants.

J 4   Cultures in the project-oriented organization

Specific values and norms characterize the culture of the project-oriented organization and enable a differentiation between the project-oriented organization and the non-project-oriented organization. The organizational culture is an essential element of the identity of a project-oriented organization. Further elements of the organization's identity are its services, the organizational structure and the business processes, the infrastructure, the budget and the environment relationships.

By defining projects and programmes sub-cultures are developed in the project-oriented organization. This development of project and programme specific cultures contributes to the success of projects and programmes, i.e. it is functional for the project-oriented organization.

J 4.1   Culture of the project-oriented organization

An organizational culture can be seen as the totality of the values, norms, behavioral patterns and artefacts which are jointly developed and used by the members of an organization. The culture of an organization is not directly tangible. It can be observed by means of symbols, skills and the tools used.

The project-oriented organization combines permanent and temporary organizations in the organization chart, applies guidelines for project and programme management, and uses specific forms and tools for project and programme management. The way a project-oriented organization sees itself can be documented in a mission statement (see Figure J4.1). Characteristics of a non-project-oriented organization are described in Figure J4.2. Non-project-oriented organizations can also perform projects, yet they do not have any explicit culture and identity as a project-oriented organization.

CHAPTER
J

---

1. We are a project-oriented organization
We use projects for business processes of small, medium and large scale. We continually develop our project management culture. We apply project management methods differently according to the project requirements

---

Fig. J4.1: Extract of a mission statement of a project-oriented organization

Characteristics of a non-project-oriented organization
• The term "project" is used for many different things, including routine tasks. A project inflation arises. Projects are not awarded adequate management attention.
• The project boundaries are defined according to the divisions and departments of the company. This results in too many small projects, which leads to sub-optimizations. The integrative work must be performed by the permanent organization which is, however, overtaxed by it.
• Nobody knows which and how many projects are being performed at any time. No information regarding the project portfolio exists. Projects arise informally, parallel projects with the same objectives are performed, and resources required for projects cannot be controlled.
• No project management methods are used in projects. This means that the transparency and the chance for efficient communication in projects are lost. This does not increase creativity, it reduces it.
• Individuals leave their mark on the working forms in projects. This means "reinventing the wheel" for each project. The professionalism in project management depends exclusively on the qualifications of individual persons.
• The objectives and the tasks are always agreed from one project meeting to the next.The lack of a "big project picture" means that the members of the project organization lack orientation.

Fig. J4.2: Characteristics of a non-project-oriented organization

The values, norms and behavioral patterns of an organization are also expressed in its management paradigm. A traditional management paradigm is characterized by hierarchy as a central integration instrument, by cooperation based on interface definitions and by operations performed in functional organizational units. Organizations with a traditional management paradigm cannot use the organizational potentials of projects and programmes.

A "new management paradigm" is required for the successful, efficient management of projects and programmes. This "new management paradigm" can be characterized on the basis of:

- Customer orientation,
- Perception of the organization as a competitive factor,
- Process-orientation and team-orientation,
- "Empowerment" of employees,
- Networking with customers, partners and suppliers, and
- (Dis)continuous change.

The "new management paradigm" is influenced by the model of the learning organization, by lean management and total quality management (see Figure J4.3).

Influences of the learning organization[a]
• Distinction between individual, collective and organizational learning
• Perception of the organization as a competitive factor
• Necessity of learning and de-learning
• Continuous and discontinuous learning
Influences of lean management[b]
• Process orientation
• Concentration on core competencies
• Flat, lean organizational structures
• Team work
• Networks and cooperations
• Continuous improvement
Influences of total quality management[c]
• Customer-orientation
• Product and process quality
• Quality control, quality assurance and quality management
(a) *Senge, P.,* (fifth discipline) The fifth discipline, 1998 (b) *Womack, J.,* (Autoindustrie) Die zweite Revolution in der Autoindustrie, 1992 (c) *Juran, J. M.,* (Qualitätsplanung) 1991.

Fig. J4.3: Influences on the "new management paradigm"

The management paradigm existing within an organization creates a cultural framework for the performance of projects and programmes. Even though a traditional management paradigm suits a traditional, method-oriented project management approach (see Chapter B2.1: Traditional project management), it results in inefficient projects. Efficient projects based on a systemic, process-oriented project management approach (see Chapter B2.2: ROLAND GAREIS Project and Programme Management®) are promoted by a new management paradigm.

The relationships between the management paradigm of a project-oriented organization and the practiced project management approach are depicted in Figure J4.4.

	Traditional management paradigm	New management paradigm
Traditional project management	• Fitting culture • Inefficient projects • Dominance of the line organization	• Cultural combination is hardly conceivable • High potential for conflicts
Systemic project management	• Cultural differences • Efficient projects are possible, yet … • High potential for conflicts	• Fitting culture • Efficient projects • Empowerment" of the projects

Fig. J4.4: Relationship between the management paradigm and the project management approach

J 4.2   Sub-cultures in the project-oriented organization

In the project-oriented organization, the cultural differentiation is promoted by the development of cultures specific to projects and programmes. Projects are not only an instrument of cultural differentiation, they are also an instrument of integration. Cultural integration occurs when members of the project organization come from various divisions and/or when the project content refers to the organization as a whole.

As a sub-system of an organization, a project can be distinguished from other sub-systems, such as departments or other projects, by means of a specific culture. A cultural differentiation of a project is functional for the project-oriented organization if the existing autonomy helps the project to achieve its objectives.

Basically, the cultural division process into sub-cultures can be performed as far as desired. However, it is a prerequisite that the relevant cultures are characterized by their own values, rules and behavioral patterns which can be learned and jointly shared by the members.

The objectives of the development of cultures specific to projects and programmes were described in Chapter D3, whereas the methods for the development of a culture were depicted in Chapter F1.9.

J 5   Management of discontinuities in the project-oriented organization

J 5.1   Continuous and discontinuous developments of organizations

The development of organizations takes place both continuously and discontinuously. On the one hand, continuous developments of strategies and objectives, services, the infrastructure, the organization and the personnel of companies take place by continuous product developments, by investments in buildings, in information and telecommunications technologies or by organizational and personnel developments. On the other hand, organizations also experience discontinuous developments.

Chances can arise, for example, from entering into a strategic alliance, from a merger with another company or as a result of the accessibility of new technologies. By means of a merger, organizations intend on using synergies, such as the reduction of competition, the development of new markets or reduced costs. The identities of the merging organizations change. The definition of new objectives and strategies, the redefinition of the services, the redesign of the organization, for example, create a new joint identity.

Crises can result from a basic shift of the market or from a necessary recall campaign of a faulty product, for example. There are various causes for crises of organizations. They can be caused by changes to the organization's relationships to relevant environments or by the dynamics of the "organization" as a social system (see Figure J5.1).

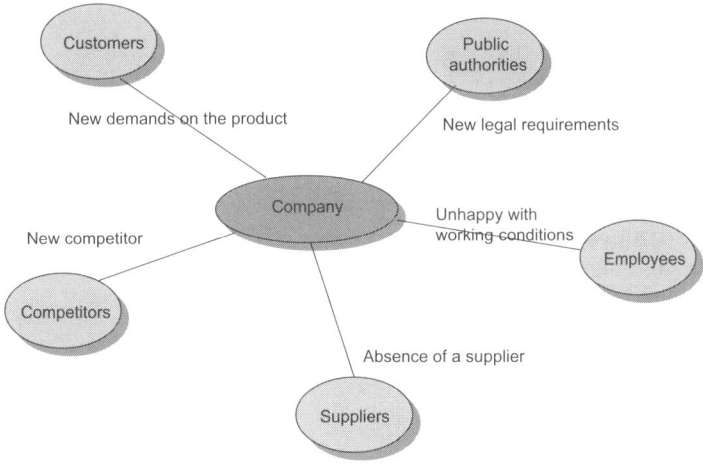

Fig. J5.1: Possible causes for the definition of a company crisis [3]

---

3)   Cp. Gareis R., (ed.), (Erfolgsfaktor Krise) 1994, p. 24.

There are also structurally caused discontinuities in organizations. The generational change in a family-run company is one example for such a change in identity.

Dynamics of the project-oriented organization

The project-oriented organization is characterized by dynamic boundaries (see Figure J5.2). The quantity, the objectives and the scope of projects and programmes being performed as well as the resources being used and cooperation partners being involved are continually changing. The annual budgets of organizations which perform customer contracts in a project form (e.g. IT, construction or engineering construction companies) can differ by more than 50%.

The context of the project-oriented organization is also dynamic. Relations with the constantly newly-established project and programme environments are to be designed, and strategic alliances are to be established according to specific requirements.

In addition to the professionalism in performing the specific business processes of project and programme management, project portfolio management, etc., project-oriented organizations need special competencies for strategic controlling, organizational learning and for the reflection of the organization's identity.

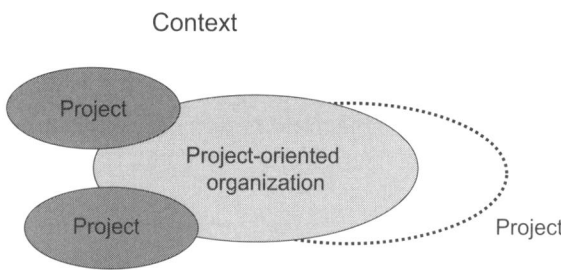

Fig. J5.2: Dynamic boundaries of project-oriented organizations

Due to their dynamics and complexity, project-oriented organizations have high discontinuity potentials. The following causes can result in crises in project-oriented organizations

- Non-perception of the dynamics of the organization,
- Inadequate project definition due to insufficient differentiation between business processes, investments and projects,
- Insufficient competencies for performing the specific business processes of the project-oriented organization,
- Projects being too closely coupled, and
- Inadequate structure of the project portfolio.

J 5.2   Projects and programmes for the management of discontinuities

Existing structures of the permanent organization are frequently relied upon for resolving discontinuities. However, the organizational structures and the business processes designed for the management of routine processes are not suitable for the performance of complex, relatively unique business processes. The hierarchy prevents unbureaucratic, fast solutions to problems in critical situations. Projects, on the other hand, promote organizational flexibility and dynamics. By using a flat project organization, the high communication demand for resolving a discontinuity can be coped with. It is therefore recommendable to use projects or programmes to resolve the discontinuities of an organization.

In larger companies and in medium to long-term processes of change, several closely-coupled projects are to be performed as a programme with the aid of programme management (see Figure J5.3).

Programmes for resolving company chances
• "Symphonie": Merger between Bawag and PSK • "LKS 2000": Reorganization of state-owned hospitals in Salzburg • "Railway stations initiative": Modernization of Austrian railway stations • "Star 2000": IT integration between AUA and Lufthansa

Fig. J5.3: Programmes for resolving company chances

Also measures for avoiding a company crisis, for promoting a company chance and for providing for discontinuities can be performed as projects or as programmes.

J 5.3   Specific demands on project management when resolving a discontinuity

Due to the ambivalence of the outcome, projects for resolving a discontinuity of an organization are characterized by a high level of uncertainty. Traditional structures of the organization have to be challenged or even fundamentally changed. New solutions are not obvious. On an almost daily basis, consequences of implemented measures, new facts or reports in the media can fundamentally change the project status and the further course of action. Creativity potential is required and there is extreme pressure with regard to time and decision-making. .

The high social complexity of projects for resolving a discontinuity in an organization results from all employees being affected by the success or failure. Professional communication within the project and with relevant environments is necessary. The differentiation of the measures implemented in the project to resolve the discontinuity from the organization's current activities presents a particular challenge.

The demands on the management of projects for resolving discontinuities exceed the requirements of "normal" projects, and demand a high level of project management competence. Since projects for resolving discontinuities are (relatively) unique , standard project plans and experience gained in similar projects cannot be used, as in repetitive projects.

A project organization adequate for these requirements is to be designed. Operational project plans which are as flexible as possible are to be developed. The establishment of alternative project plans for various project scenarios is recommended.

Design of the project organization for resolving a discontinuity

Projects for resolving a discontinuity are to be designed adequately. Necessary roles are to be defined, experts to perform these roles are to be recruited internally or externally, appropriate communication structures are to be established, incentive systems are to be developed, etc.

CHAPTER

J

Projects for resolving a discontinuity are "internal" projects, i.e. there are no external customers, but "only" internal customers. Appointing the personnel for the roles of project owner team and project manager involves strategic decisions which significantly influence the success of the project. The project manager should be part of the management board. The project owner team can be a committee made up of owners' representatives and representatives of the management board. In the case of crises in a business segment, the role of the project manager may be assumed by either a representative of the management board or by the business segment manager.

When designing the project organization, it must be minded that specialist competencies are available for analyzing, strategic planning and the implementation of measures, also decision-making competencies and relational competencies are required. Decision-making competencies are necessary in order to react to changes at short notice and to relatively autonomously determine strategies and measures. Dependencies on the decision-making structures of the hierarchy are to be minimized. Important decision-makers of the affected social system therefore have to assume roles in the project organization. The provision of relational competencies in the project is to ensure acceptance by those affected by the decisions taken.

Communication structures require special attention. Intensive information exchange, constructions of current "realities" and regular reflections on the successes and failures of the measures implemented are to be used by the project team. Representatives of relevant environments are to be provided with appropriate information.

Symbolic management is important in projects for resolving discontinuities. The significance of changes can be visualized by means of the name of the project, by events and publications to communicate the project objectives and the project results as well as by rituals when presenting the results.

The name "symphony" was chosen for the merger of the two Austrian banks BAWAG and P.S.K., which is an example of a "chance" programme. A launching event entitled "Overture" was organized for around 300 employees in the concert hall in Vienna. Furthermore, an employee magazine including progress reports and highlights was published regularly.

Project planning for resolving a discontinuity

Project plans enable a common project view of the project team members, a common project language and the establishment of commitment in the project. The project plans support communication within the project. Using professional project management for resolving discontinuities conveys confidence in a professional course of action.

The work breakdown structure is the central planning instrument. Further project plans build on this. Due to the high complexity and dynamics of projects for resolving discontinuities, the use of lists of dates and bar charts is recommended for scheduling purposes. Neither the work breakdown structure nor the schedules are to be designed in too much detail, otherwise the effort of making adaptations upon the expected changes is too great, and acceptance of the project management methods is diminished.

The communication costs, PR costs and consultancy costs are to be considered in project cost planning. Controlling of the costs and services in projects for resolving discontinuities is generally performed in short periods. In order to promote the motivation of those involved, the current results can be presented with the help of appropriate visualization tools.

Projects for resolving discontinuities require a high degree of external orientation. Great attention must therefore be paid to the analysis of the project environment relationships. Only the construction of relevant project environments and the development of new networks create the potential for resolving discontinuities.

Case Study J1 depicts the organization and the strategies of the project for resolving a crisis of the project management association Projekt Management Austria . The phases and work packages of the project for resolving the crisis are apparent from the work breakdown structure.

Project-performing organization:

- Projekt Management Austria

Reason:

- First signs of illiquidity of the association

Project organization:

- Project owner team (managing board of the association), project manager (managing director), project team members (all employees)

Strategies for resolving the crisis:

- Cost savings for the association, offer of additional services (e.g. large-scale project management events), use of the network of the association's members for marketing purposes and optimization of the organization.

Project duration:

- Approx. 8 months

Use of project management methods:

All fundamental project management methods were applied for resolving the crisis. The following work breakdown structure is depicted as an example.

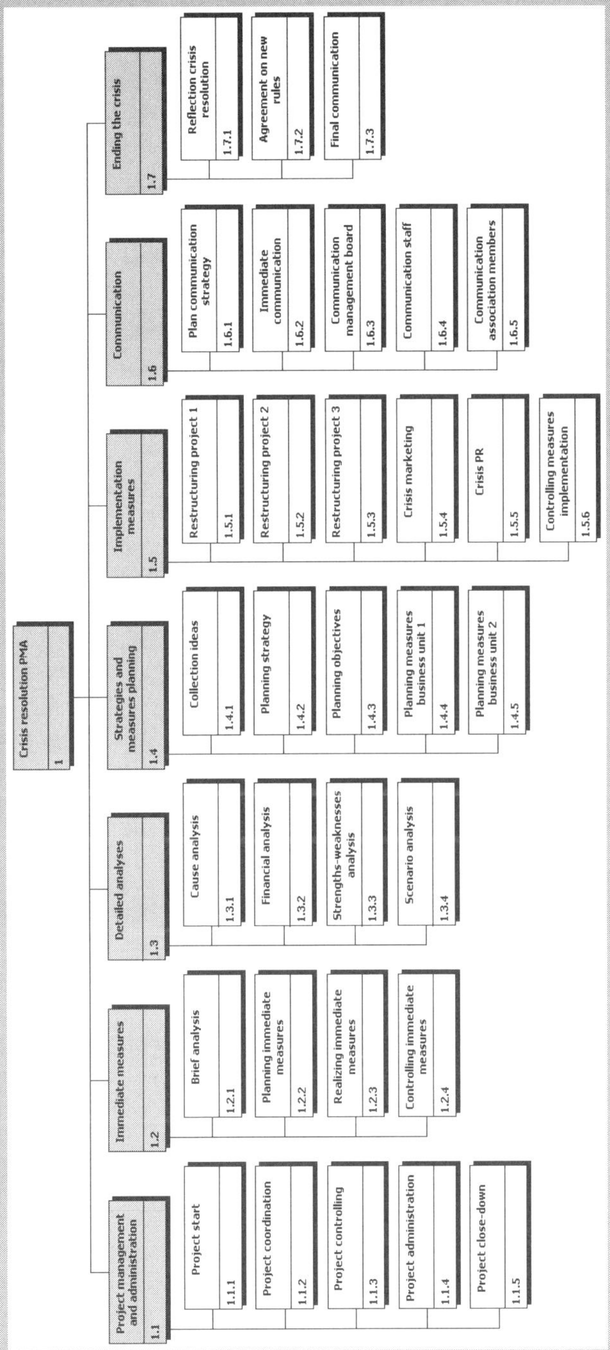

Fig. J5.4: Work breakdown structure of the project "PMA crisis"[(a)]

(a)   *Gareis, R.,* (Erfolgsfaktor Krise) 1994, p. 176

J 5.4   Projects for avoiding crises, for promoting chances and for providing for discontinuities

A project (or a programme) can be used not only to resolve discontinuities but also for avoiding crises, for promoting chances and for providing for discontinuities.

Chains of a concept and a realization project are generally needed in order to avoid crises and to promote chances. Starting from the development of crisis/chance scenarios, concepts and implementation plans are developed. The implementations themselves occur in realization projects. As an example, Case Study J2 depicts the project "SAVE" for providing for a company crisis at an insurance company. .

⋯⟩ **CASE STUDY J2: Project "SAVE"**

Project-performing organization:
* International insurance company

Problem fomulation and project objectives:
* Due to the annual review by the concern revision, the central eastern Europe division of the insurance company was asked to develop a business continuation plan in crisis situation. Possible causes for crises were, on the one hand, natural phenomena, such as fire, floods and earthquakes and, on the other hand, acts, such as arson, bomb attacks and computer viruses.
* The business continuation plan should present concrete help should a crisis arise, in addition to measures for a smooth restart of the business operations. The duration of the restart process should be minimized in order to ensure appropriate looking after of the insurance customers.
* To this end, existing plans had to be revised and integrated into an overall crisis provision plan. The plan was implemented as part of the project for the division location in Austria. In follow-up projects, implementation took place in the company's locations in eastern Europe.

Project management documentation of the project "SAVE"
* Project objectives plan·
* Objects of consideration plan
* Work breakdown structure

SAVE	PROJECT OBJECTIVES PLAN	Insurance company XY
Type of objective	Project objectives	
Main objectives	• Developing guidelines for the case of a crisis • Ensuring the necessary infrastructure in the case of a crisis in order to guarantee the continuation of business operations • Creating awareness for crisis management • Developing competencies among the employees of the insurance company in Austria for resolving crises	
Additional objectives	• Providing a basis for developing competencies for resolving crises in the eastern European units of the insurance company • Identifying any safety defects	
Non-objectives	• Implementing the crisis provision plan in the eastern European units • Developing a concept for avoiding crises	

Fig. J5.5: Objectives plan of the Project "SAVE"

SAVE	OBJECTS OF CONSIDERATION PLAN	Insurance company XY
Type of object	Object of consideration	
Definition and scenarios	Definition of crisis	
	Definition of damage	
	Scenarios	
Business processes	Crisis management manual	
	Guidelines	
	Checklists	
	Simulation	
Organizational structure	Departments	
	Branches	
	Crisis management staff	
Infrastructure	Hardware	
	Software	
	Buildings	
Creation of awareness	Training	
	Marketing	

Fig. J5.6: Objects of consideration of the project "SAVE"

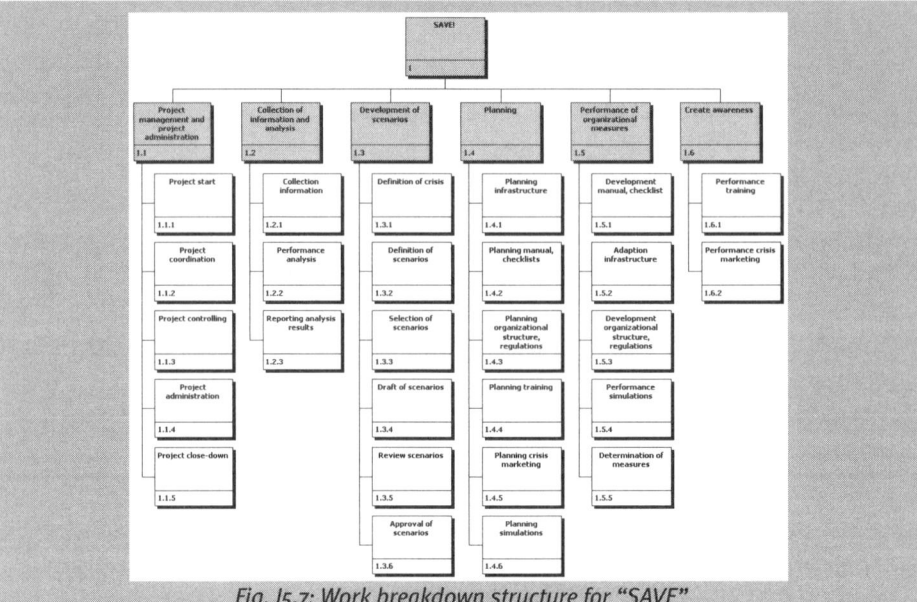

Fig. J5.7: Work breakdown structure for "SAVE"

Interpretation:

- From the point of view of the project manager of the project "SAVE", the use of project management contributed to the high quality of the project results, and enabled the creation of transparency in the project as well as a holistic view of the project.
- Furthermore, an appropriate basis in terms of content was developed for the follow-up projects in eastern Europe. Challenges to the performance of the project were the non-immediacy of the dangers and a low level of willingness amongst employees to tackle an abstract subject.

J 6   Development of project-oriented organizations

(in cooperation with Michael Stummer)

J 6.1   Objectives and objects of consideration of developing a project-oriented organization

The use of projects and programmes for performing relatively unique business processes of large scope and high strategic importance requires both individual competencies and the appropriate organizational competencies of the project-oriented organization.

The first establishment and the further development of the project-oriented organization represent phases in the development of the organizational competencies.

In accordance with a traditional view, measures of organizational development refer to changes of the organizational structure and the business processes, the organizational culture and the personnel structures. In accordance with a systemic understanding of the organization, also additional elements of the identity of an organization, i.e. the strategies, services and products, infrastructures and environment relationships, are to be considered (see Excursus B3: Identity of an organization).

Objects of consideration of the development of a project-oriented organization thus represent all these identity elements. The extent of change to these identity elements is to be planned in the design of the organization development process (see Figure J6.1). .

Identity elements	Object of consideration in the development process
Organizational strategy	• Management by Projects
Services	• Project and programme management as a differentiation characteristic for the performance of services or as a service itself
Organizational structure	• PM Office; Project Portfolio Group; Expert Pools • Consideration oof project and programme management responsibilities of line managers in role descriptions
Business processes and instruments	• Project and programme management • Assurance of management quality in projects and programmes • Project portfolio management·Instruments: Forms and checklists
Personnel	• Project management career path • Project management personnel development
Infrastructure	• Software for project and programme management • Software for project portfolio management • Moderation material, project rooms
Environment relationships	• Relationships with customers, partners, suppliers, etc.
Budget	• Budget for the PM Office, • Budget for personnel development, • Budget for project management consulting and project management auditing.

Fig. J6.1: Objects of consideration in the development process of the
project-oriented organization

Measures for the development of a project-oriented organization can lead to continuous or discontinuous changes. If development measures only affect a few identity elements, this leads to a continuous change. No fundamental change of the culture should result.

Measures such as project management training or the introduction of project management software are examples of this.

However, the development can also be discontinuous. The objective of discontinuous development is a change of the identity of the organization. This requires a consideration of many or all identity elements of the project-oriented organization. The organization develops a new self-understanding.

In the following, the two phases of first establishing and further developing a project-oriented organization are described. The option of introducing project management to a non-project-oriented organization is also considered.

The design of these identity-forming processes represents a central factor for success. The objectives and objects of consideration are to be defined, the measures to be performed and the methods and tools to be used are to be planned and the organization is to be designed.

J 6.2   First establishment of a project-oriented organization

The first establishment of a project-oriented organization assumes that no or only a few individual and few organizational competencies exist within the organization. The development of competencies in project, programme and project portfolio management is striven for by means of personnel development measures, infrastructural, environmental and budgetary measures as well as measures to restructure the organization and redesign the business processes..

The process of establishing a project-oriented organization is depicted in Figure J6.2.

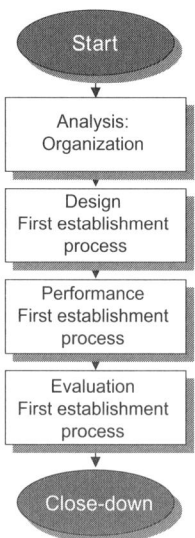

Fig. J6.2: Process of the first establishment of a project-oriented organization

Organization analysis

The organization analysis serves the purpose of collecting and analyzing information on the organization in order to create an adequate basis for designing the development process. General information to be obtained refers to the existing business processes and the organizational structure, the personnel structure, the infrastructure, the environment relationships, etc. Specific information refers to the number and the types of projects and programmes and the current competencies in project, programme and project portfolio management. Methods for organization analysis are interviews, documentation analysis, observations, reports, presentations and workshops.

The organization analysis already represents an intervention in the organization. By means of questions in interviews, by reports from the analyses of project management documentation, etc., the representatives of the organization are supplied with targeted information about the project-oriented organization. The objective of this is to promote the process of change.

CHAPTER

J

Design of the establishment process

The design of the establishment process can be implemented based on the results of the organization analysis. In addition to the planning of the objectives and the objects of consideration, the organization for the implementation of the establishment process is to be designed. Comprehensive establishment processes should be performed in the form of a project or a chain comprising a conception project and a realization project. This ensures the necessary management attention.

An adequate project owner team is to be formed and a project manager with good social competencies is to be nominated. The project team members need to have competencies about the management of the project-oriented organization, decisional authorities and relational competencies to ensure acceptance of the results of the establishment process.

Performance of the establishment process

The establishment process is performed in two phases. In the first phase, fundamental personnel, organizational and infrastructural structures are to be developed. In the second phase, these structures are to be transferred into the projects, programmes and the project portfolio management.

The following measures must be performed to create the fundamental structures of a project-oriented organization:

- Developing guidelines and forms for project, programme and project portfolio management,
- Developing standard project plans,
- Establishing personnel development plans and a career path for the project management personnel,
- Providing software for project, programme and project portfolio management,
- Possibly establishing a PM Office.

These structures can be transferred into the projects, programmes and the project port-folio management by the following measures:

- Training of owners, managers and team members of a project or a programme,
- Coaching of project and programme owners, of project and programme managers and project and programme teams,
- Exchange of experience workshops for project and programme management,
- Management consulting of projects and/or programmes,
- Project and programme management marketing.

Investment controlling

Once the measures for establishing a project-oriented organization and a pilot phase of the application have been performed, this investment is to be subjected to controlling. This can be performed only qualitatively (observations, reflections, etc.) or supported by quantitative analyses. One instrument for analyzing the competencies of a project-oriented organization is the "maturity model" (see Chapter J6.3). The results of the analysis of the competencies of the project-oriented organization enable the evaluation of the benefits of the investment and the planning of further measures.

J 6.3   Further development of a project-oriented organization

Organizations in which individual and organizational competencies with regard to project and programme management and possibly even project portfolio management exist, are considered in the further development of a project-oriented organization.

Objectives of the further development can be, on the one hand, the optimization of the established business processes of the project-oriented organization and, on the other hand, the establishment of additional business processes and the structures necessary for their performance. If a project portfolio management has not yet been established, its establishment could be the objective. The introduction of management consulting and management auditing are also objectives of further development.

The phases of the further development process are similar to those of the process for first establishing a project-oriented organization (see Figure J6.2). For defining optimization measures the analysis of the project-oriented organization is of special importance. Other further development measures are to be defined depending on the respective objectives.

Analysis of the project-oriented organization

In addition to compiling general information, the analysis of a project-oriented organization further includes compiling the existing competencies as project-oriented organization. This refers both to the individual and the organizational competencies. Methods for analyzing individual competencies of the project management personnel are described in Chapter K.

The analysis of the organizational competencies of a project-oriented organization can be performed using the model "maturity of the project-oriented company (mm-poc)". Based on a structured questionnaire, both organization-internal and -external experts can assess the maturity..

The questionnaire on the maturity model is structured based on ROLAND GAREIS Management of the Project-oriented Company® according to the specific business processes of the project-oriented organization. In addition to the questions regarding the assessment of the organizational competencies for performing these business processes, the questionnaire contains general data on the considered project-oriented organization. For example, the number of projects and programmes performed, and the types of projects and programmes are analyzed.

The "maturity" of each business process is analyzed on the basis of questions. Most questions refer to project management, since this is the central business process of the project-oriented organization. However, programme management and the assurance of the management quality in a project or a programme are also comprehensively analyzed (see Figure J6.3).

The individual questions per group of questions are equally weighted. If, for example, five questions are planned for one group of questions, each question is weighted with 20%. The weighting of the groups of questions depends on the importance of the processes in the project-oriented organization. The "project management" group of questions has a weight of 20%. All other groups of questions have a weight of 10%. This weighting is important for the calculation of the "maturity ratio". .

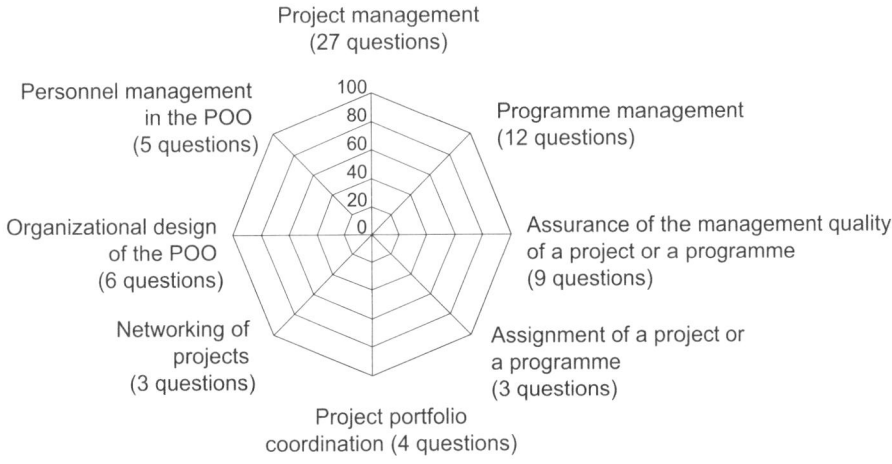

Fig. J6.3: Maturity model of a project-oriented organization

One question on project management and one question on project portfolio management are depicted as an example in Figures J6.4 and J6.5.

A1.1. Methods of project planning in the project start process	
Project objectives plan	
Objects of consideration plan	
Work breakdown structure	
Work package specifications (for selected work packages)	
Project bar chart	
CPM schedule	
Project resource plan	
Project finance plan	
Project cost plan	
Project risk analysis	
Project scenario analysis	

1 = never, 2 = seldom, 3 = sometimes, 4 = frequently, 5 = always

Fig. J6.4: Question from the model "Maturity of the project-oriented organization"

E1. Methods of project portfolio coordination	
Project portfolio database	
Project portfolio reports	
Project portfolio score card	
Investment proposals	
Analyses of alternative project portfolios	
Project/programme progress reports	

1 = never, 2 = seldom, 3 = sometimes, 4 = frequently, 5 = always

Fig. J6.5: Question from the model "Maturity of the project-oriented organization"

The maturity model is based on an algorithm which enables the depiction of a "maturity area" in the spider web and the calculation of a "maturity ratio". The interpretation of the results of the analysis is to consider not only the individual business processes, but also the relationships between the business processes. For example, project portfolio co-ordination can only be performed based on a professional project management.

In a benchmarking of project-oriented organizations, the results of the analysis can be compared with the "maturities" of other organizations. The results of the analysis are to be interpreted in relation to the results of the interviews and documentation analyses.

The analysis of an Austrian IT company as a project-oriented organization is described in Case Study J3. .

⋯❯ CASE STUDY J3: Analysis of an Austrian IT company as a project-oriented organization

Project-oriented organization:

- Austrian software development company with 260 employees, part of an IT service provider. Products: Individual developments, introduction and customizing of standard solutions in the field of banking.

Objectives of the analysis:

- Analysis of the competencies as a project-oriented organization
- Creation of a basis for the decision regarding the performance of personnel development and organizational development measures
- Analysis of the similarities/differences of the project management practice of the organization with *ROLAND GAREIS Management of the Project-oriented Company*®.

CHAPTER

J

Process of the analysis:

- Performance of interviews and a documentation analysis
- Performance of a self-assessment workshop with experts from the project-oriented organization
- Development of an analysis report
- Presentation and discussion of the analysis results

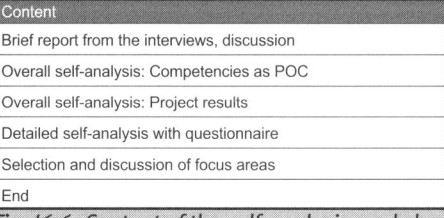

Content
Brief report from the interviews, discussion
Overall self-analysis: Competencies as POC
Overall self-analysis: Project results
Detailed self-analysis with questionnaire
Selection and discussion of focus areas
End

Fig. J6.6: Content of the self-analysis workshop

Results of the analysis:

- Maturity as a project-oriented organization as area in the spider web
- Maturity ratio.

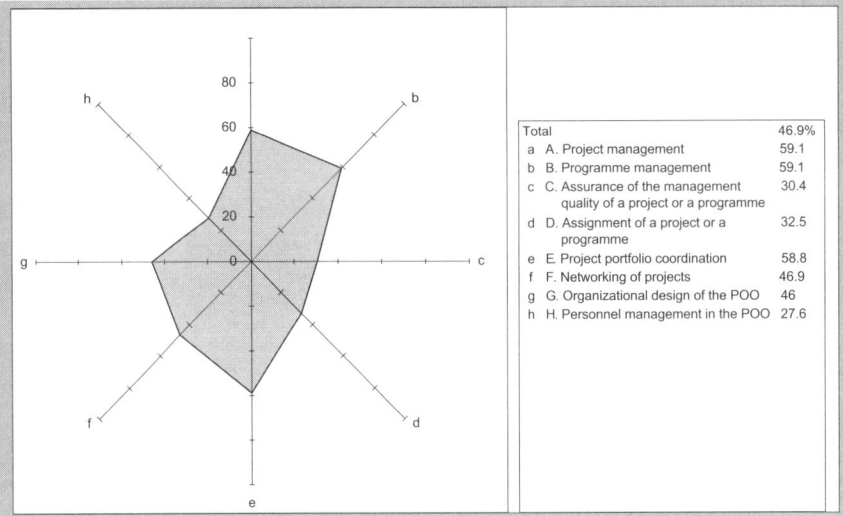

Total	46.9%
a A. Project management	59.1
b B. Programme management	59.1
c C. Assurance of the management quality of a project or a programme	30.4
d D. Assignment of a project or a programme	32.5
e E. Project portfolio coordination	58.8
f F. Networking of projects	46.9
g G. Organizational design of the POO	46
h H. Personnel management in the POO	27.6

Fig. J6.7: Maturity of the IT-company as a project-oriented organization

Interpretation:
Description of the results (extract)

- The organization has an overall maturity ratio of 47%.
- This is a relatively high maturity as a project-oriented organization.
- In particular, there are comprehensive competencies in project and programme management and in project portfolio coordination.
- There are striking deficits in personnel management in particular, in the quality management of projects and programmes and in the assignment of projects and programmes.

Suggested measures

- Project and programme management: Continuous further development based on the existing competencies,
- Assignment of projects and programmes: Rapid further development by means of defining a standardized process and developing tools (forms, etc.).
- Personnel management in the project-oriented organization: Fast cover of the call for action by means of a definition of career paths and personnel development plans (assessment, training, coaching),
- Quality management: Slow further development of the competencies, handling as a second priority,
- Benchmarking of the maturities of project-oriented organizations.

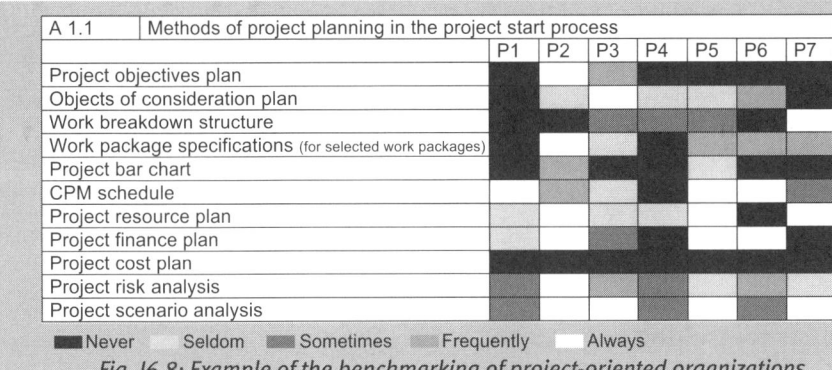

A 1.1	Methods of project planning in the project start process	P1	P2	P3	P4	P5	P6	P7
Project objectives plan								
Objects of consideration plan								
Work breakdown structure								
Work package specifications (for selected work packages)								
Project bar chart								
CPM schedule								
Project resource plan								
Project finance plan								
Project cost plan								
Project risk analysis								
Project scenario analysis								

■ Never   Seldom   ■ Sometimes   ■ Frequently   Always

Fig. J6.8: Example of the benchmarking of project-oriented organizations (considered organization: P7)

Interpretation:

Description of the results:

- The considered organization (depicted as P7) was subjected to a benchmarking with 6 other (P1-P6) organizations. The benchmarking partners were selected according to the criterion of belonging to the same industry.
- Compared to the partners, P7 is in the top third with regard to application of the methods of project planning.
- Conspicuous points:
  - Objects of consideration plan is always used,
  - Work breakdown structure is never used,
  - Lots of attention paid to schedules, finance plan and cost plan,
  - The resource plan is not broadly applied:

Suggested measures (extract):

- Introduction of the work breakdown structure as a central instrument for planning projects (also as the basis for scheduling and cost planning),
- Further development of resource planning (based on the structures for cost planning).

Design of the further development process

The further development of a project-oriented organization should also be performed in project form. The analysis of the project-oriented organization is relatively complicated and requires an appropriate organizational design. People with relevant know-how (project owner team, project and programme managers, PM Office manager, Expert Pool managers) and decision-makers are to be included in the process. The scope of the further development project depends on the extent of the further development measures.

Performance of the further development process

Measures in the further development process can be divided into optimization and establishment measures. The optimization measures refer to the established business processes of project, programme and project portfolio management. Measures for establishing additional competencies and structures can be, for example:

- Promoting the networking of projects,

- Developing guidelines for management consulting and management auditing of a project or a programme,

- Personnel development of potential project management consultants and project management auditors, and

Investment controlling

The competencies of the project-oriented organization are to be periodically subjected to an analysis and possibly an optimization. The responsibility for this lies with the PM Office.

J 6.4   Implementation of project management in non-project-oriented organizations

The objective of implementing project management in non-project-oriented organizations is usually the development of individual competencies in project management by means of training and coaching measures.

These measures lead to a continuous change in the organization. Changes are made to the personal competencies of individual employees, which does not lead to a further-reaching change within the organization. Generally, this is not desirable. The benefit for the organization comes from the increased competencies in solving problems by the participants of the project management training.

Case Study J4 describes the objectives and measures for performing project management training in a non-project-oriented organization.

⋯⋗ **CASE STUDY J4: Project management training in an Austrian administration organization**

Organization:
- Austrian administration organization
- Approx. 100 employees, high proportion of college graduates
- Few to no organizational competencies in project and programme management
- Approx. 8-10 projects per year

Objectives:
- Increase in the personal project management competencies of employees
- Creation of a basis for improving the quality in the management of projects

Measures:
- Planning of a project management seminar
- e-supported preparation of the seminar:
    - Trainer: Seminar invitation, scripts, design of the seminar (see Figure J6.9: Seminar invitation)
    - Participants: Pre-study texts, questions regarding the seminar content, self-assessment, discussion forum
- Performance of the seminar:
    - Target group: (Potential) project managers, selected project team members
    - Duration: 2 days
- e-supported follow-up to the seminar:
    - Trainer: Evaluation of the feedback questionnaires, development of a flip-chart protocol, seminar report
    - Participants: Knowledge test, reference studies

Invitation:
Seminar on project management

1) Objectives of the seminar

- Familiarization with the benefits of project management
- Familiarization with the process and the methods and of project management
- Development of a common project management understanding
- Work on concrete projects by the participants ("training on the project")

2) Schedule

Contents
- Introduction
- Project, project management
- Selection of training projects, formation of groups
- Project definition and context definition, group work
- Project environment analysis, group work
- Project scope planning, group work
- Project scheduling, group work
- Project cost/resource planning
- Project organization, group work

Working forms
- Brief inputs
- Examples and cases
- Work situations like in projects
- Group work
- Work on participants' own examples (several projects per seminar, to be worked on in small groups)

Fig. J6.9: Seminar invitation

Benefits of the training measures:
- Increased competencies in solving problems by the trained employees
- Joint project management understanding by the trained employees
- Initial analyses of project plans of the projects dealt with in the seminar

References

*Grimm, R., Kozok. B., L*afos, F., (Kommunikationsinfrastruktur) Kommunikationsinfrastruktur für virtuelle Organisationen, in Gora: W. (ed.), Virtuelle Organisationen im Zeitalter von E-Business und E-Government, Springer, Berlin, 2001

Senge, P., (fifth discipline) The fifth discipline: the art and practice of the learning organization, Century Business, London, 1998

Womack, J., (Autoindustrie) Die zweite Revolution in der Autoindustrie, Campus, Frankfurt, 1992

CHAPTER

J

K

Personnel management in the project-oriented organization

members, project contributors and members of the project owner team with project management competencies.

A distinction is made in project-oriented organizations between a management career, an expert career and a project management career. The establishment of a project management career path contributes to the establishment of the project manager as a profession in the project-oriented organization.

Business processes for the personnel management in the project-oriented organization are the recruiting, disposing, leading, developing and releasing of project personnel. Specific methods for personnel management are methods for the assessment of project management competencies, the project-related appraisal interview, the training and further education in project management and the project management coaching and mentoring.

Specific incentives are used in order to ensure personnel loyalty and motivation in the project-oriented organization. Incentives for individuals are to be differentiated from team incentives.

K Personnel management in the project-oriented organization
(in cooperation with Martina Huemann)

Contents

K 1   Roles and careers in the project-oriented organization

K 1.1   Roles in the project-oriented organization

A distinction can be made between permanent and temporary roles within project-oriented organizations.

Permanent management roles are, for example, the roles of member of the managing board, division manager, Expert Pool manager, member of a Project Portfolio Group and PM Office manager (see Figure K1.1 and role descriptions in Chapter J). Temporary roles in a project or a programme are: Member of a project owner team, project manager, project team member and project contributor (see role descriptions in Chapter C), management consultant and management auditor (see role descriptions in Chapter H).

Not only project and programme managers but all roles in projects and programmes require project management or programme management competencies.

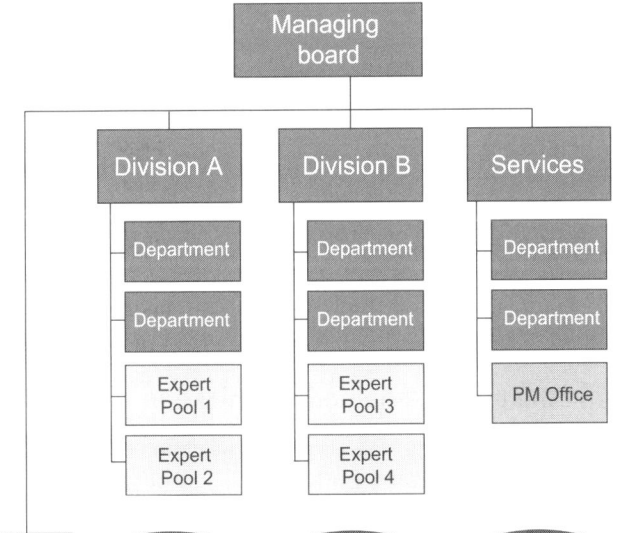

Fig. K1.1: Organization chart of the project-oriented organization (according to ROLAND GAREIS Management of the Project-oriented Company®)

K 1.2   Career paths in the project-oriented organization

"Career" is a hierarchical promotion of a person both within an organization and over several organizations.[1] Due to the flat organization structure of the project-oriented organization, career movements are not necessarily associated with achieving a higher position within the hierarchy.

Schein[2] differentiates between vertical, horizontal and centripetal career movements. In the case of vertical career movements, promotion is associated with a hierarchical advancement. In the case of horizontal career movements, no hierarchical advancement is performed. Centripetal career movements indicate changes toward the inner core of the organization. An example of a centripetal career movement in the project-oriented organization is, for example, the assumption of a membership in the Project Portfolio Group.

The career in the project-oriented organization is not defined according to the increased number of subordinate employees, but rather as a further personal development process and the attaining of competencies.

CHAPTER

K

A career path is defined as standardized stages in the process of individual professional development.[3] Career paths give the employees of an organization orientation with regard to the planning of their professional development. In project-oriented organizations, a distinction can be made between a management career, an expert career and a project management career.

Management career in the project-oriented organization

The management career is to be understood as the assuming of various permanent roles in the project-oriented organization with personnel and management responsibilities.[4]

The realization of a traditional management career requires that the organization has a hierarchical structure. A typical management career path includes the career stages of "Junior manager", "Manager", "Senior manager" and "Management executive". The management career in the project-oriented organization requires also the performance of owner roles in projects and programmes.

Expert career in the project-oriented organization

Since the specialist knowledge of experts is an important competitive factor of project-oriented organizations, experts must be offered personal development opportunities. The expert career can be structured by the stages of junior expert, expert, senior expert and, possibly, executive expert.

A devaluation of experts by means of a one-sided promotion of the management and the project management careers is to be avoided in the project-oriented organization. The balance between expert and (project) management careers is to be ensured by means of appropriate status and payment.

1) *Spiesshofer, U.,* (Ingenieure) 1991, p. 35
2) *Schein, E.,* (Career) 1978, p. 37 ff
3) *Lehnert, C.,* (Karriereplanung) 1996, p. 178
4) *Walker, J.W.,* (Human Resource Strategy) 1992, p. 211

Project management career in the project-oriented organization

Figure K1.2 depicts the project management career path with the career stages of "Junior project manager", "Project manager", "Senior project manager" and "Project management executive".

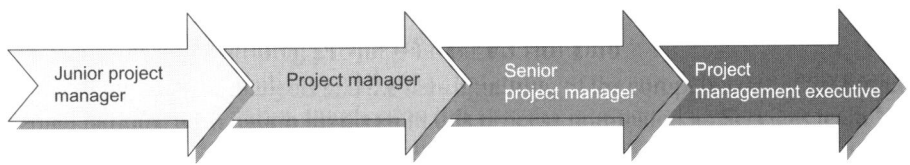

Fig. K1.2: Project management career path

The opportunities for making a career in project management depend on the "maturity" of the project-oriented organization. The performance of projects of various degrees of complexity, the definition of programmes, the offer of services for management consulting and management auditing of a project or a programme, the existence of a PM Office and a Project Portfolio Group determine the opportunities for performing various roles in the project-oriented organization.

The stage on the career path of a person is to be differentiated from the roles that the person performs. Figure K1.3 depicts the relationship between the project management career stages and possible roles within the project-oriented organization.

Project management career stage	Role
Project management executive	• Programme manager • PM Office manager • Member of the Project Portfolio Group
Senior project manager	• Project manager of complex projects • Project management coach • Management consultant of a project or a programme • Management auditor of a project or a programme
Project manager	• Project manager • Employee of the PM Office
Junior project manager	• Project manager of small projects • Project controller • Project management assistant

Fig. K1.3: Relationship of the project management career stages and possible roles within the project-oriented organization

The management, the expert and the project management career paths are parallel to each other and complement each other.

Turner and *Keegan*[5] coin the term "staircase career" for the project-oriented organization. Career movements can advance up the career ladder, but also along the career ladder. This prevents the "Peter principle", which says that persons are promoted until they reach their level of their incompetence. If an employee finds himself/herself in a position not befitting his/her competence, he/she can make a sideways career movement which – compared to a step down the career ladder, is not negative. The path to a higher career stage can also be associated with a change-over to another project-oriented organization.[6]

In order to guarantee the flexibility in the personnel management of the project-oriented organization, the permeability between the individual career paths is to be ensured. Figure K1.4 depicts relationships, ideal change-over opportunities and restrictions between the career paths of a project-oriented organization.

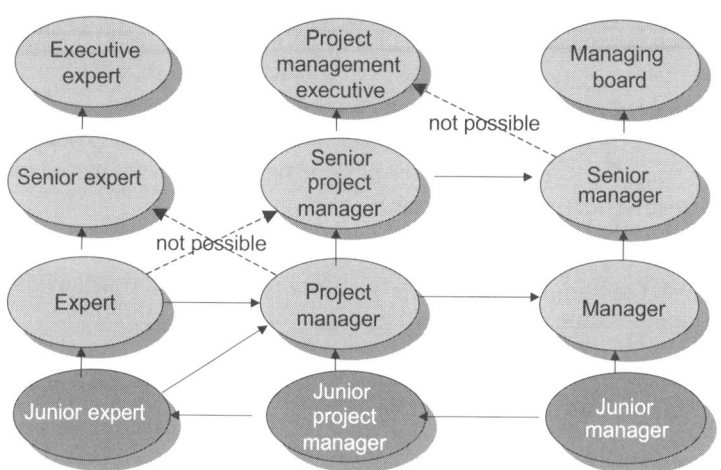

Fig. K1.4:Relationships between the career paths in the project-oriented organization

K 1.3   Competencies of the project manager

A person needs project management competencies in order to perform the role of project manager. Individual project management competence can be defined as a potential based on project management knowledge and project management experience for performing a project role.[7] Project management knowledge and project management experience are not independent of each other, but rather mutually influence each other. Knowledge presents the foundation on which experience can be gained, which, in turn, extends the knowledge.[8]

---

5)   *Keegan, A.E., Turner, J.R.,* (Managing Human Resources) 2003, p. 6
6)   *Kessler, H., Hönle, C.* (Karriere) 2002, p. 110
7)   *Huemann, M.,* (Individuelle Projektmanagement-Kompetenzen) 2002, p. 82
8)   For example, it is necessary to be familiar with the method of work breakdown structuring in theory, in order to be able to gain experience in this project management method.

To ensure the appropriate competence of the project manager , minimum requirements regarding project management knowledge and project management experience can be determined for the various stages of the project management career. As an example, the requirement profiles for the career stages of "Project manager" and "Senior project manager" are depicted in Figure K1.5. The knowledge and experience elements of the project management used to describe the requirements are basically structured according to subprocesses of project management.

Elements	Project management knowledge					Project management experience				
	5 Great deal	4 Good level	3 Average	2 Minimal	1 None	1 None	2 Minimal	3 Average	4 Good level	5 Great deal
1 Project definition, project management process										
2 Methods for the project start Project context										
3 Methods for the project start: Design project organization										
4 Methods for the project start: Project planning										
5 Methods for project coordination										
6 Methods for project controlling										
7 Methods for project closedown										
8 Methods for resolving a project discontinuity										
9 Programme definition and programme management										

Legend:
Project manager
Senior project manager

Fig. K1.5: Minimum degree of project management competencies of project managers and senior project managers

To perform his/her role, a project manager requires not only project management competence, but also:

- Technical competence, i.e. knowledge and experience in the technologies and products used in the project,

- Organizational competence, i.e. knowledge and experience in the organizations performing the project, and

- Cultural competence, i.e. knowledge and experience with various national cultures of any partners cooperating in the project.

K 1.4 Project manager – a profession

Requirements of the profession of project manager

Professions are subjected to a change in meaning over the course of time. New professions arise, such as IT specialist, solar technician or project manager. Professions go through a life cycle, starting with a phase of rank growth, through a phase of orderliness until the profession is formally established. Professions can also become obsolete, such as the professions of master of ceremonies or lady of the court.

The requirement for a profession of project manager results from the increased project orientation of organizations. There arises an increased requirement for project managers.

It is no longer only organizations which perform contracting projects, such as construction, engineering construction and IT companies, which employ members of staff exclusively in the role of project managers, but also banks, insurance companies, hospitals, etc. The creation of the profession of project manager contributes to the formalization of the status of the project manager and to the marketing of project management in the organization.

Establishment of the profession of project manager

The question of establishing the profession of project manager arises on the one hand in society and, on the other hand, in organizations. The task of establishing the profession of project manager in society is the responsibility of project management associations in cooperation with organizations as well as training and research institutes. This effort is describes in the excursus K1: Establishment of the profession of project manager in society.

Professions recognized in society and in organizations are based on[9]:

- A joint knowledge basis,
- Defined entry barriers,
- A code of ethics, and
- An established professional association.

At the level of a single project-oriented organization this basis is to be created in order to establish the profession of project manager: The guidelines for project and programme management represent the joint knowledge base in a project-oriented organization with regard to project management. Entry barriers are created by a project management career

---

9) *Kuwan, H., Waschbüsch, E.,* (Zertifizierung) 1996

path and appropriate requirement profiles. The formal evidence that these requirements are being fulfilled can be produced in the form of project management certifications.

A draft of a code of ethics for project managers is supplied in Figure K1.6. This can be adapted in accordance with the organization's specifications. The affiliation of project managers to one or several project management associations is to be organized.

Code of Ethics for Project Managers

- Placing the project's interests over those of individual project partners
- Consolidation of short-term project objectives and the long-term objectives of the business case of the investment initialized by the project
- Project management is a service to the project, not a means to exercise power
- Application of professional project management methods
- Realistic depiction of project management costs
- Avoidance of misinformation on projects, the development of truthful project progress reports and project close-down reports
- Avoidance of manipulations of project results
- Promotion of development of the project team members
- Assurance of contributions for learning by the project-oriented organization

---

Fig. K1.6: Code of ethics for project managers

···❯ **EXCURSUS K1: Establishment of the profession of project manager in the society**

At society level, it is the task of project management associations to establish the profession of project manager. This task can be performed in cooperation with training and research institutions.

Internationally, there are two globally-active project management associations, i.e. the International Project Management Association IPMA (www.ipma.ch) and the Project Management Institute PMI (www.pmi.org). Whereas PMI is centrally organized and active in many countries in the form of "Chapters", IPMA, currently comprising 33 independent national project management associations, is a federalist organization.

PMA (Project Management Austria), the Austrian project management association (www.p-m-a.at) belongs to IPMA. The objective of PMA is the promotion and the formal establishment of the profession of project manager in Austria.

Internationally, the profession of project manager is establishing itself. The status of the establishment of the profession can be assessed as a transition from rank growth to the phase of orderliness.

There exist internationally-varying standards for project management certifications and, thus, various entry barriers for project managers. Standardization of the project management knowledge bases and project management certifications has not yet been achieved.

PMI has formulated a code of ethics for project managers, which defines the conduct of a professional project manager in the performance of his/her profession.[a]

The national and international project management associations perform marketing activities and lobbying for the profession of project manager. Knowledge bases on project

management have been developed, published and defined as the basis for the certifications of project managers. The ICB-IPMA Competence Baseline[b] is the knowledge basis published by IPMA, and *A Guide to the Project Management Body of Knowledge* (*PMBoK® Guide*)[c] is published by PMI. The national project management associations of IPMA develop national "baselines" based on the ICB. In Austria, the "pm baseline"[d] forms the basis for project management certifications by PMA.

CHAPTER

K

[a]   *Project Management Institute*, (Fact Book) 2001, p. 93 ff
[b]   *Caupin, G.* (et al.), (ICB) 1999
[c]   *Project Management Institute*, (Body of Knowledge) 2000
[d]   *Gareis, R.*, (pm baseline) 2001

K 1.5   Certifications for project managers

The certifications of project managers prove the individual project management competencies. The certifications are to guarantee a project management standard to be followed by the project management personnel of a project-oriented organization.

Project management certifications can be performed either within the organization or externally. International IT concerns combine internal and external certifications. The globally-active project management associations IPMA and PMI offer project management certifications in accordance with the associations' relevant standards. The advantage of the certification according to IPMA lies in a 4-stage model, whereas certification according to PMI is limited to the certification stage of the Project Management Professional PMP. In Austria, certification according to IPMA is performed by PMA.

The advantage of external project management certifications lies in the national/international validity of the certificates. They represent independent confirmations of the project management competencies. The certification is an external confirmation of quality for project managers.

Project-oriented organizations can structure their project management career path according to the project management certification levels. Figure K1.7 depicts the relationship between project management competencies, project management certification levels according to PMA and possible professional roles.

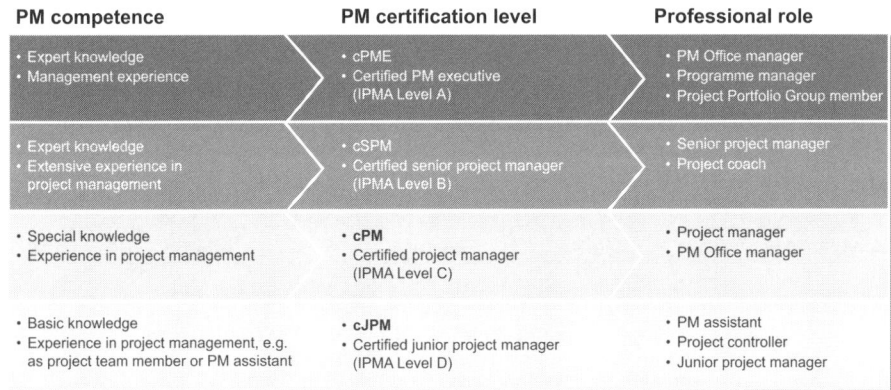

PM competence	PM certification level	Professional role
• Expert knowledge • Management experience	• cPME • Certified PM executive (IPMA Level A)	• PM Office manager • Programme manager • Project Portfolio Group member

• Expert knowledge
• Extensive experience in
  project management
	• cSPM
	• Certified senior project manager
	  (IPMA Level B)
		• Senior project manager
		• Project coach

• Special knowledge • Experience in project management	• cPM • Certified project manager (IPMA Level C)	• Project manager • PM Office manager

• Basic knowledge
• Experience in project management, e.g.
  as project team member or PM assistant
	• cJPM
	• Certified junior project manager
	  (IPMA Level D)
		• PM assistant
		• Project controller
		• Junior project manager

Fig. K1.7: Project management career path according to PMA

In many project-oriented organizations, remuneration of the project management personnel is connected to the certification levels. Figure K1.8 depicts the project management career path of an international IT company. The career stages are consultant (IT expert), project manager, senior project manager and project director. The consultant has 2-5 years' project experience and manages small teams. The project manager manages projects with a low degree of complexity. The senior project manager manages complex projects. The highest level is project director. The project director performs the role of programme manager and is responsible for the further development of project management within the organization.

The years specified for the individual career stages are generally shorter in Europe than in the USA, where the IT concern has its headquarters. The scope of the project management training which a person must undergo in order to attain a certain career stage is precisely defined. Training is generally held internally and is based on the internal project management guidelines and tools.

In this project-oriented organization, the career path is structured according to the model of IPMA certifications. The PMI-PMP is equated with the IPMA certification Level C (certified project manager).

Career Progression	Consultant (IT-Expert)	PM Level 3 Project Manager	PM Level 2 Senior Project Manager	PM Level 1 Project Director
Total Years	2 - 5	4 - 9	6 - 14	6 - 14
Training & Testing	12 Courses	9 Courses	3 Courses	2 Courses
Experience	2–5 years project team experience; lead small teams	2–4 years managing L3 projects; participate in L2 project	2–5 years managing L2 projects; participate in L1 projects	2–3 years mangaging L1 projects
External Certification	- IPMA Level D	- IPMA Level C - PMI PMP	- IPMA Level B	Continuing Professional Development

Fig. K1.8: Project management career path of an IT company

K 2 Business processes for personnel management in the project-oriented organization

Business processes for personnel management in the project-oriented organization are the recruiting, disposing, leading, developing and releasing of project personnel. An overview of these business processes is depicted in Figure K2.1.

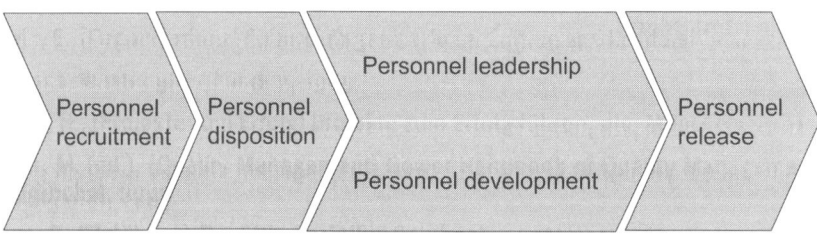

Fig. K2.1: Business processes for personnel management

If the recruiting, disposing, developing and releasing of project personnel are performed within projects, these tasks as well as leading project personnel are integrated functions of project management. The objectives and methods for performing these project management tasks are described in Chapter F.

The recruiting, the developing and releasing of personnel in project-oriented organizations, however, is also performed independent from the management of a single project. Future project managers can be recruited without a certain project already being planned, the further education of project personnel can serve the purpose of general personnel development objectives and project managers can be released from the project-oriented organization should their competencies no longer be required. These general personnel management business processes of the project-oriented organization are described in the following.

K 2.1 Business process: Recruiting project managers

"Recruitment" is the procurement and selection of personnel[10]. The procurement and selection of project managers can be performed for future projects of a project-oriented organization. The objective of such a recruitment measure is the expansion of the Expert Pool of project managers.

Company-internal personnel procurement is to be differentiated from external personnel procurement. In the case of internal procurement of potential project managers, the Expert Pools of the project-oriented organization represent the procurement market. Experts from various Expert Pools who, due to their work as project team members, have also attained project management competences, may be interested in a career in project management and apply for such positions.

In the case of external personnel procurement, the external procurement market is accessed. Project managers working as freelancers frequently belong to a network of experts with which a project-oriented organization can maintain relations. The chance of accessing such networks increases the personnel flexibility of project-oriented organizations.

The choice of personnel is made using the competencies described in a profile for a project manager. To support recruitment, databases with descriptions of the employees' project management competencies can be used.

The business process for recruiting a project manager from the external procurement market is described in Figure K2.2.

---

10) *Lueger, G.* (Beschaffung und Auswahl von Mitarbeitern) 1996, p. 338

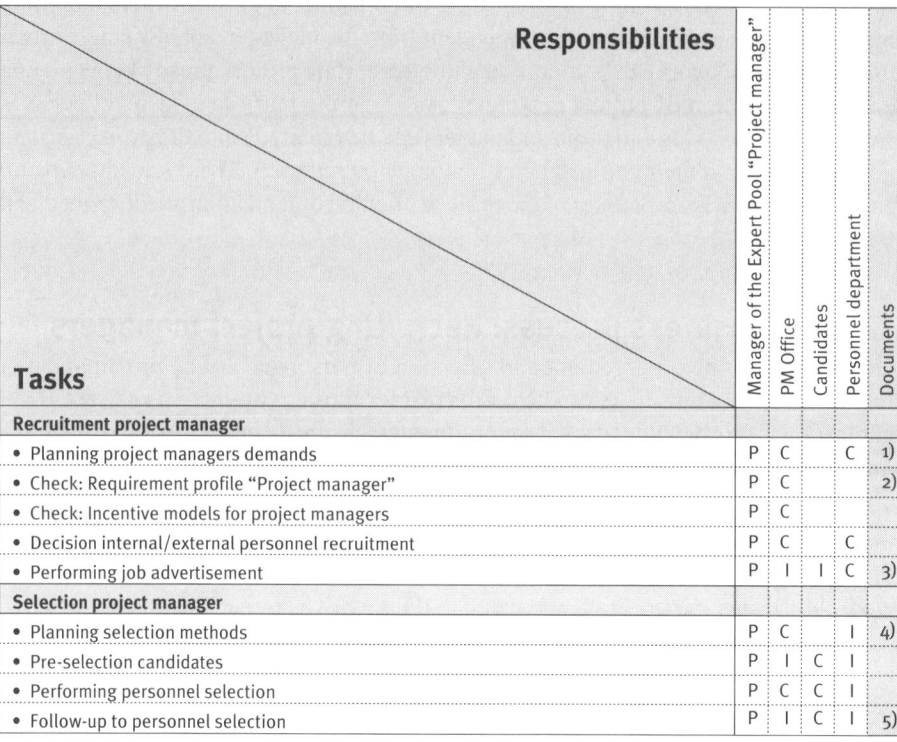

Tasks	Manager of the Expert Pool "Project manager"	PM Office	Candidates	Personnel department	Documents
Recruitment project manager					
• Planning project managers demands	P	C		C	1)
• Check: Requirement profile "Project manager"	P	C			2)
• Check: Incentive models for project managers	P	C			
• Decision internal/external personnel recruitment	P	C		C	
• Performing job advertisement	P	I	I	C	3)
Selection project manager					
• Planning selection methods	P	C		I	4)
• Pre-selection candidates	P	I	C	I	
• Performing personnel selection	P	C	C	I	
• Follow-up to personnel selection	P	I	C	I	5)

Legend:
P ... Performance
C ... Contribution
I ... Information

Results/documents:
1) Requirement plan "Project managers"
2) Competencies Profile "Project manager"
3) Advertisement text
4) List: Selection methods

Fig. K2.2: Business process: Recruitment of a project manager from the external procurement market

The requirement for additional project managers may arise from the requirement planning of the Expert Pool of project managers. Before starting recruitment activities, the competencies profile for the project manager is to be examined and a decision must be taken whether to recruit from the internal or the external personnel market.

The selection methods to be used, such as letters of application, interviews, tests or assessment centers for project managers, are to be planned.

K 2.2   Business process: Developing project managers

The objective of the development of project managers is the improvement of their project management competencies. Development measures can be training/further education, coaching and mentoring. The development measures can be offered only to the employees of the considered project-oriented organization, or encompass freelance project managers, who network with the project-oriented organization. The business process of developing project managers is depicted in Figure K2.3.

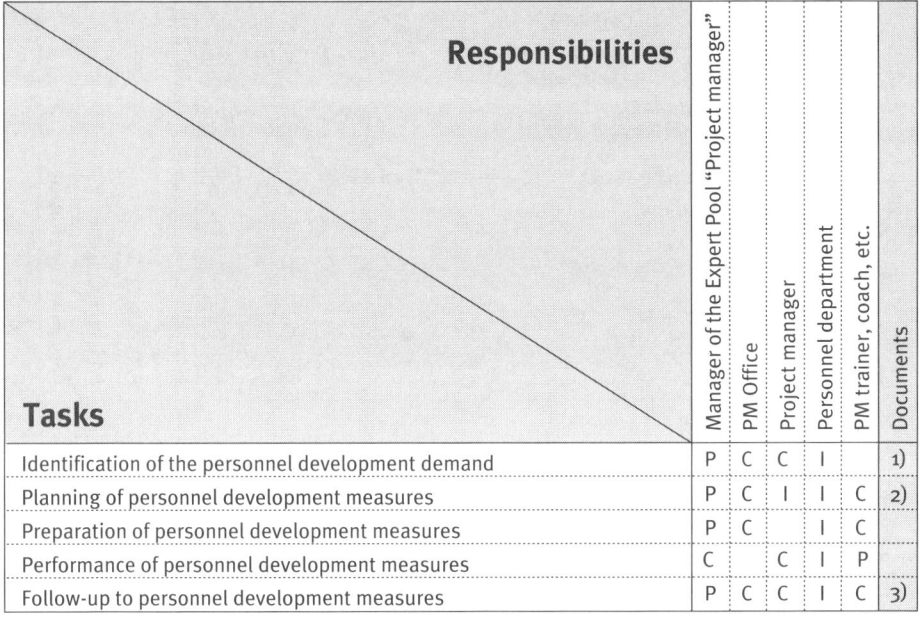

Tasks \ Responsibilities	Manager of the Expert Pool "Project manager"	PM Office	Project manager	Personnel department	PM trainer, coach, etc.	Documents
Identification of the personnel development demand	P	C	C	I		1)
Planning of personnel development measures	P	C	I	I	C	2)
Preparation of personnel development measures	P	C		I	C	
Performance of personnel development measures	C		C	I	P	
Follow-up to personnel development measures	P	C	C	I	C	3)

Legend:
P ... Performance
C ... Contribution
I ... Information

Results/documents:
1) Development requirements "Project managers"
2) Personnel development measures
3) Evaluation: Personnel development measures

Fig. K2.3: Business process: Developing project managers

K 2.3   Business process: Releasing project managers

The release of personnel encompasses all measures by which a surplus of personnel – from the aspects of quantity, quality, locality and chronology – are reduced. The business process of releasing project managers from the project-oriented organization is depicted in Figure K2.4.

The requirement to release project managers from a project-oriented organization results from the planning of the requirement for project managers for the planning period. Before the release of project managers is planned and performed, alternative employment opportunities within in the organization are to be considered. These highly-qualified resources may be used short-term, for example as temporary PM Office employees. Following the release of freelance project managers, it is important to remain in contact with them in order to enable cooperation in the future.

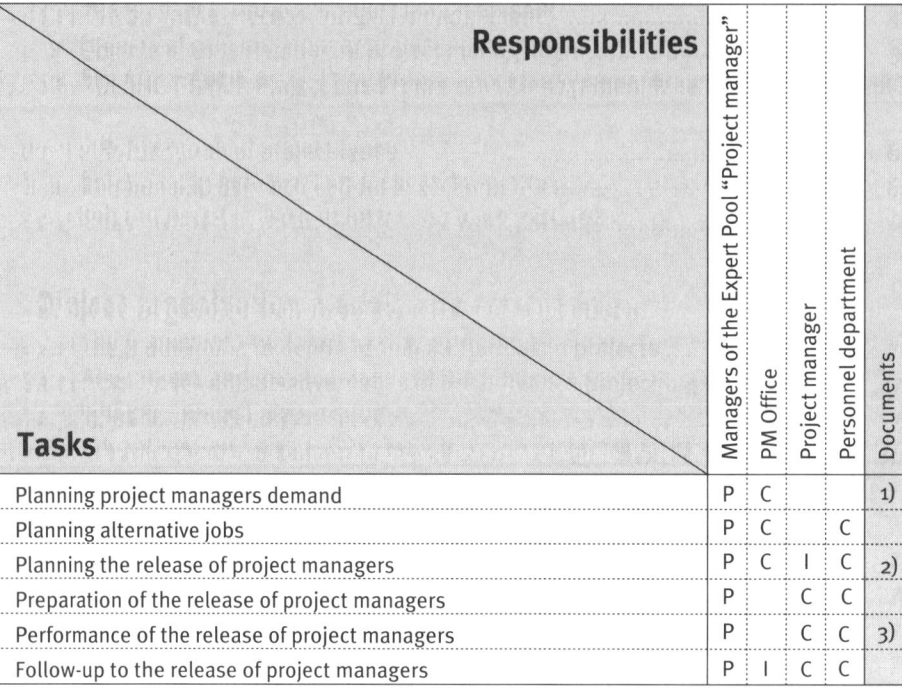

Tasks	Managers of the Expert Pool "Project manager"	PM Office	Project manager	Personnel department	Documents
Planning project managers demand	P	C			1)
Planning alternative jobs	P	C		C	
Planning the release of project managers	P	C	I	C	2)
Preparation of the release of project managers	P		C	C	
Performance of the release of project managers	P		C	C	3)
Follow-up to the release of project managers	P	I	C	C	

Legend:
P ... Performance
C ... Contribution
I ... Information

Results/documents:
1) Requirement plan "Project managers"
2) Plan: Release of project managers
3) Personnel releasing measures

Fig. K2.4: Business process: Releasing project managers

K 3 Methods for personnel management in the project-oriented organization

Specific methods for personnel management in the project-oriented organization, such as methods for the assessment of the project management competencies, the project-related appraisal interview, the training and further education in project management, and the project management coaching and mentoring, are described in the following.

K 3.1 Methods for assessing project management competencies

The following methods can be used for assessing the project management competencies of employees:

- Test for assessing project management knowledge,

- Self-assessment and external assessment of individual project management competencies, and
- Assessment center for project managers and programme managers.

Test for assessing project management knowledge

An assessment of knowledge can be performed by either an oral or a written test.

In order to test project management knowledge in accordance with *ROLAND GAREIS Project and Programme Management*®, the "*pm test*" can be used. The "*pm test*" is a multiple choice test with around 450 questions in German and English. The questions are arranged in accordance with the business processes of the project-oriented organization. The questions on project management are structured in accordance with the project management sub-processes. There are four possible answers to each question. One, several, all or none of the possible answers could be correct. Figure K3.1 depicts a question in the "*pm test*" and its possible answers.

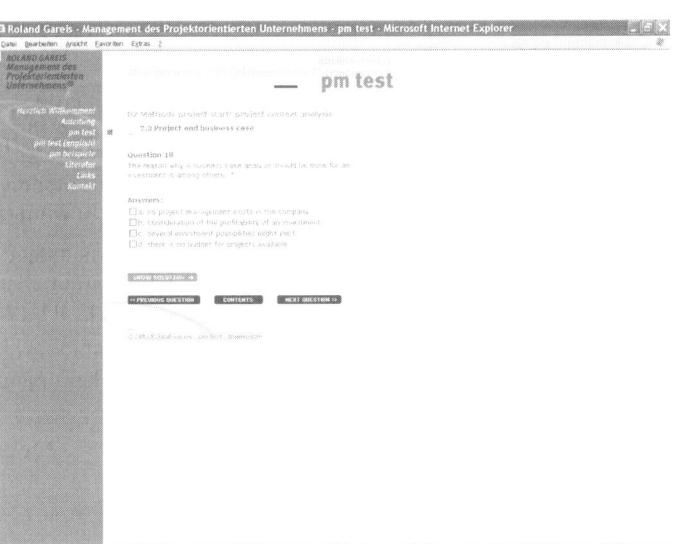

Fig. K3.1: A question from the "pm test"

The "*pm test*" is used by the University of Economics and Business Administration, Vienna, and by other training institutions as well as by project-oriented organizations..

Self-assessment and external assessment of individual project management competencies

The objective of the assessment of individual project management competencies is the evaluation of the project management knowledge and the project management experience of an individual. The assessment can either be performed by the person himself/herself as a self-assessment and/or by a third party as an external assessment.

An extract of the questionnaire for self-assessment is depicted in Figure K3.2. An external assessment should be based on the results of the self-assessment. In a meeting, the external assessor questions the self-assessment and analyzes the project management documents drawn up by the assessment candidate. Using concrete documents, the project management knowledge and the relevant experience can thus be assessed. Interviews with people who cooperate with the assessment candidate can also contribute to this assessment.

3. Project management knowledge and experience of the project manager		
3.1 Project start process		
3.1.1. Methods of project planning		
1=none, 2=minimal, 3=average, 4=good level, 5=great deal	Knowledge	Experience
Project objectives plan		
Objects of consideration plan		
Work breakdown structure		
Work package specifications		
Project resource plan		
Project finance plan		
Project milestone plan		
Project bar chart		
CPM schedule		
Project cost plan		
Business case analysis		
Business Case Analyse		
3.1.2. Methods of project risk management and discontinuity management		
1=none, 2=minimal, 3=average, 4=good level, 5=great deal	Knowledge	Experience
Project risk analysis		
Project scenario analysis		
Alternative project plans		
3.1.3. Methods for designing the project context relationships		
1=none, 2=minimal, 3=average, 4=good level, 5=great deal	Knowledge	Experience
Project environment analysis		
Analysis: Pre- and post-project phase		
Analysis: Relationship between the project and other projects		
Analysis: Relationship between the project and the company strategies		
Project marketing		

Fig. K3.2: Extract of the questionnaire on self-assessment of individual project management competencies

Figure K3.3 depicts the summary of the results of an assessment of the individual project management competencies of project managers of a project-oriented IT company. The objective of the assessment was to gain an overview of the project management competencies of the project managers. The depicted table, on the one hand, compares the competencies of the individual project managers and, on the other hand, it enables a comparison with the minimum project management knowledge and minimum project management experience required for the certifications to as certified project manager (cPM) and as certified senior project manager (cSPM) by the Austrian project management associtaion. Based on the results, differentiated measures are initiated for the further development of the project management competencies of each project manager.

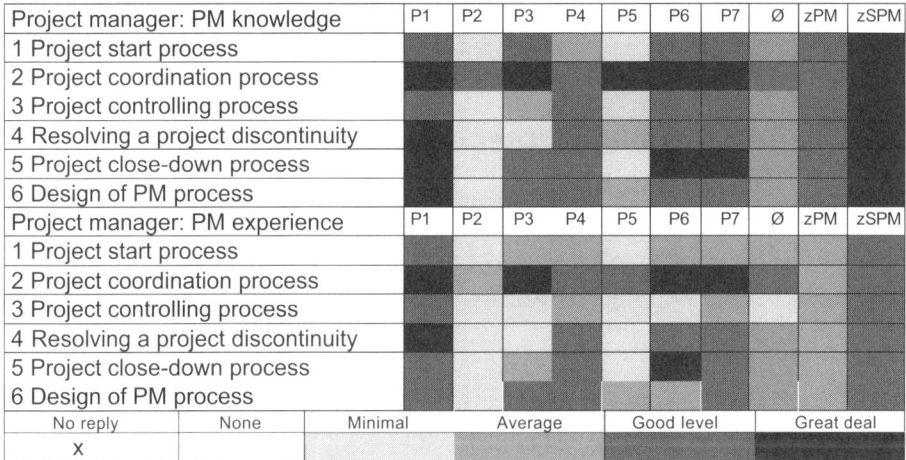

Fig. K3.3: Summarized results of the assessment of the individual project management competencies of project managers of an IT company

Self-assessments and external assessments of individual project management competencies can also be used in management auditing (see Chapter H).

Project management assessment center

In an assessment center, candidates for a management position are placed in a simulated working situation in order to be able to assess how competent they are to perform the appropriate role.[11]

In project-oriented organizations, assessment centers are becoming increasingly popular for the selection or for further development of project managers and programme managers.

At the assessment center the roles of assessor, moderator, candidate and observer are to be differentiated. Assessment centers for project managers generally last two days.

---

11) *Kompa, A.,* (Assessment Center) 1999, p. 31 ff

The general methods used in assessment centers must be adapted specifically to the working situations in projects. Examples of this are depicted in Figure K3.4.

Assessment center method	Project management-related application
Presentation	Project start workshop: Presentation of a project
Group discussion	Discussion about the role of the project manager compared to the role of the project owner team
Role play	Project owner meeting in a project crisis
Analysis	Interpretation of a project progress report and a project score card

Fig. K3.4: Methods for the project management assessment center

K 3.2   Training and further education in project management

Target groups for the training and further education measures in the project-oriented organization are not only project managers but also project team members, project contributors and members of project owner teams. The content and the durations of training and further education measures are to be designed specifically for these different target groups.

Training and further education measures can be organized "on the job" or "off the job", company internally or externally .

The training and further education of project managers "on the job" can be performed in the form of internships, job rotation measures and by individual coaching measures. Training and further education options "off the job" are visits to lectures, seminars and courses.

The internal performance of lectures, seminars and courses has the advantage that the content can be adapted specifically to the project-oriented organization. The advantage of open trainings is in learning the practices of different organizations and in the opportunity to network with project managers from other project-oriented organizations.

K 3.3   Coaching and mentoring of project managers

A coach helps somebody to help himself/herself by giving the coached person advice for resolving concrete work situations. In comparison to consulting, coaching relates to the individual, not the organization. It is used in work situations in which an individual requires a solution to a problem.

Coaching of a project manager can, for example, be used following project management training in order to help the project manager to implement the imparted content into a current project.

Mentoring serves the purpose of longer-term instruction and advice of young employees by means of regular meetings with a mentor. Experienced colleagues within the organization can act as mentors. Support can take the form of communicating basic information for an increased understanding of the structures and cultures of the organization. The mentor assumes the role of advisor, friend and role model[12]. In project-oriented organizations, experienced project managers frequently support "young" project managers as mentors.

K 4   Incentive models in the project-oriented organization

K 4.1   Design of incentive models

The objectives of the use of incentive models in the project-oriented organization are the recruitment of competent employees, the motivation of employees to perform especially well, and employees' commitment to the organization.

Motives, incentives and motivation are closely connected with each other. These terms are defined in the following excursus K2.

···⟩ EXCURSUS K2: Motives, incentives and motivationMotive

A motive is a desired state. Intrinsic motives, whose satisfaction is achieved in the work itself, are to be differentiated from extrinsic motives, which cannot be satisfied by the job itself but rather by the consequences of the work or its attendant circumstances.[a]

Intrinsic motives are performance, competence and relational motives. Extrinsic motives are financial, security and prestige motives.

The performance motive is expressed by the fact that the individual experiences satisfaction by achieving performance objectives set by himself/herself. Performance-motivated people are motivated to perform mainly by difficult tasks. Money, for these people, plays only a minor role as an incentive. It is merely a means of assessing their own performance in comparison to the performance of others.

The competence motive is expressed in the striving for professional development, high performance and the desire to be able to influence future developments. Should competence-motivated employees be confronted with routine or heavily-supervised tasks, this can quickly lead to frustration.

The relational motive results from the desire to make contact with others. There are team-oriented people who enjoy teamwork with others, and there are those who prefer to work alone.

---

12) *Olfert, K., Steinbuch, P.,* (Personalwirtschaft) 1995, p. 464

The money motive is probably the most obvious motive for working. However, money tends to lose its predominant importance once an employee's income increases.

The security motive finds expression in the person's endeavors to eliminate any potential dangers and obstacles which impair the fulfillment of one's needs. This covers the conscious security motives through to the subconscious security motives which influence any decisions taken by people averse to risks.

The prestige motive is based on the desire to distinguish oneself from others. Prestige-motivated employees primarily strive for careers which enable them a high income and respected positions.

Incentives activate motives

An incentive is that component of the perceived situation which activates motives[b]. Incentives can be classified according to:[c]

- The sources of the incentives: Intrinsic and extrinsic incentives,
- The incentive object: Material and immaterial incentives, and
- The number of recipients of the incentive: Individual incentives, group incentives and organization-wide incentives.

Intrinsic incentives arise from jobs themselves in the form of personal successes and failures. They go hand-in-hand with intrinsic work motives (performance, competence and relational motive). Extrinsic incentives are dependent more on the environment and less on the job. Extrinsic incentives can be both material and immaterial.

Material incentives are financial rewards. Immaterial incentives are rewards which cannot be (directly) measured in financial terms. Immaterial incentives include power, responsibility/job content, autonomy/, promotion and career opportunities, recognition, status, image, information, communication, security.[d]

In individual incentives, the individual person is the center of attention, while in group incentives the team is considered.

Motivation from a systemic point of view

Motivation can be defined as a complex interaction of various activated motives.[e]

From a systemic point of view, social systems can only control themselves. Thus, implemented "incentives" first of all have to be perceived as incentives by the individual or by the team in order to be effective. Therefore, it is not possible to motivate anybody "from the outside". Motivation is always self-motivation.

[a]   *Jung, H.,* (Personalwirtschaft) 1999
[b]   *Rosenstiel, L.,* (Organisationspsychologie) 1975, p. 320
[c]   *Becker, F.,* (Anreizsysteme) 1990, p. 10
[d]   *Guthof, P.,* (Strategische Anreizsysteme) 1995, p. 24-30
[e]   *Jung, H.,* (Personalwirtschaft) 1999, p. 359

In the project-oriented organization incentive models for employees and teams of the permanent organization are to be differentiated from incentive models for members and teams of project and programme organizations. It must be ensured that different incentive systems do not compete against each other.

Project-specific incentives are to be agreed in the project start process. The agreement is to be made jointly between the project owner team and the project team. The effectiveness of the incentives is to be increased by offering the members the chance to participate in shaping the incentives.

Possible project-related incentives are the project work itself, project bonuses and other incentives, such as gifts or esteem as rewards for special achievements. These incentives are described in the following:

K 4.2 Project work as incentive

Project work itself is an intrinsic incentive which addresses peoples' performance motive, competence motive and relational motive.

Projects are usually new and challenging, they require teamwork, grant autonomy and promote creativity. The agreement of project objectives provides the individual with orientation and enables to receive feedback regarding his/her performance. In the case of positive project results, the project member can be proud of himself/herself. Many projects are exciting and emotional. They offer the individual opportunities to learn and develop. The temporary character of projects enables the individual to accept new challenges upon close-down of the project.

It is the responsibility of managers of project-oriented organizations to create and to offer challenging and exciting projects to the employees. These are an incentive to work in the organization.

K 4.3 Project bonuses

Project-specific incentive models are frequently reduced to the agreement of project bonuses.

Project bonuses can either be agreed with the individual members of the project organization, e.g. with the project manager, or with the project team overall. The results to be achieved in order for the project bonuses to be paid out, their amount and the division of the project bonuses between the members of the project team, are to be agreed in the project start process.

The payment of project bonuses is to be made during the project close-down process based on the evaluation of the project success (see Chapter F5).

K 4.4 Other project-related incentives

Instead of project bonuses or in addition to project bonuses, further incentives of a more symbolic nature can be used in projects.

Here, too, individual incentives are to be differentiated from team incentives. Individual incentives are rewards or esteem for individual members of the project organization. Following an intensive work period in a project, a project manager can be awarded a weekend away with his/her family. Or, for example, a project team member known as a wine connoisseur may receive a special bottle of wine for his/her exceptional performance in the project. However, this requires the project owner team and/or project manager to be on familiar terms with the employees in order to be able to express such personal recognition.

The expression of personal esteem, e.g. by giving praise in a project team meeting, can have a positive influence on the motivation of an employee.

Possible team incentives are, for example, a "Dinner for the project team" or a visit to a musical by the whole project team.

K 5   Organization of the personnel management in the project-oriented organization

The project-related tasks of recruiting, disposing, leading, developing and releasing of project personnel in the project-oriented organization are performed, on the one hand, by Expert Pool managers, by the PM Office, the Project Portfolio Group and the human resource department and, on the other hand, by project owner teams and project managers.

General personnel management tasks, such as the establishment of the profession of project manager, the development of a project management career path, the organization of training and further education in project management and the promotion of project management certifications, are to be performed jointly by the PM Office and the manager of the Expert Pool "Project managers".

It is the responsibility of the manager of the Expert Pool "Project managers" to ensure that adequately-trained project managers are available. Project-oriented organizations can cooperate with universities in order to recruit future project managers. Excursions, workshops and internships can be performed with project management students. Contacts with a network of freelance project managers must also be maintained.

References

Becker, F., (Anreizsysteme) Anreizsysteme für Führungskräfte – Möglichkeiten zur strategisch-orientierten Steuerung des Managements, Poeschel, Stuttgart, 1990

Caupin, G., Knöpfel, H., Morris, P., Motzel, E., Pannenbäcker O., (ICB), IPMA Competence Baseline, IPMA, Bremen, 1999

Deneke, J.F., (Berufsbild) Berufsbild des Vermögensberaters, Poeschel, Stuttgart, 1998

Eskerod, P., (Human Resource Allocation) The Human Resource Allocation Process When Organizing by Projects, in: *Lundin, R.A., Midler, C.* (eds.), Projects as Arenas for Renewal and Learning Processes, Kluwer Academic Publishers, Boston, 1998

Gareis, R., (pm baseline) pm baseline – Wissenselemente zum Projekt- und Programmmanagement sowie zum Management projektorientierter Unternehmen, PMA, Version: July 2nd, 2001

Guthof, P., (Strategische Anreizsysteme) Strategische Anreizsysteme – Gestaltungsoptionen im Rahmen der Unternehmensentwicklung, Gabler, Wiesbaden, 1995

Huemann, M., (Individuelle Projektmanagement-Kompetenzen) Individuelle Projektmanagement-Kompetenzen im projektorientierten Unternehmen, Peter Lang Europäischer Verlag der Wissenschaften, Frankfurt am Main, 2002

Jung, H., (Personalwirtschaft) Personalwirtschaft, Oldenbourg, München, Wien, 1999

Keegan, A.E.; Turner, J.R., (Managing Human Resources) Managing human resources in the project-based organization, in: *Turner, J.R.,* (ed.), People in Project Management, Gower, Aldershot, 2003

Kessler, H., Höhnle, C., (Karriere) Karriere im Projektmanagement, Springer, Berlin, Heidelberg, 2002

Kompa, A., (Assessment Center) Assessment Center – Bestandsaufnahme und Kritik, Rainer Hampp, München, 1999

Kuwan, H., Waschbüsch, E., (Zertifizierung), Zertifizierung und Qualitätssicherung in der beruflichen Weiterentwicklung; Bundesinstitut für Berufsbildung, Berlin, 1996

Lauer, B., (Berufsbild) Projektmanager/in – das Berufsbild in der Zukunft, in: *Haarbeck, S.,* (ed.), Szenarien der Arbeitswelt von morgen, Dt. Wirtschaftsdienst, Köln, 2000

Lehnert, C., (Karriereplanung) Neuorientierung der betrieblichen Karriereplanung, Dt. Univ-Verlag, Wiesbaden, 1996

Lueger, G., (Beschaffung und Auswahl von Mitarbeitern), in: *Kasper, H.* (ed.), Personalmanagement – Führung – Organisation, Ueberreuter, Wien, 1996

Olfert, K., Steinbuch, P., (Personalwirtschaft) Personalwirtschaft, Kiehl, Ludwigshafen, 1995

Project Management Institute, (Fact Book) Project Management Fact Book, Project Management Institute, Pennsylvania, 2001

Project Management Institute, (Body of Knowledge) Project Management Body of Knowledge, Project Management Institute, Pennsylvania, 2000

Rischar, K., (Optimale Personalauswahl) Optimale Personalauswahl, TÜV Rheinland, Köln, 1990

Rosenstiel, L., (Organisationspsychologie) Organisationspsychologie, Kohlhammer, Stuttgart, 1975

Schein, E., (Career) Career Dynamics: Matching individual and organizational needs, Addison-Wesley, Reading, Mass., 1978

Spiesshofer, U., (Ingenieure) Ingenieure im europäischen Management: Karrieren von Ingenieuren im Topmanagement von europäischen Industrieunternehmen, VDI, Düsseldorf, 1991

Vogt, G., (Arbeitswelt), Nomaden der Arbeitswelt: virtuelle Unternehmen, Kooperation auf Zeit, Versus, Zürich, 1999

Walker, J. W., (Human Resource Strategy), Human Resource Strategy, MacGraw-Hill, New York, 1992

Wildenmann, B., (Professionell führen) Professionell führen – Empowerment für Manager, die mit weniger Mitarbeitern mehr leisten müssen, Luchterhand, Neuwied, 1996

CHAPTER

K

L

The project-oriented society

A society which frequently uses projects and programmes for the performance of relatively unique business processes of medium to high complexity in profit and non-profit areas can be perceived as a project-oriented society.

In the project-oriented society there are organizations that have competencies in project and programme management, in project portfolio management, in personnel management and in the organizational design of project-oriented organizations.

In the project-oriented society, there are institutions, that perform project management-related training, research and marketing services.

L The project-oriented society

Contents

L 1 Model of the project-oriented society

L 1.1 Project orientation as a macro-economic pheno-menon

In many national societies, projects and programmes are performed not only in companies but also in other organizations, such as municipal administrations, associations, schools and even in families. "Management by Projects" becomes an organizational strategy of societies in order to deal with the increasing complexity and dynamics of the social environment. The globalization of the economy, new technologies with ever-shorter product life cycles and the application of a new management paradigm, characterized by virtual organizations, "empowerment" and knowledge management, promote the use of project and programme management.

Not only industry, but also non-profit organizations, see projects and programmes as adequate organization forms in order to perform relatively unique business processes of medium to high complexity. In addition to contracting projects, new types of projects such as marketing, product development and organization development projects, are gaining in importance.

The significance of projects and programmes in society, the structure of the society, its history and expectations regarding its future, influence the development of the maturity as a project-oriented society. The significance of project-orientation in a society can be measured by the number of project-oriented branches and project-oriented organizations.

L 1.2 Construct: The project-oriented society

A society which uses projects and programmes to perform relatively unique processes of medium to high complexity, can be perceived as a project-oriented society.

The perception of a society as a project-oriented society is a construct. It requires the observation of the society through "special glasses", i.e. the glasses of project-orientation. All communications by the society connected with the management of projects, programmes and project portfolios, are analyzed.

The model of the project-oriented society considers, on the one hand, the practices of project-oriented organizations in project management, programme management, project portfolio management, management of the project personnel and in the organizational design as a project-oriented organization.[1] On the other hand, the model describes project management-related services by institutions which promote the application of pro-

---

1) Instead of "project-oriented company" the term "project-oriented organization" is used, since the project orientation may relate not to the company as a whole, but to individual divisions, such as departments or profit centers, and also to organizations, such as schools, municipal administrations, associations and families.

ject, programme and project portfolio management . Project management-related services are performed by training, research and marketing institutions.

Maturity model of the project-oriented society

The project-oriented society can be visualized in a "maturity model" by means of a spider web model (see Figure L1.1). The axes of the spider web represent the dimensions of the practice of project-oriented organizations and the project management-related services of training, research and marketing institutions. The "maturity" of a project-oriented society can be assessed based on this model.

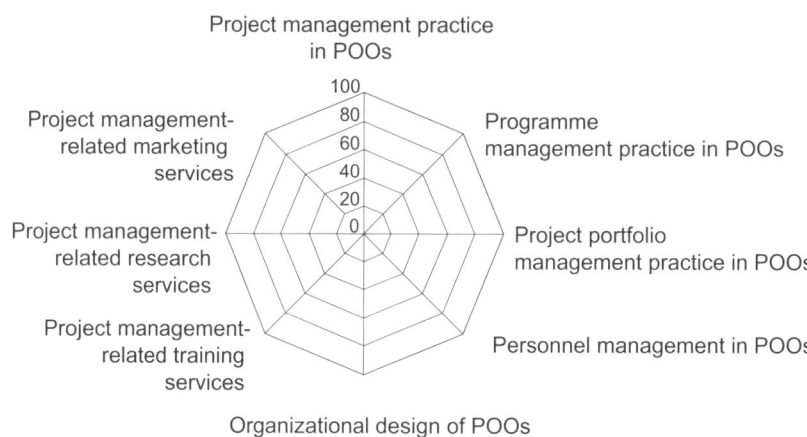

Fig. L1.1: Maturity model of the project-oriented society

The "practice" axes of the maturity model of the project-oriented society correspond to the axes of the maturity model of the project-oriented organization depicted above (see Chapter J2), with the processes "Assignment of a project or a programme", "Project portfolio coordination" and "Networking of projects" being summarized as the "Project portfolio management practice in POOs". The content in the "practice" axes to be considered were described in detail in Chapters E to K.

In the following those axes of the maturity model are described which refer to the project management-related services offered in the project-oriented society.

Project management-related services

Project management-related services are differentiated in training, research, and marketing services.

Formal project management training programmes can be offered by private and public training institutes. Universities, universities of applied science, and training and consultancy companies all offer project management courses. These can be part of a general academic education (e.g. a component of a university course in economics or civil engineering), can lead to an academic degree in project management, or can only be con-firmed by means of a certificate.

The quality of project management training programmes can be differentiated with re-gard to the amount of courses offered and the communicated project management ap-proach.

Project management-related services by research institutes can take the form of research projects and research programmes, publications and research events on project manage-ment subjects. Specific financing may be allocated for the research in project manage-ment.

Project management marketing tasks in the project-oriented society are performed main-ly universities and by national project management associations. Services provided by national project management associations are: Member service, certification of project managers, performance of project management events, etc. The application of project management is also promoted by developing project management-related standards and by defining formal requirements for project management in performing public contracts.

CHAPTER

L

The service dimensions of the model of the project-oriented society are described in de-tail with the aid of the questionnaire for analyzing the maturities of these dimensions of project-oriented societies (see e.g. Figure L1.2).

How many of the following institutions are offering formal project management training programmes?	
Secondary schools (such as high schools, trade schools …)	
Colleges/technical colleges of education	
Universities	
Institutions of further education	
Consulting companies	
Other educational institutions (please state)	

1…none of them, 2…few of them, 3…some of them, 4…many of them, 5…all of them

How many of the following institutions perform project management-related research?	
Colleges/technical colleges of education	
Universities	
Institutions of further education	
Consulting companies	
Other educational institutions (please state)	

1…none of them, 2…few of them, 3…some of them, 4…many of them, 5…all of them

Which services are performed by the national project management associations/chapters?		
	IPMA association	PMI chapters
Certification of project management personnel		
Project management marketing		
Lobbying for the promotion of project management as a profession		
Project management research		
Project management training		
Project management events		
Development of project management standards		
Awards for project manager or project of the year		
Project management publications		
Project management newsletters		
National project management initiative		
Project management newsgroups or discussion platforms		
Others (please state)		

1…Yes, 2…No

Fig. L1.2: Selected questions from the questionnaire for analyzing project-oriented societies

L 2 Analysis and benchmarking of project-oriented nations

L 2.1 Analysis of Austria as a project-oriented nation

The model of the project-oriented society was developed as part of a research project by the International Project Management Association IPMA by the author to analyze and benchmark project-oriented nations. This model was first used for analyzing Austria as a project-oriented nation in 2000.

Objectives of the analysis of Austria as a project-oriented nation were:

- The assessment of the practice of project-oriented organizations in Austria in project management, programme management, project portfolio management, in personnel management and in organizational design,
- The assessment of the project management-related services of Austrian training, research and marketing organizations, and

- The definition of strategies for the further development of Austria as a project-oriented nation.

Process of the analysis of Austria as a project-oriented nation

The analysis process included the following phases:

- Analysis of the project management training, the project management research and the project management marketing in Austria,
- Preparation of a workshop to analyze the practice of project-oriented organizations in Austria in project management, programme management, project portfolio management, in personnel management and in organizational design,
- Performance of the analysis workshop, and
- Follow-up to the analysis workshop.

The analysis of the project management-related services of training, research and marketing institutes was performed by means of documentation analyses, Internet research and interviews. The questionnaire on the analysis of the maturity of a project-oriented nation formed the basis for the analysis workshop. Around 60 Austrian project management experts and 20 project management students participated in the analysis workshop at the University of Economics and Business Administration, Vienna. The participants were divided into 8 groups. The questions from the questionnaire were discussed and answered consensually. This was not a quantitative but rather a qualitative analysis based on expert opinions.

In the follow-up to the analysis workshop, the achieved results were summarized and given to the experts for final feedback. After the feedback, the results were adapted, and reports were developed and published. In the spider web the maturity of Austria as a project-oriented nation was visualized. Additionally, the maturity ratio for Austria was calculated and tables comprising the results were developed and interpreted.

The "maturity" of Austria as a project-oriented nation in the year 2002

The "maturity" of Austria as a project-oriented nation in 2003 is depicted as the area in the spider web model of the project-oriented nation (see Figure L2.1). The overall "maturity ratio" for Austria was 37%. The maturity ratios for the individual dimensions of the project-oriented nation can be seen from Figure L2.1. The highest maturity was in project management marketing at 75%, followed by 38% and 37% for the project management practice in project-oriented organizations and for project management training.

The strengths and weaknesses of Austria as a project-oriented nation were expressed by these values. These analyses formed the basis for the definition of the objectives of "programm I austria – the Austrian project management initiative" (see Chapter L3.1).

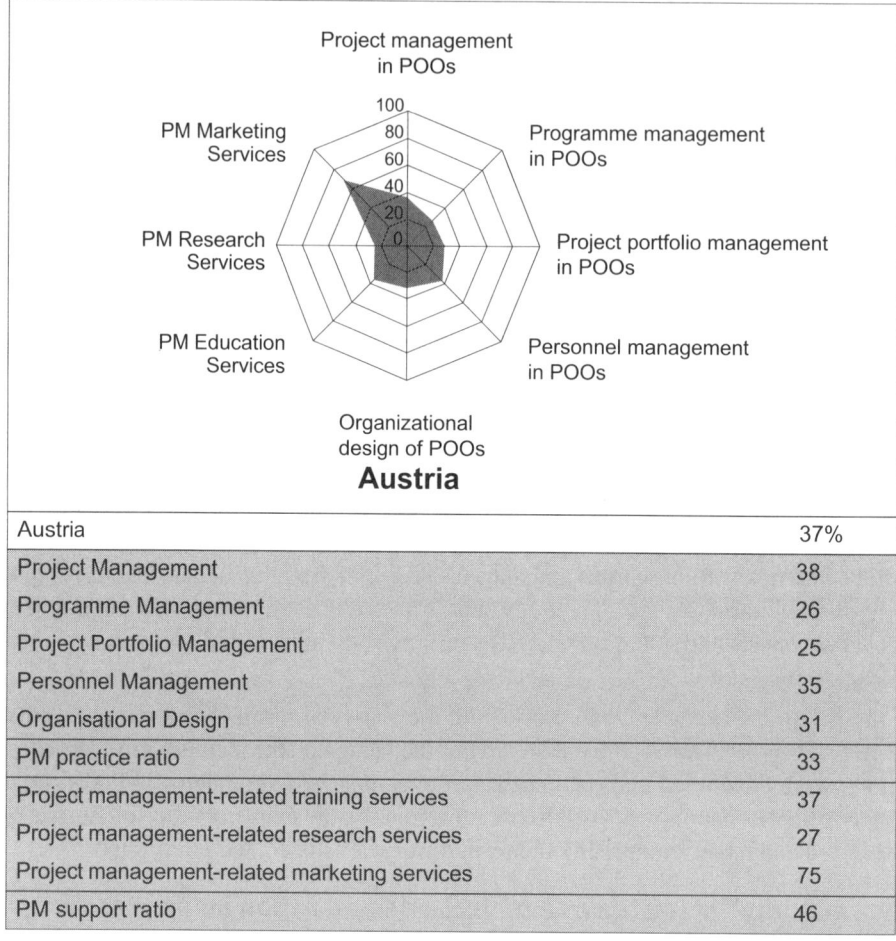

Austria	37%
Project Management	38
Programme Management	26
Project Portfolio Management	25
Personnel Management	35
Organisational Design	31
PM practice ratio	33
Project management-related training services	37
Project management-related research services	27
Project management-related marketing services	75
PM support ratio	46

Fig. L2.1: "Maturity" of Austria as a project-oriented nation in 2002

L 2.2   Benchmarking project-oriented nations and regions

Between 1999 and 2002, the IPMA (International Project Management Association) performed the research initiative "Project-oriented society" under the management of the author. It was the objective of two research projects to develop a model of the project-oriented society and to apply it to nations. In a project entitled "Conception of the project-oriented society ", the model of the project-oriented society was developed.

Based on this model, the "Benchmarking of project-oriented nations" was performed. The objectives of this project were the assessment of the maturities of project-oriented nations and regions, the analysis of common features and of differences between these project-oriented nations, and the development of strategies for the further development of the maturities of project-oriented nations.

CHAPTER

L

An initial group of project-oriented nations, i.e. Denmark, UK, Austria, Romania, Sweden and Hungary, performed a benchmarking in the period between October 2000 and March 2001. A second group, i.e. Ireland, Latvia, Norway and, once again Austria, performed the benchmarking in the period between 2001 and 2002. An analysis of the project-oriented region of "Moscow" followed in January 2003.

Process of benchmarking project-oriented nations

In order to perform the analysis and the benchmarking of the competencies of the project-oriented nations, each nation involved in this process nominated a "Services team" and a "Practices team" comprised of representatives of the national project management associations, managers of project-oriented organizations, researchers from universities and project management trainers and consultants.

The significance of projects and programmes, and the project management-related services were identified and described. This analysis was performed with the aid of the questionnaire for analyzing project-oriented organizations.

The analysis of practices of the project-oriented organization was carried out by the "Practices team" in the form of a workshop. The "Practices team" invited along to this workshop experts from various project-oriented branches and from non-profit organizations. In the analysis workshop, the questions from the questionnaire for analyzing project-oriented nations were discussed and answered.

The assessments of the results of the analyses of the individual nations were performed by the "Coordination team". These assessments were presented at a benchmarking workshop with representatives of all nations involved in the process. Common features and differences between the project-oriented nations were discussed and interpreted. These formed the bases for the formulation of strategies and measures for the further development of the individual project-oriented nations.

The benchmarking results were summarized in a benchmarking report which was then given to all national associations of the IPMA. Summaries of the results were published in various media.[2]

Maturity ratios of the project-oriented nations

The maturities of the project-oriented nations are expressed by maturity ratios. The project-oriented nations and the project-oriented region of Moscow are depicted in Figure L2.2, ordered according to the maturity ratio. It can be seen that in 2002 Sweden and the UK had the highest ratios. Latvia had the lowest maturity as project-oriented nation.

---

[2] *Gareis, R., Huemann, M.,* (Benchmarking Projektorientierter Gesellschaften) 2002
Gareis, R., Huemann, M., (Assessing and Benchmarking) 2001

	SWE	UK	NOR	MOS	IRL	DEN	AUT	HUN	ROM	LAT
POS Ratio	**56**	**55**	**46**	**45**	**44**	**42**	**37**	**34**	**23**	**22**
Project Management	63	56	51	52	54	46	39	42	31	28
Programme Management	70	30	55	40	44	37	26	37	17	16
Project Portfolio Management	57	30	45	28	39	43	25	37	10	0
Personnel Management	62	58	47	31	46	50	35	46	26	25
Organisational Design	65	40	45	28	45	55	31	35	30	22
PM Practice Ratio	**63**	**47**	**46**	**40**	**46**	**46**	**33**	**40**	**25**	**21**
PM Education	40	80	30	51	34	30	36	30	30	11
PM Research	40	70	44	38	16	30	28	20	10	11
PM Marketing	40	70	38	70	67	40	71	15	10	50
PM Service Ratio	40	73	37	53	39	33	45	22	17	24

Fig. L2.2: Comparison of the "maturities" of the project-oriented societies in 2002

L 3   National project management initiatives

L 3.1   "programm I austria" – the Austrian project management initiative

Context and strategies of "programm I austria"

Inspired by project management initiatives in Norway and Finland and the "Benchmarking POS" project by the International Project Management Association IPMA, "programm I austria – the Austrian project management initiative" programme was launched in October 2000.

The planning of the programme was performed in the project "Conception: programm I Austria". Programme objectives, programme strategies and the organizational structures were planned, and the projects to be started in the first two years of the programme were defined. Ideas were gathered for projects to be performed in the following years. However, their realization depended on the necessary cooperation partners and financing opportunities.

The overall topics of the programme were "Project management for everybody", "Further development of project-oriented organizations" and "Development of the model of the project-oriented society". The objective of "Project management for everybody" was to contribute to the establishment of project management not only as an essential maturity of companies but also of administrations of small municipalities, associations, schools and families.

The objective of "Further development of the competencies of project-oriented organizations" was to contribute to the professionalism of project and programme management, project portfolio management, personnel management and organizational design in project-oriented organizations.

The objective of the "Development of the model of the project-oriented society" was, on the one hand, to communicate the economic relevance of project-orientation to the public, and on the other hand to contribute to the establishment of the profession of "Project manager".

"A student-driven initiative" was defined as the organizational strategy of "programm I austria". The PROJEKTMANAGEMENT GROUP from the University of Economics and Business Administration, Vienna, and the Austria project management association were the supporting organizations of the programme.

Programme bar chart of "programm I austria"

The scope of "programm I austria" and the timing of the projects are depicted in the bar chart in Figure L3.1.

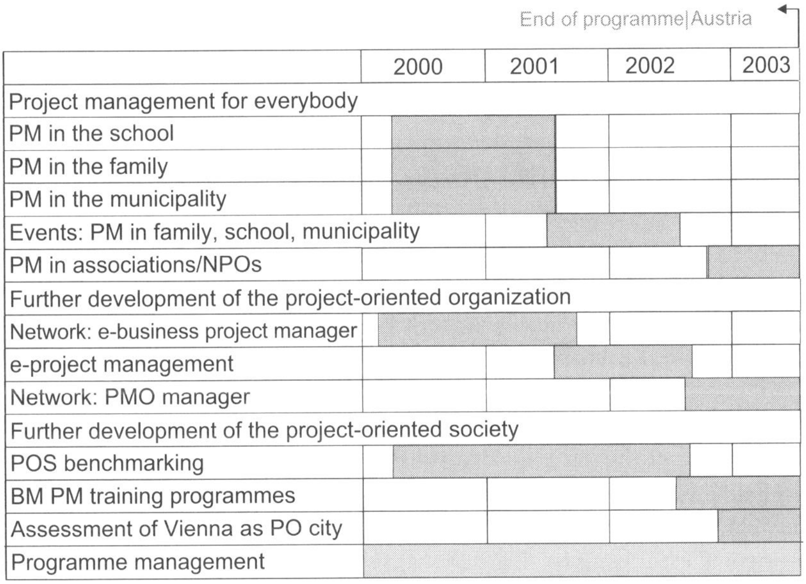

End of programme|Austria

	2000	2001	2002	2003
Project management for everybody				
PM in the school	▓			
PM in the family	▓			
PM in the municipality	▓			
Events: PM in family, school, municipality		▓		
PM in associations/NPOs				▓
Further development of the project-oriented organization				
Network: e-business project manager	▓	▓		
e-project management		▓	▓	
Network: PMO manager				▓
Further development of the project-oriented society				
POS benchmarking	▓	▓		
BM PM training programmes				▓
Assessment of Vienna as PO city				▓
Programme management	▓	▓	▓	▓

CHAPTER

L

Fig. L3.1: Bar chart of "programm I austria"

It can be seen from the programme bar chart that the topics "Project management for everybody", "Further development of project-oriented organizations" and "Development of the model of the project-oriented society" were the basic criteria for the structuring of the programme.

Due to the student culture in the programme and the resulting high fluctuation of personnel, an attempt was made to define relatively short-term projects. A high differentiation by projects enabled not only the formation of project teams (with students) active over short periods, but also increased the potential for finding various cooperation partners and sponsors for various projects.

Chains of projects were created in order to ensure continuity in terms of content and to organize the transfer of knowledge between projects accordingly. One example of a chain of projects is the sequence of the projects "Project management at school", "Event: Project management at school" and "Marketing: Project management for everybody".

The results of "programm I austria" were communicated in workshops, events, brochures and on the Internet. This developed a sort of network for the participating research partners, which were project-oriented companies,, public administration institutions and universities.

The success of "programm I austria" depended on whether Austrian companies and political institutions recognized the necessity of using and further developing Austria as a project-oriented society. This was the prerequisite for ensuring the necessary financial resources for the wide-scale realization of the vision "Project-oriented nation Austria".

In retrospect, it can be said that the objective of the rapid further development of Austria as a project-oriented nation could not be realized. However, important research results were achieved, and concrete contributions toward the further development of Austria as a project-oriented nation were achieved, such as the establishment of a network of PM Office managers.

L 3.2   Other national project management initiatives

On the one hand, the Austrian project management initiative was geared to other national initiatives and, on the other hand, influenced such initiatives, e.g. the Romanian project management initiative.

Recently further national project management initiatives were performed in Finland, Norway, Sweden and Romania.

Finland: Global Project Business

- **Objectives of the "Global Project Business" initiative (GPB)**

The Finnish "Global Project Business" initiative was a national research programme with the objective of increasing the competitiveness of internationally-active, Finnish project-oriented organizations. The "Global Project Business" was performed between 1998 and 2001.

The objectives of GPB were:

- The development of new methods and procedures regarding product, technology and process management in the international project business, and
- The promotion of the exchange of information and of cooperation between project-oriented organizations, including at international level.

- **Organization and infrastructure of the GPB initiative**

The strategies of the Finnish initiative were determined by a team in which representatives of project-oriented organizations worked together. With representatives from the telecommunications sector, forestry/the timber industry and the energy sector, the three most important branches of Finnish industry were represented (NOKIA, Valmet, Jaakko Pöyry, Andritz-Ahlstrom, Fortum Power&Heat).

The project management research was performed by the Helsinki University of Technology. The tasks of coordination and project management of the "Global Project Business" initiative were assumed by the RAMSE Consulting Oy consultants in cooperation with Tekes.

- **Resources of the GPB initiative**

A budget of approximately € 15 million was allocated to the four-year initiative.

Strategies of the GPB initiative

The initiative was available exclusively to Finnish companies. Small and medium-sized companies were particularly interested in the results, since their future depends a great deal on successful operations on the global market.

Norway: Project 2000 and Norwegian Center of Project Management

- Objectives of the "Project 2000" initiative

The "Project 2000" was performed between 1994 and 1999. The basic vision was to improve the identification, evaluation, planning and performance of projects in order to increase the competitiveness of the Norwegian industry.

A renewal of the applied project management methods and procedures in order to then be able to use them in industry and public administration, was aimed for.

The initiative was performed principally in the form of research projects and training programmes at universities and other training centers. Norwegian industry assumed a decisive role in determining the main thematic points.

- **Organization and infrastructure of the "Project 2000" initiative**

Structurally, the initiative was performed over several years as a cooperation between Norwegian universities (Norwegian University for Science and Technology, BI Norwegian School of Management), The Foundation for Scientific and Industrial Research and Norwegian companies. The cooperation partners played a decisive role in selecting and concretizing the research activities to be performed.

50 researchers and 11 PhD students were involved in "Project 2000". In total, 28 organizations participated in "Project 2000".

- **Resources of the "Project 2000" initiative**

The budget for the "Project 2000" initiative amounted to approximately € 5.5 million. Of this, approximately 50% were allocated to research programmes, 30% to the dissertation and other training programmes, approximately 10% to the administration, and the rest to workshops and seminars.

- **Strategies of the "Project 2000" initiative**

A central strategy was the establishment of the "Norwegian Center of Project Management". This "center" was intended to ensure the long-term further development of Norway as a project-oriented nation.

Sweden: Project Sweden

- **Objectives of the "Project Sweden" initiative**

Once various research groups in Sweden had been united under the "Project Sweden" initiative, an attempt was made, in cooperation with the most successful Swedish project-oriented organizations to produce research results on the topic of project management.

Workshops and conferences for researchers and practitioners, and joint research projects, are being performed (currently approximately 20 projects). In addition, so-called "Experience days" are being held. The objective of such is the exchange of information regarding project management topics.

- **Organization and infrastructure of the "Project Sweden" initiative**

A "Steering group" was established in order to manage the initiative. This group consists of researchers and people from the cooperating companies. The main task of this group is to report to the financing companies about the current research projects. This communication is performed primarily via a web community.

- **Resources of the "Project Sweden" initiative**

Financial resources were provided for the employed coordinator of "Project Sweden". There were about 25 people working as researchers on the "Project Sweden", of which around 10 were entrusted exclusively with topics pertaining to "Project Sweden".

The annual expenses amounted to approximately € 100,000 for central organization activities and € 800,000 for other expenses accrued in the network. These costs were financed by cooperation partners from industry (ABB, Volvo, Vinnova, F AS, Riksbankens jubileumsfond).

Romania: EPROM – European Union Compatible Postgraduate Course on Project Management

- **Objectives of the project "EPROM"**

The "EPROM" initiative was performed as an EU-sponsored project between 1998 and 2001.

An important reason for the project was the increased cooperation between Romanian companies and companies from the EU, especially with regard to Romania's planned accession to the EU. The industrial region of Romania was to become more active by means of project management competencies.

The following results were achieved as part of the project:

- Master degree programmes in project management were established at three Romanian universities.
- Romania was analyzed as a project-oriented nation. Based on this, strategies for the further development of Romania as a project-oriented nation were developed.
- A competence center for project management was established at the Academy of Economic Studies in Bucharest.
- A project management guidebook was developed as the basis for project management training programmes.

- A national project management association was established.
- The profession of "Project manager" was authorized and registered by the Romanian Ministry of Labor.

- **Organization and infrastructure of the project "EPROM"**

The project "EPROM" was performed as a cooperation between four Romanian universities (Academy of Economic Studies in Bucharest, Transylvania University of Brasov, University of Craiova, University of Constanta) and five universities (University of Athens in Greece, University of Limerick in Ireland, University of Bradford in the UK, the Stockholm Royal Institute of Technology in Sweden and the University of Economics and Business Administration in Austria).

- **Resources of the project "EPROM"**

A development fund for central and eastern Europe (Tempus Fonds) amounting to approximately € 230,000 was provided by the EU in order to finance the expenses of the approximately 30 people working on the project.

CHAPTER

L

References

Gareis, R., Huemann, M., (Benchmarking projektorientierter Gesellschaften) Benchmarking projektorientierter Gesellschaften: Dänemark, Österreich, Rumänien, Schweden, Ungarn und UK im Vergleich, Projekt Management, GPM, 13th vol., 2/2002

Gareis, R., Huemann, M., (Assessing and Benchmarking) Assessing and Benchmarking Project-oriented Societies, in: Project Management – International Project Management Journal, Project Management Association Finland, Norwegian Project Management Forum, Vol. 7, No. 1., 2001

References

Books

Aarto, K., Martinsuo, M., Aalto, T., (Strategic Management) Strategic Management through
Projects, 1st edition, PMA Finland, Helsinki, 2001

Ashby, W.R., (Process of model-building) The process of model-building in the behavioral sciences, Ohio State Univ. Press, Columbia, OH, 1970

Bateson, G., (Geist und Natur) Geist und Natur, Suhrkamp, Frankfurt am Main, 1990

Becker, F., (Anreizsysteme) Anreizsysteme für Führungskräfte – Möglichkeiten zur strategisch-orientierten Steuerung des Managements, Poeschel, Stuttgart, 1990

Bortz, J., Döring, N. (Forschungsmethoden) Forschungsmethoden und Evaluation, Springer, Berlin, 1995

Bounds, G.M., Dobbins, G.H., Fowler, O.S. (Quality Perspective) Management – A total Quality Perspective, South-Western Publication, Cincinnati, OH, 1995

Bullinger, H.J. (ed.), (Lernende Organisationen) Lernende Organisationen, Schäffer-Poeschel, Stuttgart, 1996

Čamra, J.J., (REFA-Lexikon) REFA-Lexikon, 2nd edition, Beuth, Berlin, 1976

Caupin, G., Knöpfel, H., Morris, P., Motzel, E., Pannenbäcker O., (ICB), IPMA Competence Baseline, IPMA, Bremen, 1999

Daft, R.L., (Symbols in Organizations) Symbols in Organizations: A Dual-Content Framework for Analysis, in: Pondy, L.R., Frost, P.J., Morgan G. (eds.), Organizational Symbolism, JAI Press Inc., Greenwich, CT, 1983

Dandridge, T.C., (Symbols) Symbols' Function and Use, in: Pondy, L.R., Frost, P., Morgan, G. (eds.), Organizational Symbolism, JAI Press Inc., Greenwich, CT, 1983

Deming, W.E., (Crisis) Out of the Crisis, Cambridge University Press, Cambridge, 1992

Doujak, A., Doujak, G., (Krise ist wie Krieg) Krise ist wie Krieg, in: Gareis, R. (ed.), Er- folgsfaktor Krise, Signum, Wien, 1994

Endruweit, G. (ed.), (Soziologie) Wörterbuch der Soziologie, dtv, München, 1989

Eschenbach, R., Kunesch, H., (Prozessmanagement) Prozessmanagement: Instrumente des Prozessmanagements, Wien, 1993

Exner, A., Königswieser, R., Titscher, S., (Unternehmensberatung) Unternehmensberatung

– systemisch, in: Das systemisch-evolutionäre Management, Orac, Wien, 1992

Fischer, T.M., (Sicherung unternehmerischer Wettbewerbsvorteile) Sicherung unternehmerischer Wettbewerbsvorteile durch Prozess- und Schnittstellen-Management, in: Zeitschrift für Organisation 5/1993

Gaitanides, M., (Prozessmanagement) Prozessmanagement – Konzepte, Umsetzungen und

Erfahrungen des Reengineering, Hanser, München, Wien, 1994

Gareis R., (ed.), (Erfolgsfaktor Krise) Erfolgsfaktor Krise. Konstruktionen, Methoden, Fall- studien zum Krisenmanagement, Signum, Wien, 1994

Gareis, R., (pm baseline), pm baseline – Wissenselemente zum Projekt- und Programmmanagement sowie zum Management Projektorientierter Unternehmen, PMA, Version 2, July, 2001

Gareis, R., (Handbook) Handbook of Management by Projects, Manz, Wien, 1990

Gareis, R., (Non-Profit) Projekte und Projektmanagement in Non-Profit-Organisationen, in: Badelt, C. (ed.), Handbuch der Non-Profit Organisationen, Schäffer-Poeschel, Stuttgart, 1997

Gester P., (Systemische Gesprächs- und Interviewgestaltung) Warum der Rattenfänger von Hameln kein Systemiker war? Systemische Gesprächs- und Interviewgestaltung, in: Schmitz, C., Gester P., Heitger B., (eds.), Managerie – Systemisches Denken und Handeln im Management, 1st yearbook, Carl Auer, Heidelberg, 1992

Goldman, S.L., Nagel, R.N., Preiss, K., Warnecke, H.J., (Agil im Wettbewerb) Agil im Wettbewerb: Die Strategie der virtuellen Organisation zum Nutzen des Kunden, Springer, Berlin, Heidelberg 1996

Grimm, R., Kozok. B., Lafos, F., (Kommunikationsinfrastruktur) Kommunikationsinfrastruktur für virtuelle Organisationen, in: Gora, W. (ed.), Virtuelle Organisationen im Zeitalter von E-Business und E-Government, Springer, Berlin, 2001

Guthof, P., (Strategische Anreizsysteme) Strategische Anreizsysteme – Gestaltungsoptionen im Rahmen der Unternehmensentwicklung, Gabler, Wiesbaden, 1995

Hill, W., Fehlbaum, R., Ulrich, P., (Organisationslehre 1) Organisationslehre 1: Ziele, Instrumente und Bedingungen der Organisation sozialer Systeme, 5th edition, Paul Haupt (UTB), Stuttgart, Bern, 1994

Hillier, F.S., Lieberman, G.J., (Operations Research) Introduction to operations research, McGraw-Hill, Boston, 2001

Hoffmann, F., (Unternehmenskulturen) Erfassung, Bewertung und Gestaltung von Unternehmenskulturen – Von der Kulturtheorie zu einem anwendungsorientierten Ansatz, in: zfo 3/1989

Huemann, M., (Individuelle Projektmanagement-Kompetenzen) Individuelle Projektmanagement-Kompetenzen im Projektorientierten Unternehmen, Peter Lang Europäischer Verlag der Wissenschaften, Frankfurt am Main, 2002

Huemann, M., (Improving Quality) Improving Quality in Projects and Programs, in: Morris, P.W., Pinto J. (eds.), The Handbook of Managing Projects, Wiley & Sons, Somerset, NJ, 2004

Ishikawa, K., Lu, D., (Quality Control) What is Total Quality Control? The Japanese Way, Prentice Hall, Englewood Cliffs, NJ, 1985

Jann, B., (Statistik) Einführung in die Statistik, Oldenbourg, München, Wien, 2002

Jones, M.O., (Organizational Symbolism) Studying Organizational Symbolism, Sage Publications, Thousand Oaks, CA, 1996

Jung, H., (Personalwirtschaft) Personalwirtschaft, Oldenbourg, München, Wien, 1999

Juran, J.M., (Qualitätsplanung), Handbuch der Qualitätsplanung, 3rd revised edition, Moderne Industrie, Landsberg am Lech, 1991

Kaplan, R., Norton, P., (Balanced Scorecard) Balanced Scorecard, Strategien erfolgreich umsetzen, Schäffer-Poeschel, Stuttgart, 1997

Kasper, H., (Organisierte Sozialsysteme) Die Handhabung des Neuen in organisierten Sozialsystemen, Springer, Wien 1990

Keegan, A.E., Turner, J.R., (Managing Human Resources) Managing human resources in the project-based organization, in: Turner, J.R. (ed.), People in Project Management, Gower, Aldershot, 2003

Kepper, G., (Qualitative Marktforschung) Qualitative Marktforschung: Methoden, Einsatz-möglichkeiten und Beurteilungskriterien, 2nd revised edition, Dt. Univ. Verl., Wiesbaden, 1996

Kessler, H., Höhnle, C., (Karriere) Karriere im Projektmanagement, Springer, Berlin, Heidelberg, 2002

Klimecki, R., Probst, G., Eberl, P., (Systementwicklung als Managementproblem) Systementwicklung als Managementproblem, in: Staehle, W., Sydow, J. (eds.), Managementforschung, Volume 1, Berlin, 1991

Kloock, J., (Flexible Prozesskostenrechnung) Flexible Prozesskostenrechnung und Deckungsbeitragsrechnung, in: KRP, 2/1993

Kompa, A., (Assessment Center) Assessment Center – Bestandsaufnahme und Kritik, Rainer Hampp, München, 1999

Königswieser, R., Exner, A., (Systemische Intervention) Systemische Interventionen, Architekturen und Designs für Berater und Veränderungsmanager, Klett-Cotta, Stuttgart, 1998

Konrad, L., (Strategische Früherkennung) Strategische Früherkennung – eine kritische Analyse des „Weak Signal"-Konzeptes, Brockmeyer, Bochum, 1991

Krczal, A., (Von Qualitätskontrolle zu Qualitätsverbesserung), Von der Qualitätskontrolle zur kontinuierlichen Qualitätsverbesserung, in: Eckardstein, D., Kasper, H. Mayerhofer, W. (eds.), Management, Theorien – Führung – Veränderung, Schäffer-Poeschel, Stuttgart, 1999

Krüger, W., (Organisation der Unternehmung), Organisation der Unternehmung, Kohlhammer, Stuttgart, 1993

Krystek, U., Redel, W., Reppegather, S., (Virtualität) Erfolgsfaktoren und Elemente der Virtualität, in: Gablers Magazin 3/97

Kuwan, H., Waschbüsch, E., (Zertifizierung), Zertifizierung und Qualitätssicherung in der beruflichen Weiterentwicklung, Bundesinstitut für Berufsbildung, Berlin, 1996

Lamnek, S., (Qualitative Sozialforschung 1) Qualitative Sozialforschung, Volume 1, Methoden und Techniken, Psychologie s-Union, Weinheim, 1995

Lamnek, S., (Qualitative Sozialforschung 2) Qualitative Sozialforschung, Volume 2, Methoden und Techniken, Psychologie s-Union, Weinheim, 1996

Lauer, B., (Berufsbild) Projektmanager/in – das Berufsbild in der Zukunft, in: Szenarien der Ar- beitswelt von morgen, Haarbeck, S. (ed.), Dt. Wirtschaftsdienst, Köln 2000

Lehnert, C. (Karriereplanung) Neuorientierung der betrieblichen Karriereplanung, Dt. Univ. Verlag, Wiesbaden, 1996

Linde, F., (Virtualisierung von Unternehmen) Virtualisierung von Unternehmen, Dt. Univ. Verl., Wiesbaden, 1997

Lipnack, J., Stamps, J., (Networking) Networking – The first report and directory, Doubleday, New York, 1982

Lueger, G., (Beschaffung und Auswahl von Mitarbeitern) Beschaffung und Auswahl von Mitarbeitern, in: Kasper, H. (ed.), Personalmanagement – Führung – Organisation, Ueberreuter, Wien, 1996

Luhmann, N., (Funktionen) Funktionen und Folgen formaler Organisation, Duncker und Humblot, Berlin, 1964

Luhmann, N., (Komplexität) Komplexität, in: Grochla, E. (ed.), Handwörterbuch der Organisation, 2nd edition, Poeschel, Stuttgart, 1980

Luhmann, N., (Soziale Systeme) Soziale Systeme: Grundriss einer allgemeinen Theorie, Suhrkamp, Frankfurt am Main, 1984

Mayntz, R., Holm, K., Hübner, P., (empirische Soziologie) Einführung in die Methoden der empirischen Soziologie, Westdeutscher Verlag, Opladen, 1974

Mertens, P., Faisst, W., (Virtuelle Unternehmen) Virtuelle Unternehmen – eine Organisationsstruktur für die Zukunft? (Internet), 1997

Motzel, E., (Projektmanagement Kanon) Projektmanagement Kanon – der deutsche Zugang zum Project Management Body of Knowledge, TÜV, Köln, 1998

Neuberger, O., (Führen und geführt werden) Führen und geführt werden, Ferdinand Enke, Stuttgart, 1990

Olfert, K., Steinbuch, P., (Personalwirtschaft) Personalwirtschaft, Kiehl, Ludwigshafen, 1995

Paslack, R., (Urgeschichte der Selbstorganisation) Urgeschichte der Selbstorganisation, Vieweg, Braunschweig, 1991

Project Management Institute, (Body of Knowledge) A Guide to The Project Management Body of Knowledge, Project Management Institute, Upper Darby, PA, 2000

Project Management Institute, (Fact Book) Project Management Fact Book, Project Management Institute, Upper Darby, PA, 2001

Pondy, L.R., Frost, P., Morgan, G. (eds.), (Organizational Symbolism) Organizational Symbolism, JAI Press Inc., Greenwich, CT, 1983

Reibnitz, U. von, (Szenario-Technik) Szenario-Technik: Instrumente für die unternehmerische und persönliche Erfolgsplanung, Gabler, Wiesbaden, 1992

Reschke, H., (Aufbauorganisation) Formen der Aufbauorganisation in Projekten, in: Reschke, H., Schelle, H., Schnopp, R. (eds.), Handbuch Projektmanagement, Volume 2, TÜV Rheinland, Köln, 1989

Rischar, K., (Optimale Personalauswahl) Optimale Personalauswahl, TÜV Rheinland, Köln, 1990

Rosenstiel, L., (Organisationspsychologie) Organisationspsychologie, Kohlhammer, Stuttgart, 1975

Schein, E., (Career) Career Dynamics: Matching individual and organizational needs, Addison-Wesley, Reading, MA, 1978

Schein, E., (Organizational Culture) Organizational Culture and Leadership, Jossey Bass, San Francisco, Washington, London, 1985

Schelle, H., (Projekte zum Erfolg) Projekte zum Erfolg führen, dtv, München, 2001

Seaver, M. (ed.), (Quality Management) Gower Handbook of Quality Management, Gower, Aldershot, 2003

Senge, P., (Fieldbook) The fifth discipline fieldbook: strategies and tools for building a learning organization, Doubleday, New York, 1994

Senge, P., (fifth discipline) The fifth discipline: the art and practice of the learning organization, Century Business, London, 1998

Simon, D., (Schwache Signale) Schwache Signale – die Früherkennung von strategischen Diskontinuitäten durch Erfassung von „weak signals", Service Fachverlag an der Wirtschaftsuniversität, Wien, 1986

Spiesshofer, U., (Ingenieure) Ingenieure im europäischen Management: Karrieren von Ingenieuren im Topmanagement von europäischen Industrieunternehmen, VDI, Düsseldorf, 1991

Steinle, H., Bruch, H., Lawa, D. (eds.), (Instrument moderner Dienstleistung) Projektmanagement: Instrument moderner Dienstleistung, Edition Blickbuch Wirtschaft, Frankfurt, 1995

Steyrer, J., (Theorien der Führung) Theorien der Führung, in: Kasper, H., Mayrhofer, W., Personalmanagement, Führung, Organisation, Linde, Wien, 2002

Sydow, J., (Erfolg als Vertrauensorganisation) Erfolg als Vertrauensorganisation, Virtuelle Unternehmung, in: Business Review 7-8, 1996

Titscher, S., (Professionelle Beratung) Professionelle Beratung – Was beide Seiten vorher wissen sollten..., Ueberreuter, Wien, 1997

Van Wieren, H. D., (Alliance) Alliance, an Excellent Solution to Meet Project Execution Challenges, in: 16th IPMA World Congress on Project Management, Making the Vision Work, 4-6 June 2002, Berlin, 2002

Vogt, G., (Arbeitswelt), Nomaden der Arbeitswelt: virtuelle Unternehmen, Kooperation auf Zeit, Versus, Zürich, 1999

Walker, J. W., (Human Resource Strategy), Human Resource Strategy, MacGraw-Hill, New York, 1992

Watzlawick, P., (Wirklichkeit) Wie wirklich ist die Wirklichkeit?, Piper & Co., München, 1976

Weibler, J., (Personalführung) Personalführung, Franz Vahlen, München, 2001

Weibler, J., (Symbolische Führung) Symbolische Führung, in: Kieser, A., Reber G., Wunderer, R. (eds.), Handwörterbuch der Führung, 2nd edition, Schäffer-Poeschel, Stuttgart, 1995

Weick, K. E., (Systematic Observation Methods) Systematic Observation Methods, in: Gardner, L., The Handbook of Social Psychology, Volume 2: Research Methods, Addison-Wesley, Reading, MA, 1968

Wildenmann, B., (Professionell führen) Professionell führen – Empowerment für Manager, die mit weniger Mitarbeitern mehr leisten müssen, Luchterhand, Neuwied, 1996

Wilke, H., (Interventionstheorie) Systemtheorie II: Interventionstheorie, Lucius & Lucius, Stuttgart, 1996

Witt, F. J., (Aktivitätscontrolling) Aktivitätscontrolling und Prozesskostenmanagement, Poeschel, Stuttgart, 1991

Womack, J., (Autoindustrie), Die zweite Revolution in der Autoindustrie, Campus, Frankfurt, 1992

Papers

Enzenhofer, D., Semper, W., (Controlling von Projekte-Netzwerken) Controlling von Projekte-Netzwerken in projektorientierten Unternehmen, in: Projekt Journal

Eskerod, P. (Human Resource Allocation) The Human Resource Allocation Process When Organizing by Projects, in: Lundin, R.A., Midler, C. (eds.), Projects as Arenas for Renewal and Learning Processes, Kluwer Academic Publishers, Boston, MA, 1998

Gareis, R., (Programmmanagement) Programmmanagement und Projektportfolio-Management, in: Deutsche Gesellschaft für Projektmanagement (ed.), Projekt Management 1/2001, TÜV, Köln, 2001

Gareis, R., (Virtuelle Organisationen) Projekte und Virtuelle Organisationen, PROJEKTMA-NAGEMENT GROUP, Wirtschaftsuniversität Wien, 1997

Gareis, R., Huemann, M., (Assessing and Benchmarking) Assessing and Benchmarking Project-oriented Societies, in: Project Management – International Project Management Journal, Project Management Association Finland, Norwegian Project Management Forum, Vol. 7, 1/2001

Gareis, R., Huemann, M., (Benchmarking Projektorientierter Gesellschaften) Benchmarking Projektorientierter Gesellschaften: Dänemark, Österreich, Rumänien, Schweden, Ungarn und UK im Vergleich, Projekt Management, GPM, Vol. 13, 2/2002

Thatcher J. R., (New Age Managers for Projects) New Age Managers for Projects – Dealing with a "World Turned Upside Down", in: Project Management Network, Vol. IV, No. 4, Project Management Institute, Drexel Hill, PA, 1990

Further reading

Knutson, J. (ed.), (Project Management for Business Professionals) Project Management for Business Professionals: A Comprehensive Reference Guide, Wiley & Sons, Somerset, NJ, 2000

Morris, P.W., Pinto J. (eds.), (Handbook) The Handbook of Managing Projects, Wiley & Sons, Somerset, NJ, 2004

Figures

C   Project organization models and project roles

D Teams, leadership and project culture

E Project management

F   Methods of project and programme management

H   Assurance of the management quality in projects and programmes

I Project portfolio management

J   Organizational design of the project-oriented organization

K   Personnel management in the project-oriented organization

L   The project-oriented society

Abbreviations

AFSCM	Air Force System Command Manual
BI	Norwegian School of Management
BM	Benchmarking
C/SCSC	Cost/Schedule Control System Criteria
EEE	European Environment Education-project
EPC	engineering, procurement and construction
EPROM	European Union Compatible Postgraduate Course on Project Management
GMP	Gesellschaft für Projektmanagement (Association for Project Management)
GPB	Global Project Business
ICB	International Competence Baseline of the International Project Management Association
IPMA	
ILO	International Labour Organization
IPMA	International Project Management Association
NSP	Norwegian Center of Project Management
NTNU	Norwegian University of Science and Technology
PERT	Programme Evaluation and Review Technique
PM	Project management
PMA	PROJEKT MANAGEMENT AUSTRIA
PM BoK®	Project Management Body of Knowledge
PMG	PROJEKTMANAGEMENT GROUP
PMI	Project Management Institute
PMP	Project Management Professional
POC	Project-oriented company
POO	Project-oriented organization
POS	Project-oriented society
WBS	Work breakdown structure
WP	Work package
RGC	ROLAND GAREIS CONSULTING
SINTEF	Foundation for Scientific and Industrial Research
UNIDO	United Nations Industrial Development Organization

Index

R

S

T

U

The author and the team

Author: Roland Gareis

Roland Gareis was born in Vienna in 1948. He has two children with his wife Haldis: Luisa and Lorenz. Once a professional soccer player, today he plays tennis and goes skiing. He likes to spend his spare time in Reichenau an der Rax, Lower Austria.

He studied at the University of Economics and Business Administration, Vienna, and habilitated at the University of Technology, Vienna. He was a professor at the Georgia Institute of Technology, Atlanta, a visiting professor at the ETH, Zurich, at the Georgia State University, Atlanta and at the University of Quebec, Montreal. He is a university professor for project management at the PROJEKTMANAGEMENT GROUP from the University of Economics and Business Administration, Vienna, and director of the university programme "International Project Management".

From 1986–2002 he was chairman of the board of the project management association Project Management Austria. In 1990 he organized the IPMA world congress on the topic "Management by Projects" in Vienna. He is a former research director of the International Project Management Association IPMA.

His research activities:

- 1988–1991: Research programme "Management by Projects";
- 1992–1993: Research programme "Crisis Management";
- 1995–1998: Research programme "Project Management – A Business Process of the Project-oriented Company";
- 1999–2002: Research programme "Project-oriented Society" and "programm I austria";
- 2003–2007: Research programme *project and process [orientation]*".

His consulting services:

- 1982: Formation of ROLAND GAREIS CONSULTING; consulting national and international project-oriented companies in further developing their competencies.
- 2001: Development of licences for *ROLAND GAREIS Project and Programme Management*® and *ROLAND GAREIS Management of the Project-oriented Company*®.
- 2002: Development of the "Maturity model for a project-oriented organization".

Co-author of individual chapters: Martina Huemann

Martina Huemann was born in Vienna in 1969. Her sons, Daniel and Florian, keep her busy. In the past, extensive travels led her to various corners of this earth, today, project management conferences take her to exotic places. She studied business administration at the University of Economics and Business Administration, Vienna, at the University of Prague, Czech Republic and at the University of Lund, Sweden.

Since 1995 she is a university assistant for project management at the PROJEKTMANAGEMENT GROUP. She wrote her PhD dissertation on the topic "Individual project management competencies in project-oriented companies". She is a trainer in the university programme "International Project Management". Since 2003 she is Research Fellow at the University of Technology Sydney, Australia. She has gained experience in research, marketing and organizational development projects and is a certified project manager. She is a member of the board of Project Management Austria, assessor of the IPMA Award and network partner of ROLAND GAREIS CONSULTING.

Co-author of individual chapters: Michael Stummer

Michael Stummer was born in Neunkirchen, Lower Austria, in 1974. He has two sons, Alexander and Benjamin. He is an enthusiastic runner, soccer player and tennis player. He is a graduate in business administration and of the university programme "International Project Management". He has gained experience as a project manager in IT, organizational development, research and product development projects, and is a certified project manager. Since 1999 he is a trainer of the university programme "International Project Management".

He is a member of the board of Project Management Austria and in this position he is also facilitator of the PM Office network. Since 1998 he works as trainer and consultant at ROLAND GAREIS CONSULTING and since 2003 he is its authorized representative.

624